Multi-physical and Non-orthogonal Geotechnical Plasticity Modeling and Algorithms

Dechun Lu · Xin Zhou · Jingyu Liang · Xiuli Du

Multi-physical and Non-orthogonal Geotechnical Plasticity Modeling and Algorithms

Science Press
Beijing

Springer

Dechun Lu
College of Architecture and Civil Engineering
Beijing University of Technology
Beijing, China

Xin Zhou
College of Architecture and Civil Engineering
Beijing University of Technology
Beijing, China

Jingyu Liang
School of Civil and Transportation Engineering
Beijing University of Civil Engineering and Architecture
Beijing, China

Xiuli Du
College of Architecture and Civil Engineering
Beijing University of Technology
Beijing, China

ISBN 978-981-96-2368-6 ISBN 978-981-96-2369-3 (eBook)
https://doi.org/10.1007/978-981-96-2369-3

Jointly published with Science Press
The print edition is not for sale in China mainland. Customers from China mainland please order the print book from: Science Press.

This Springer imprint is published by the registered company Springer Nature Singapore Pte Ltd.
The registered company address is: 152 Beach Road, #21-01/04 Gateway East, Singapore 189721, Singapore

Preface

The coupling effect of multi-physics has emerged as a core scientific issue in modern geotechnical engineering, which is widely present in complex engineering practices, including tunnel excavation in water-rich strata, degradation of municipal solid waste, nuclear waste disposal, and the treatment of roadbed disease in permafrost regions. Research in this field has attracted considerable attention from both academia and industry. The main characteristic of these effects is that the multi-physics coupled environments in which geotechnical media are situated significantly alter the physical and mechanical properties of materials. This is manifested in the nonlinear evolution of key material parameters such as strength, stiffness, and permeability. The degradation of material properties induced by the multi-physics coupling effect has become a major cause of progressive failure of engineering structures.

The main bottleneck in current theoretical research lies in the limitations of traditional geotechnical plasticity theory, which is confined to the stress–strain constitutive description within a purely mechanical framework. This framework struggles to capture the dynamic evolution characteristics of geomaterial properties under unsteady environments and fails to analyze the nonlinear response of engineering structures under the coupling action of environmental factors and mechanical loads. This book builds on emerging challenges in modern geotechnical engineering. It systematically studies multi-physics coupled geotechnical plasticity theory, constitutive models, and numerical algorithms, providing the foundational theory and analytical methods that combine both theoretical depth and engineering applicability for analyzing the performance evolution of geotechnical engineering structures under the environmental-mechanical coupling action.

This book innovatively introduces environmental factors as independent constitutive variables from the perspective of material parameters. It proposes the unsteady consistency condition and the multi-physical loading/unloading criterion to determine the magnitude of plastic strain increment. Additionally, it proposes a novel three-dimensional non-orthogonal plastic flow rule, which directly utilizes the non-orthogonal gradient of the yield function to determine the direction of the plastic strain increment, without the need for a plastic potential function. The “stress–strain” incremental relationship under steady environments is extended to the "stress–strain-environment" incremental relationship under unsteady environments, thus achieving a leap from traditional mechanical constitutive theory to multi-physical constitutive theory.

In addition to the theoretical innovations, this book also presents the development and validation of non-orthogonal elastoplastic models for soil and concrete under both steady and unsteady environments. These models have been validated through extensive experimental data. To address the technical challenges of numerical implementation of elastoplastic models, the book proposes efficient implicit stress update algorithms. These algorithms integrate advanced mathematical tools such as smooth function, line search/trust region methods, hyper-dual number derivative approximation, and nonlocal theory, effectively solving computational issues related to non-smoothness, non-linearity, and mesh dependence. These innovative geomaterial models and stress update algorithms lay a solid foundation for analyzing and solving large-scale complex geotechnical engineering problems, bridging the gap between fundamental theories and engineering applications.

This research is funded by the National Natural Science Foundation of China (52025084; 52438005; 52408374) and the National Key Research and Development Program of China (2022YFC3800901). Special thanks are extended to colleagues and students, including Xiaoqiang Li, Fanping Meng, Cancan Su, Guosheng Wang, Yaning Zhang, Anyu Shi, and Yun Chen, for their substantial contributions throughout the research process. Despite our best efforts to ensure rigor, due to the inherent complexity of the research field, some oversights are inevitable. We sincerely welcome constructive feedback from peers and readers to jointly promote the continued development of this field.

Beijing, China
January 2024

Dechun Lu
Xin Zhou
Jingyu Liang
Xiuli Du

Contents

Chapter 1
Introduction

In recent years, with the expansion of human activities and the development of industry, multiphysical geotechnical engineering problems in complex environments have become the focus of attention for engineers, researchers, and policymakers, including but not limited to the storage of high-temperature radioactive waste, the climate change effects on slope stability or permafrost evolution, the collapse of unsaturated soil foundation, the settlement of landfill caused by the degradation of municipal solid waste, the corrosion of chloride and sulfate on concrete structures in the marine environment, the submarine landslide induced by combustible ice mining the reinforcement for soils based on microbial-induced carbonate precipitation, etc. The unsteady environmental conditions and the interaction between the environmental factors and geomaterials are common characteristics of these multiphysical geotechnical engineering problems. Under unsteady environmental conditions, geomaterial behavior is no longer controlled by mechanical loads alone. Local environments such as temperature changes, freezing–thawing, wetting–drying, groundwater table changes, pollution intrusion, etc. will change geomaterial characteristics. These characteristics, in turn, dictate the resulting mechanical behavior of geomaterials, as manifested by a possible change in strength, deformation, or hydraulic conductivity. Engineering structures in unsteady environments will show more complex structural responses, which urgently need new basic theories and analysis tools to analyze the performance evolution of engineering structures under the coupling action of mechanical loads and environmental factors.

Geotechnical plastic mechanics is the fundamental theory for analyzing deformation and strength problems in geotechnical engineering. Its main task is to determine the plastic strain increment, both in magnitude and direction, for a given stress increment or strain increment. The magnitude of plastic strain increment can be determined by the consistency condition together with the yield function and the hardening function. These two characteristic functions encompass the model parameters closely related to the mechanical properties of materials. In the classical geotechnical plasticity theory, the mechanical properties of the material are assumed to

D. Lu et al., *Multi-physical and Non-orthogonal Geotechnical Plasticity Modeling and Algorithms*, https://doi.org/10.1007/978-981-96-2369-3_1

remain unchanged during the loading process, which means that the material is a definite analysis object. The constitutive relationship of materials is established in the form of incremental stress–strain. This assumption holds for geomaterials in a steady environment. In unsteady environments, however, while some properties remain constant, others are subject to change as a function of environmental factors (Fang & Daniels, 2017). Plastic modeling should incorporate environmental factors as constitutive variables, akin to stress and strain, to formulate the multi-physical constitutive relationship in the form of incremental stress–strain-environment, which is necessary for analyzing geotechnical engineering problems under the coupling effect of mechanical loads and environmental factors. The absence of environmental variables is the first modeling limitation of classical geotechnical plasticity theory. On the other hand, classical geotechnical plasticity theory determines the direction of plastic strain increment via the orthogonal gradient of the plastic potential function, employing the orthogonal plastic flow rule. However, geomaterials are notably sensitive to hydrostatic pressure due to its inherent friction deformation mechanism. The direction of plastic strain increment direction is influenced by the stress state, displaying complex dilatancy behavior. It is very difficult to construct a reasonable potential function directly from experimental data, which is also the second modeling limitation of classical geotechnical plasticity theory. These limitations are challenges at the theoretical model level faced by analysis methods for geotechnical engineering problems.

At the numerical calculation level, the nonlinear elastoplastic model is generally defined in the form of ordinary differential equations (ODEs), which need to be integrated in the time domain to convert into the algebraic equations for the solution. The numerical algorithm for solving elastoplastic models is called the stress update algorithm, also known as the stress integration algorithm or constitutive integration algorithm. Corresponding to the complex mechanical behavior of geomaterials, researchers have to face some challenging computational difficulties in the numerical implementation of models, e.g. the non-smoothness corresponding to the loading/unloading inequalities, the nonlinearity that may result in the iterative divergence, and the mesh sensitivity caused by the softening constitutive law. Especially in unsteady environments, the coupling effect between environmental factors and geomaterials can make these computational difficulties more prominent. The appropriate numerical algorithms, a combination of multiple numerical algorithms, or even a new computational framework are required to ensure the computational efficiency and accuracy of numerical simulation. In the following, the research of geomaterials elastoplastic models and stress update algorithms in recent years is summarized.

1.1 Geotechnical Plasticity Modeling

The classical plasticity theory for metal materials can be traced back to the maximum shear stress yield criterion proposed by Tresca in 1864. He was probably influenced by the earlier work of Coulomb on soil mechanics. In 1870, Saint–Venant proposed a fundamental coaxial assumption, that is, the principal axis of plastic strain increment coincides with the principal axis of stress. Then, Levy in 1870 and von Mises in 1913 independently proposed that with negligible elastic strain, the total strain increment is proportional to the deviatoric stress. In 1928, von Mises (Mises, 1928) proposed the concept of plastic potential function, which determines the direction of plastic strain increment $\mathrm{d}\boldsymbol{\varepsilon}^{\mathrm{p}}$ by the gradient of the plastic potential function g. From then on, the plastic strain increment can be expressed as the product of magnitude and direction.

$$\mathrm{d}\boldsymbol{\varepsilon}^{\mathrm{p}} = \mathrm{d}\phi \frac{\partial g}{\partial \boldsymbol{\sigma}}$$

where $\boldsymbol{\sigma}$ is the stress tensor. $\mathrm{d}\phi$ is a non-negative plastic multiplier.

In 1951, the Drucker postulate (Drunker, 1951) for stable materials was proposed, which states that incremental internal energy can only increase. Soon, the Ilyushin postulate (Il'Iushin, 1961) applicable to unstable materials was also proposed one after another. These two postulates have made important contributions to the improvement of classical plasticity theory. In the late 1950s, the geotechnical plasticity theory became an independent subject developed from classical plasticity theory. In 1957, Drucker (Il'Iushin, 1961) pointed out that the mean stress or strain could cause the volumetric yield of geomaterial and suggested that a hat-style yield surface should be added in Mohr–Coulomb cone-shaped yield space. Furthermore, Roscoe et al. (1958) submitted the paper "On the yielding of soils" in 1958, which laid the foundation for critical state soil mechanics. Then, Roscoe and his co-workers proposed the famous Cam-clay model (Roscoe et al., 1958) and its modified version (Roscoe & Burland, 1968). Subsequently, a series of elastoplastic models for geomaterials have been proposed and further developed, focusing on two common issues: how to determine the magnitude and direction of plastic strain increments.

1.1.1 Magnitude of Plastic Strain Increment

In the plasticity theory, the yield function and hardening function are the keys to determining the increment of plastic strain increment. The yield function describes the stress conditions of the material when plastic deformation occurs, while the hardening function, also known as the hardening parameter, records the loading history of the material. For a specific elastoplastic model, the two functions collectively characterize the deformation mechanism and mechanical properties of materials.

To reasonably describe the yield and hardening behaviors of soil, researchers have proposed various forms of yield functions, and the hardening parameters that match the yield functions are used to describe the mechanical behaviors of soils. The classic Modified Cam-Clay (MCC) model (Roscoe et al., 1963) uses the plastic volumetric strain as the hardening parameter and an elliptical yield function, which reasonably describes the hydrostatic pressure yield, compressive hardening, and shear contraction of normally consolidated soil. Hashiguchi (1989) proposed the concept of the sub-loading surface to reasonably reflect the influence of the over-consolidated ratio (OCR) on the deformation characteristics of soils. Asaoka et al. (1997) proposed the concept of the super-loading surface by combining the concept of the sub-loading surface with the MCC model. The OCR can be described by the relative position relationship between the super-loading surface and the sub-loading surface. The structural characteristics and stiffness degradation law of soil during the loading process can be described by introducing the OCR into the hardening parameter (Nakano et al., 2005). In the elastoplastic models of soil proposed by Dafalias (1980), the OCR was introduced into the hardening modulus. The influence of OCR is considered by a mapping relationship between the loading surface and the boundary surface. Yao et al. (2008) proposed the unified hardening parameter independent of stress paths by revealing the similarity in deformation patterns between sand and clay. Then, the overconsolidation parameter R is defined based on the relative relationship between the current yield surface and the reference yield surface. Furthermore, R is incorporated into the unified hardening parameter to uniformly describe the strain hardening/softening, shear dilatancy/contraction, and stress-path-dependency behaviors of soil (Yao et al., 2009). To reflect the state-dependent of granular materials, i.e., the deformation characteristics of granular materials are affected not only by the stress state but also by the current void ratio state. Yao et al. (2019) transformed R into the state variable that quantifies the current state of sand in the e-ln$p$ coordinate system and established a unified hardening model for clays and sands (CSUH). Some other scholars have also defined various forms of state variables based on the relative position of the current state point (p, e) to the critical state line (CSL). Li and Dafalias (2000) and Wang et al. (2002) introduced the state variables into the hardening parameters of granular materials, thereby describing the variation of the state-dependent magnitude of plastic strain increment of granular materials. In addition, by constructing different forms of yield functions or hardening parameters, the elastoplastic models can further describe the intermediate principal stress effect, anisotropic characteristics, structure–property, etc. of soil.

For the elastoplastic models of concrete, there are two types of yield surfaces: the open-form yield surface and the closed-form yield surface. In the past, the open-form yield surface has been widely used in the elastoplastic constitutive models of concrete (Cervenka & Papanikolaou, 2008; Lu et al., 2016; Sánchez et al., 2012). Almost all of these models can reasonably characterize the plastic deformation behavior under shear loading. For the open-form yield surface, however, there is no limit to the elastic domain along the hydrostatic pressure-increasing direction. Therefore, these models are unsuitable to describe the plastic strain of concrete caused by hydrostatic loading. The closed-form yield surface has piqued increasing interest in establishing the elastoplastic constitutive models of concrete (Etse & Willam, 1994; Grassl et al., 2013; Paliwal et al., 2017). These models can reasonably reflect material hardening caused by hydrostatic loading by utilizing the closed-form yield surface.

The hardening and softening behavior of concrete during loading can be described by the hardening/softening function. The specific form of the hardening/softening function can be constructed based on the stress–strain curve under a certain loading path. For example, Wang et al. (2018) construct a hardening/softening function with equivalent plastic shear strain as the plastic internal variable based on the uniaxial compressive stress–strain curve of concrete. In the plastic damage model proposed by Lee and Fenves (1998), the two hardening/softening functions are calibrated by uniaxial tensile and uniaxial compression tests of concrete. Lu et al. (2022) developed cohesion and internal friction angle into hardening functions suitable for multiaxial stress conditions based on the Mohr–Coulomb strength criterion and conventional triaxial test data of concrete. Currently, most of the existing hardening/softening functions are piecewise expressions, namely, the hardening function and softening function expressions are different. For example, the hardening/softening functions proposed by Grassl et al. (2002), Imran and Pantazopoulou (2001), and Chi et al. (2014) are piecewise expressions. When a piecewise hardening function is used in the constitutive model, the single expression of the hardening function or softening function is relatively simple, and the hardening and softening evolutions are independent. However, selecting which of the two functions should be utilized during the loading process results in a lower computing efficiency. A unified hardening function can be used to avoid this drawback of the piecewise hardening function, increasing the computing efficiency of the constitutive model, and there are fewer parameters in the unified hardening function than those in the piecewise hardening/softening function.

Compared to the vigorous development of elastoplastic models of geomaterials in the steady environment, the plastic modeling of geomaterials in unsteady environments has been preliminarily explored, but no rigorous theoretical method has been formed (Li et al., 2023; Lu et al., 2020, 2023a). The classic Barcelona model (Alonso et al., 1990) adopts constitutive variables such as net stress, strain, and suction. The latter is used to describe the effects of the water environment on the mechanical behavior of soil, including the plastic strain of soil during moisture absorption and dehumidification processes, as well as the variation of apparent cohesive strength of soil with suction. The experimental results show that temperature variation can also induce volume change in saturated soils, even under constant stress conditions. Yao and Zhou (2013) developed an elastoplastic model for clays in a non-isothermal environment. In their model, the hardening process of the yield surface is driven by both plastic volumetric strain and temperature. In the dynamic thermal, elasto-plastic damage model of concrete by Ma et al. (2022), the yield function was taken into account as a function of plastic internal variables, stress, total strain rate, and temperature. The plastic deformation caused by changes in strain rate and temperature could be described since environmental factors were considered constitutive variables. Microbially induced calcite precipitation (MICP) can improve the mechanical behavior of soils by inducing cementation within originally loose and potentially collapsible soils. Based on sub-loading concepts, Gai and Sánchez (2019) developed an elastoplastic constitutive model for microbially mediated cemented soils, in which the mass content of the precipitated calcite was chosen as the constitutive variable in the hardening function to reflect the cementation level. In the above models, the environmental factors were introduced into yield functions or hardening parameters as the constitutive variables. The contribution of the change in environmental factors to plastic deformation is reflected from the perspective of the magnitude of plastic strain increment.

1.1.2 Direction of Plastic Strain Increment

The direction of plastic strain increment, also known as the plastic flow direction, defines the proportional relationship between the components of plastic strain increment. The plastic flow direction can be determined by the orthogonal flow rule and the non-orthogonal flow rule. The orthogonal flow rule uses the gradient of the plastic potential function as the plastic flow direction. For the convenience of constitutive modeling, the plastic potential function is sometimes chosen to be the same as the yield function, as shown in Fig. 1.1a. In this case, the plastic flow direction is orthogonal to both the plastic potential surface and the yield surface. However, the research results have shown that the plastic flow direction of geomaterials is not orthogonal to the yield surface (Abdelmoula et al., 2017; Collins, 2002). The dilatancy behavior will be overestimated when the orthogonal gradient of the yield function is used to describe the plastic deformation behavior of geomaterials (Zhao & Cai, 2010).

To this end, researchers try to construct plastic potential function independently to determine the plastic flow direction of geomaterials, as shown in Fig. 1.1b. In this case, the plastic flow direction is orthogonal to the plastic potential surface, but not orthogonal to the yield surface. The non-orthogonality between plastic flow direction and yield surface is realized by the difference between plastic potential function and yield function. The Drucker-Prager (D-P) type plastic potential function has been frequently used in the constitutive modeling of geomaterials due to its simplicity

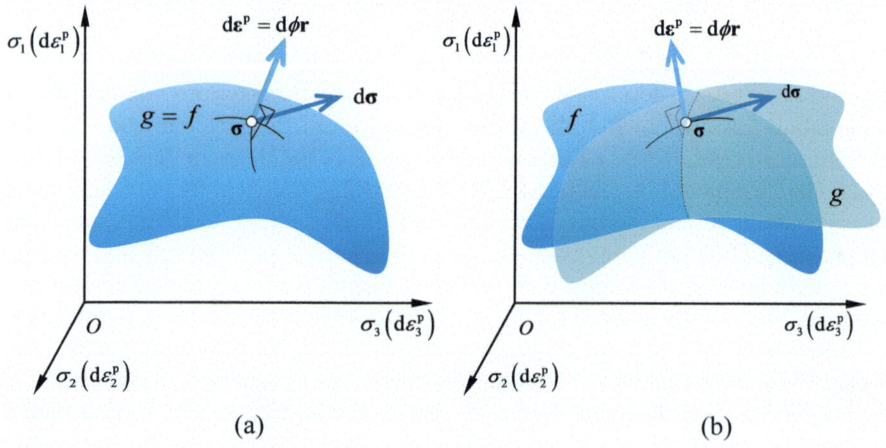

Fig. 1.1. Orthogonal flow rule: **a** $g = f$; **b** $g \neq f$

(Wu et al., 2006). The dilatancy behavior can be captured by adjusting a dilatancy parameter that defines the slope of the plastic potential curve on the meridian plane. The D-P plastic potential function is a straight line on the meridian plane, and thus it cannot describe the variations of plastic flow direction with stress states. In Yu et al.'s work (Yu et al., 2010), the dilatancy parameter of the D-P plastic potential function was developed into a function related to the pressure to describe the effect of stress state on the dilatancy behavior of concrete. Some researchers have improved the plastic damage model in ABAQUS by incorporating a hyperbolic plastic potential function (Bhartiya et al., 2021) or a polynomial one (Papanikolaou & Kappos, 2007). These modifications relate the plastic flow direction to the stress states. In the works by Grassl et al. (2013), Paliwal et al. (2017), and Zheng and Teng (2022), the plastic potential functions of the cap type were developed to describe the plastic volume shrinkage of concrete under load conditions dominated by the hydrostatic pressure. However, as the form of plastic potential function becomes more and more complex, the geotechnical plastic modeling based on the orthogonal flow rule will gradually encounter the expected bottleneck. Some additional parameters without clear physical meanings might be needed, which also will increase the difficulty of parameter calibration. A simpler and more effective plastic modeling method is eagerly desired.

In the non-orthogonal flow rule, the direction of plastic strain increment can also be directly determined by the non-orthogonal gradient of the yield function. The plastic potential function is no longer necessary for determining the plastic flow direction. Sumelka (2014) suggested that the fractional derivative can be used as a suitable mathematical work to obtain the orthogonal gradient of the yield function, which is also the fractional gradient of the yield function. Then, the fractional derivative was used by Sun and Xiao (2017) to establish the elastoplastic model of granular soil under conventional triaxial compression conditions. Furthermore, Sun et al. (2018) originally used the current stress state in the definition of the fractional derivative to model the state-dependent behavior of granular soils. However, to obtain the explicit solutions of the fractional derivative of the yield function, the yield functions in the above models have to be directly expressed in terms of the principal stresses or the stress components and established models could not reasonably describe the plastic deformation behavior of geomaterials under general stress conditions, because there is no unique definition for the chain rule for fractional derivatives. By combining fractional derivatives with covariant transformations, Lu et al. (2019a) proposed a standard method for determining the non-orthogonal gradient of the yield function in general stress space coordinates, which is not limited by the coordinate basis and form of the yield function. Based on Lu et al.'s work (Lu et al., 2019a, 2019b), the non-orthogonal flow rule is widely used to establish elastoplastic models of geomaterials. For example, Zhou et al. (2020) combined the non-orthogonal flow rule with the closed yield surface to establish a 3D non-orthogonal plastic damage model for concrete, which can reproduce the volume change law of concrete from the initial shear contraction to the following shear dilation. Liang et al. (2019, 2020) introduced the non-orthogonal flow rule into the critical state theory, and further established the 3D non-orthogonal elastoplastic models of clay and sand. Shen et al. (2022) established a 3D non-orthogonal plastic damage model for porous materials, which successfully captured the transition from volumetric shrinkage to dilatancy for clay rock using the non-orthogonal flow rule. Furthermore, researchers such as Jiang and Sun (2021), Xiao et al. (2023), and Zhou et al. (2021) have also established non-orthogonal elastoplastic models for crushed sand, rockfill, and soft soil, respectively. The applicability of the non-orthogonal flow rule for different geomaterials has been further verified.

Comparing two plastic flow rules, it can be found that the orthogonal flow rule first constructs the plastic potential function and then determines the plastic flow direction. This step of constructing the additional function is skipped in the non-orthogonal flow rule. The establishing process of the elastoplastic model will be simple, and fewer material parameters are required. Within the framework of the non-orthogonal plastic model, once the yield function and the hardening function are given, a complete plasticity model without the additional plastic potential function can be obtained. This simplicity makes non-orthogonal flow rules highly promising for the plastic modeling of geomaterials.

1.2 Stress Update Algorithms for Elastoplastic Models

The numerical calculation of nonlinear finite elements can be divided into two levels. At the local level, also known as the integration point level, the ODEs of the elastoplastic model are discretized into algebraic equations over the time domain to solve. The computational difficulties here are the non-smoothness caused by the loading/unloading inequality constraints and the nonlinearity that may lead to the divergence of the iteration. At the global level, also known as the structural level, the governing PDEs of the boundary value problem are discretized into algebraic equations over both the time and space domains to solve. When the strain softening model is used, the calculation results at the global level may encounter the problem of mesh dependence after the occurrence of strain localization. In what follows, we summarize the research work on handling the above computational difficulties in recent years.

1.2.1 Numerical Techniques for Non-smooth Problems

The non-smoothness refers to that the govern equations of elastoplastic models are constrained by the loading/unloading inequalities, also known as the *Karush–Kuhn–Tucker* (*KKT*) conditions in the numerical optimization theory. Geomaterials will exhibit vastly different mechanical behaviors during loading and unloading, such as recoverable elastic strain and unrecoverable plastic strain. In most stress update methods, the return mapping method is the most commonly used treatment for the loading/unloading inequality constraints (Simo & Hughes, 2006). In this method, the trial stress obtained by the elastic predictor is used to estimate the stress behavior (i.e., elasticity or plasticity) under the current increment step. The integration formulas that match the stress behavior are then chosen to calculate the new state variables satisfying the *KKT* conditions. The return mapping method has been incorporated into several typical stress updating algorithms, e.g., the full-implicit return-mapping algorithm (Ortiz & Simo, 1986), the cutting-plane algorithm (Simo & Ortiz, 1985; Starman et al., 2014), and the semi-implicit algorithm (Moran et al., 1990). However, this splitting inevitably leads to increased algorithm complexity since the loading/unloading estimations based on the elastic predictor have to be executed in the calculation of each step. There are also some other methods to address the constrained optimization problems in numerical optimization (Nocedal & Wright, 2006). One is to directly solve the constrained optimization problem, e.g., the projection gradient method and the Zoutendijk feasible direction method (Nocedal & Wright, 2006). The search direction of each iteration is both the descending direction of the merit function and the feasible direction of the constraint functions. For example, Zheng et al. (2020) have used the projection-contraction method for the implementation of the Mohr–Coulomb plasticity model. Few model implementations have been, however, developed along this line, probably due to its tedious constraint correction process (Arora, 2016). Another more popular idea is to use the penalty function or the smoothing function to convert inequality constraints into equality constraints, which are then added to the merit function. Then all that remains is to solve an unconstrained optimization problem. The penalty function-based optimization methods mainly include the Lagrangian method (Contrafatto & Cuomo, 2005), the multiplier method (Contrafatto & Cuomo, 2005), and the interior method (Krabbenhoft et al., 2007). The Lagrangian method and multiplier method applications to the mixed Hellinger–Reissner functional governing the elastoplasticity problem are explored by Contrafatto and Cuomo (2005). The important contributions of this study are that the *KKT* conditions are equivalently replaced by an equality constraint with the Max function and the multiplier method affects the entire equilibrium iterations. However, it is worth noting that the non-smoothness of the elastoplastic problem is not eliminated in essence but shifted to the Max function. Krabbenhoft et al. (2007) presented a detailed application of the primal–dual interior-point method in the perfect plasticity, hardening multisurface plasticity, and softening plasticity, in which the elastoplastic problem's finite element scheme is recast into a second-order cone scheme to solve. A more cost-effective and promising method to eliminate the non-smoothness of elastoplastic problems is to directly use a single smoothing function to replace two inequality constraints and one equality constraint in *KKT* conditions (Areias & Rabczuk, 2010). The updating of state variables in both the elastic and elastoplastic loading cases can be accomplished using integral equations with a unified form. Estimations for loading/unloading are also unnecessary. This method has great potential to streamline the stress update procedure (Scalet & Auricchio, 2018). Schmidt-Baldassari(2003), and Areias and Rabczuk (2010) reported the application of smooth function in crystal plasticity and finite strain plasticity, respectively. Zhou et al. (2022a) first introduced smooth function into the elastoplastic calculation of geomaterial models and developed an unconstrained stress update strategy, in which the non-smooth elastoplastic problem is transformed into an unconstrained minimization problem which can be further solved efficiently by numerical optimization algorithms. There is no doubt that such a smooth function provides a concise and valid computational framework for the implementation of the elastoplastic models.

1.2.2 Numerical Methods for Nonlinear Problems

The nonlinearity refers to that the stress–strain behavior of materials no longer follows linear laws. The nonlinear stress integral equations needed to be iteratively solved when the implicit stress integral scheme is employed, in which the Newton method has been widely used due to its asymptotic quadratic convergence speed (Potts et al., 2021). The researchers have noted, however, that the Newton method's convergence is significantly dependent on the proximity between the initial value and the final solution (Brannon & Leelavanichkul, 2010). The optimal convergence property may be lost when the iteration point exceeds the convergence radius of the Newton method or the Taylor series used in the Newton method is difficult to approximate the original problem well in the vicinity of the solution due to the presence of strong nonlinearity (Contrafatto & Cuomo, 2005). Various efforts have been made to close the gap, including the proposal of some corrective measures, e.g., optimizing the initial iteration point (Hernández et al., 2011), sub-stepping schemes (Wang et al., 2006), and multi-stage iteration (Homel & Brannon, 2015). From the perspective of mathematical programming, the nonlinear equation problem is equivalent to an unconstrained minimization problem (Nocedal & Wright, 2006), which can be solved effectively by some mature optimization methods, e.g., the line search method (LSM) (Scherzinger, 2017) and the trust region method (Lester & Scherzinger, 2017; Shterenlikht & Alexander, 2012). In some sense, the search strategies for the two methods are dual to each other. The former optimizes the step size with the given search direction, while the latter first chooses the step size (i.e., trust region radius) and then determines the search direction. Nocedal and Wright (2006) pointed out that trust region algorithms may have certain theoretical advances in solving nonlinear equations problems compared with the LSM in their monograph on numerical optimization methods. From the perspective of algorithm complexity, the LSM may be simpler and it only needs to determine an optimal step size additionally under a given search direction to obtain a larger convergence radius. The optimal step size can be obtained through a simple iteration formula that does not require the calculation of the Jacobian matrix (Scherzinger, 2017). Based on the computational framework of the return-mapping stress update strategy, the performance of the two numerical optimization methods has been thoroughly tested in the stress update problems of elastoplastic models of metal materials, e.g., the associated and non-associated plasticity (Scherzinger, 2017), the isotropic and anisotropic plasticity (Lester & Scherzinger, 2017; Scherzinger, 2017), and so on. Compared to metal materials, geomaterials have stronger nonlinear characteristics, e.g., strain hardening/softening, volume expansion/contraction, pressure dependency, etc. By combining the unconstrained stress update strategy with the two methods, respectively, Zhou et al. (2022a, 2022b) proposed two stress update algorithms with strong convergence to solve the classical MCC model. The numerical results show that the load step allowed by the trust region method and LSM is much larger than that of the standard Newton method. By using the LSM to solve the nonlinear stress integral equations in the effective stress space, Su et al. developed an implicit stress update algorithm for the plastic non-local damage model of concrete. More valuable studies in this field can be found in the literature (Lu et al., 2023b; Shterenlikht & Alexander, 2012).

1.2.3 Regularization Methods for Mesh Dependence Problem

The mesh dependence generally occurs in the structure failure analysis based on the strain softening models. With the refinement of the mesh, the shear band formed by strain localization will become narrower and the descending branch of load–displacement curves will become steeper. The underlying mathematical reason is that the loss of the ellipticity of the governing partial differential equations (PDEs) leads to the ill-posedness of the boundary-value problem. The physical explanation is that the constitutive model lacks the characteristic length to reflect the microscopic effect of materials. Various regularization methods have been explored to preserve the well-posedness of the governing PDEs, including the integral nonlocal method (Huang et al., 2018), strain-gradient or damage-gradient methods (Marigo et al., 2016; Sarkar et al., 2022), and Cosserat method (Adhikary & Dyskin, 1997). The Cosserat method can ensure that simulation results remain independent of the mesh topology by introducing rotational degrees of freedom and couple stress related to the material internal length, but the additional degrees of freedom also significantly increase the computational cost. The gradient methods require the introduction of strain gradient (Lam et al., 2003), damage gradient (Marigo et al., 2016), or plastic internal variable gradient (Jirásek & Rolshoven, 2009), also leading to the more complex and expensive model calculations. Comparatively, the integral nonlocal method may be more attractive for users of commercial finite element software, because it does not involve the modification to governing PDEs, resulting in a lower computing cost. In particular, the concept of over-nonlocal integration that was suggested by Strömberg and Ristinmaa (1996) was validated to be able to better maintain the well-posedness of the governing equations than the purely nonlocal formulation.

1.3 Main Research Work

This book presents systematically the authors' research on plasticity modeling for geomaterials in unsteady environments and the numerical algorithms for the elastoplastic model in recent years. In the first half of the book, the authors first introduced a rigorous multi-physical & non-orthogonal plastic modeling method, including the non-orthogonal flow rule without plastic potential function, the multi-physical loading/unloading criterion in unsteady environments, and the incremental stress–strain-environment constitutive relationship. This modeling method provides a rigorous theoretical framework for describing the changes in the state of material properties in unsteady environments. Then, a series of non-orthogonal elastoplastic models for soils and concrete were developed in steady and unsteady environments. The performance of these models is fully verified by a large number of experimental data, which are from the literature and the authors. In the second half of this book, by using some advanced numerical methods, such as numerical optimization methods, high-precision numerical differentiation methods, regularization methods, etc., the authors developed some efficient and robust stress updating algorithms to deal with the common computational difficulties in the numerical implementation of elasto-plastic models. These algorithms demonstrate good performance in the numerical implementation of elastoplastic models for geotechnical materials, including orthogonal and non-orthogonal ones. This book is organized into ten chapters and the main contents are introduced briefly as follows:

Part I: Multi-physical and non-orthogonal geotechnical plasticity modeling in unsteady environments

In Chap. 2, the authors elaborated on the multi-physical & non-orthogonal plastic modeling method for geomaterials in detail in unsteady environments. In the presented modeling method, the direction of the plastic strain increment is directly determined by the non-orthogonal gradient of the yield function, without the need for a plastic potential function. When executing consistency conditions, the environmental factors that affect the mechanical properties of the material are considered as the constitutive variable to determine the magnitude of plastic strain increment. The incremental stress–strain relationship in steady environments is developed into the incremental stress–strain-environment relationship in unsteady-state environments. Based on the proposed plastic modeling method, we further develop non-orthogonal elastoplastic models for geomaterials in steady-state and unsteady-state environments, including those for soils and concrete. In Chap. 3, we propose a characteristic stress method to describe the 3D failure and yield characteristics of geomaterials. Based on the characteristic stress method and non-orthogonal flow rule, the non-orthogonal elastoplastic models for clay and sand are established in sequence. In Chap. 4, saturation is considered as the constitutive variable to describe the influence of water environment changes on the deformation and strength behavior of soil. Chapter 5 establishes a non-orthogonal elastoplastic damage model for concrete in the steady environment. The model consists of the damage part driven by the plastic strain and the plastic part based on the effective stress, in which the plastic evolution is independent of the damage evolution. In Chap. 6, the strain rate is considered as the environmental factor and the constitutive variable. We propose a dynamic elastoplastic model to describe the stress–strain-strain rate behavior of concrete under dynamic loads. Particularly, constant- and variable-strain rate loading tests are performed to evaluate the performance of the established model.

Part II: Numerical Algorithms for Elastoplastic Models

From Chaps. 7 to 10, we focus on the computational difficulties in the numerical implementation of elastoplastic models, including the non-smoothness, the nonlinearity, and the mesh dependence. Chapter 7 develops an unconstrained stress update strategy, in which the *KKT* conditions in the elastoplastic constitutive equations are equivalently replaced by the *Fischer–Burmeister* smooth functions. Then, the non-smooth elastoplastic problem can be transformed into a smooth nonlinear equation system to solve, avoiding loading/unloading judgments caused by non-smoothness. In Chap. 8, we have developed two unconstrained stress update algorithms with strong convergence. The unconstrained minimization problem equivalent to nonlinear stress integral equations is solved through the non-monotonic trust region method and inexact LSM. Both two algorithms allow for iterative calculations under a larger load step size by improving the direction and step size of the search step, respectively. To avoid tedious analytical derivative operations, Chap. 9 developed an implicit stress update algorithm based on the numerical differentiation method. The high-precision hyperdual step derivative approximation is used to calculate the derivatives of nonlinear stress integral equations and consistent tangent matrix, ensuring the quadratic convergence speed of global and local iterations. Chapter 10 is dedicated to addressing the mesh dependence issue of strain softening models in structural failure analysis. We propose a non-local implicit stress update algorithm for the plastic damage model. In plastic calculation, the nonlinear implicit stress integral equations in the effective stress space are solved by the LSM. In damage calculations, the damage variable is developed into a nonlocal variable to guarantee the ellipticity of the PDEs after the occurrence of strain localization. Numerical examples under different mesh sizes show that the proposed algorithm can obtain computational results with mesh objectivity.

References

Abdelmoula, N., Harthong, B., Imbault, D., & Dorémus, P. (2017). A study on the uniqueness of the plastic flow direction for granular assemblies of ductile particles using discrete finite-element simulations. *Journal of the Mechanics and Physics of Solids, 109*, 142–159.

Adhikary, D. P., & Dyskin, A. V. (1997). A cosserat continuum model for layered materials. *Computers and Geotechnics, 20*(1), 15–45.

Alonso, E. E., Gens, A., & Josa, A. (1990). A constitutive model for partially saturated soils. *Géotechnique, 40*(3), 405–430.

Areias, P., & Rabczuk, T. (2010). Smooth finite strain plasticity with non-local pressure support. *International Journal for Numerical Methods in Engineering, 81*(1), 106–134.

Arora, J. S. (2016). *Introduction to optimum design*. Elsevier.

Asaoka, A., Nakano, M., & Noda, T. (1997). Soil-water coupled behavior of heavily overconsolidated clay near/at critical state. *Soils and Foundations, 37*(1), 13–28.

Bhartiya, R., Sahoo, D. R., & Verma, A. (2021). Modified damaged plasticity and variable confinement modelling of rectangular CFT columns. *Journal of Constructional Steel Research, 176*, 106426.

Brannon, R. M., & Leelavanichkul, S. (2010). A multi-stage return algorithm for solving the classical damage component of constitutive models for rocks, ceramics, and other rock-like media. *International Journal of Fracture, 163*, 133–149.
Cervenka, J., & Papanikolaou, V. K. (2008). Three dimensional combined fracture–plastic material model for concrete. *International Journal of Plasticity, 24*(12), 2192–2220.
Chi, Y., Xu, L. H., & Yu, H. S. (2014). Constitutive modeling of steel-polypropylene hybrid fiber reinforced concrete using a non-associated plasticity and its numerical implementation. *Composite Structures, 111*, 497–509.
Collins, I. F. (2002). Associated and non-associated aspects of the constitutive laws for coupled elastic/plastic materials. *International Journal Geomechanics, 2*(2), 259–267.
Contrafatto, L., & Cuomo, M. (2005). A globally convergent numerical algorithm for damaging elasto-plasticity based on the multiplier method. *International Journal for Numerical Methods in Engineering, 63*(8), 1089–1125.
Dafalias, Y. F. (1980). A bounding surface soil plasticity model. In *International symposium on soils under cyclic trans, Swansea, Balkema* (pp. 335–345).
Drucker, D. C., Gibson, R. E., & Henkel, D. J. (1957). Soil mechanics and work-hardening theories of plasticity. *Transactions of the American Society of Civil Engineers, 122*(1), 338–346.
Drunker, D. C. (1951) A more fundamental approach to plastic stress-strain relations. In *Proceedings of the first U. S. National congress of applied mechanics*. The American Society of Mechanical Engineers, Chicago (pp. 487–491).
Etse, G., & Willam, K. (1994). Fracture energy formulation for inelastic behavior of plain concrete. *Journal of Engineering Mechanics, 120*(9), 1983–2011.
Fang, H. Y., & Daniels, J. L. (2017). *Introductory geotechnical engineering: An environmental perspective*. CRC Press.
Gai, X. R., & Sánchez, M. (2019). An elastoplastic mechanical constitutive model for microbially mediated cemented soils. *Acta Geotechnica, 14*, 709–726.
Grassl, P., Lundgren, K., & Gylltoft, K. (2002). Concrete in compression: A plasticity theory with a novel hardening law. *International Journal of Solids and Structures, 39*(20), 5205–5223.
Grassl, P., Xenos, D., Nyström, U., Rempling, R., & Gylltoft, K. (2013). CDPM2: A damage-plasticity approach to modelling the failure of concrete. *International Journal of Solids and Structures, 50*(24), 3805–3816.
Hashiguchi, K. (1989). Subloading surface model in unconventional plasticity. *International Journal of Solids and Structures, 25*(8), 917–945.
Hernández, J. A., Oliver, J., Cante, J. C., & Weyler, R. (2011). A robust approach to model densification and crack formation in powder compaction processes. *International Journal for Numerical Methods in Engineering, 87*(8), 735–767.
Homel, M. A., & Brannon, R. M. (2015). Relaxing the multi-stage nested return algorithm for curved yield surfaces and nonlinear hardening laws. *International Journal of Fracture, 194*(1), 51–57.
Huang, M. S., Qu, X., & Lü, X. L. (2018). Regularized finite element modeling of progressive failure in soils within nonlocal softening plasticity. *Computational Mechanics, 62*(3), 347–358.
Il'Iushin, A. A. (1961). On the postulate of plasticity. *Journal of Applied Mathematics and Mechanics, 25*(3), 746–752.
Imran, I., & Pantazopoulou, S. J. (2001). Plasticity model for concrete under triaxial compression. *Journal of Engineering Mechanics, 127*(3), 281–290.
Jiang, J. S., & Sun, Y. F. (2021). Experimental evaluation and stress–fractional modeling of the state-dependent behavior of rockfill. *International Journal of Geomechanics, 21*(7), 4021109.
Jirásek, M., & Rolshoven, S. (2009). Localization properties of strain-softening gradient plasticity models. Part II: Theories with gradients of internal variables. *International Journal of Solids and Structures, 46*(11–12), 2239–2254.
Krabbenhoft, K., Lyamin, A. V., Sloan, S. W., & Wriggers, P. (2007). An interior-point algorithm for elastoplasticity. *International Journal for Numerical Methods in Engineering, 69*(3), 592–626.

Lam, D. C., Yang, F., Chong, A., Wang, J., & Tong, P. (2003). Experiments and theory in strain gradient elasticity. *Journal of the Mechanics and Physics of Solids, 51*(8), 1477–1508.

Lee, J., & Fenves, G. L. (1998). Plastic-damage model for cyclic loading of concrete structures. *Journal of Engineering Mechanics, 124*(8), 892–900.

Lester, B. T., & Scherzinger, W. M. (2017). Trust-region based return mapping algorithm for implicit integration of elastic-plastic constitutive models. *International Journal for Numerical Methods in Engineering, 12*(3), 257–282.

Li, X. Q., Lu, D. C., Lin, Q. T., Dong, X. Q., & Du, X. L. (2023). A hydro-mechanical constitutive model for unsaturated soils over a wide saturation range. *Computers and Geotechnics, 159*, 105475.

Li, X. S., & Dafalias, Y. F. (2000). Dilatancy for cohesionless soils. *Géotechnique, 50*(4), 449–460.

Liang, J. Y., Lu, D. C., Du, X. L., Wu, W., & Ma, C. (2020). Non-orthogonal elastoplastic constitutive model for sand with dilatancy. *Computers and Geotechnics, 118*, 103329.

Liang, J. Y., Lu, D. C., Zhou, X., Du, X. L., & Wu, W. (2019). Non-orthogonal elastoplastic constitutive model with the critical state for clay. *Computers and Geotechnics, 116*, 103200.

Lu, D. C., Du, X. L., Wang, G. S., Zhou, A. N., & Li, A. K. (2016). A three-dimensional elastoplastic constitutive model for concrete. *Computers & Structures, 163*, 41–55.

Lu, D. C., Liang, J. Y., Du, X. L., Ma, C., & Gao, Z. W. (2019a). Fractional elastoplastic constitutive model for soils based on a novel 3d fractional plastic flow rule. *Computers and Geotechnics, 105*, 277–290.

Lu, D. C., Meng, F. P., Zhou, X., Zhuo, Y. H., Gao, Z. W., & Du, X. L. (2023a). A dynamic elastoplastic model of concrete based on a modeling method with environmental factors as constitutive variables. *Journal of Engineering Mechanics, 149*(12), 4023102.

Lu, D. C., Su, C. C., Zhou, X., Wang, G. S., & Du, X. L. (2022). A cohesion-friction combined hardening plastic model of concrete with the nonorthogonal flow rule: Theory and numerical implementation. *Construction and Building Materials, 325*, 126586.

Lu, D. C., Zhang, Y. N., Zhou, X., Su, C. C., Gao, Z. W., & Du, X. L. (2023b). A robust stress update algorithm for elastoplastic models without analytical derivation of the consistent tangent operator and loading/unloading estimation. *International Journal for Numerical and Analytical Methods in Geomechanics, 47*(6), 1022–1050.

Lu, D. C., Zhou, X., Du, X. L., & Wang, G. S. (2019b). A 3D fractional elastoplastic constitutive model for concrete material. *International Journal of Solids and Structures, 165*, 160–175.

Lu, D. C., Zhou, X., Du, X. L., & Wang, G. S. (2020). 3D dynamic elastoplastic constitutive model of concrete within the framework of rate-dependent consistency condition. *Journal of Engineering Mechanics, 146*(11), 4020124.

Ma, J. J., Chen, J. J., Chen, W. X., & Huang, L. C. (2022). A coupled thermal-elastic-plastic-damage model for concrete subjected to dynamic loading. *International Journal of Plasticity, 153*, 103279.

Marigo, J., Maurini, C., & Pham, K. (2016). An overview of the modelling of fracture by gradient damage models. *Meccanica, 51*, 3107–3128.

Mises, R. V. (1928). Mechanik der plastischen formänderung von kristallen. *Zamm-Journal of Applied Mathematics and Mechanics/zeitschrift Für Angewandte Mathematik und Mechanik, 8*(3), 161–185.

Moran, B., Ortiz, M., & Shih, C. F. (1990). Formulation of implicit finite element methods for multiplicative finite deformation plasticity. *International Journal for Numerical Methods in Engineering, 29*(3), 483–514.

Nakano, M., Nakai, K., Noda, T., & Asaoka, A. (2005). Simulation of shear and one-dimensional compression behavior of naturally deposited clays by super or subloading yield surface cam-clay model. *Soils and Foundations, 45*(1), 141–151.

Nocedal, J., & Wright, S. (2006). *Numerical optimization.* Springer Science & Business Media.

Ortiz, M., & Simo, J. C. (1986). An analysis of a new class of integration algorithms for elastoplastic constitutive relations. *International Journal for Numerical Methods in Engineering, 23*(3), 353–366.

Paliwal, B., Hammi, Y., Moser, R. D., & Horstemeyer, M. F. (2017). A three-invariant cap-plasticity damage model for cementitious materials. *International Journal of Solids and Structures, 108*, 186–202.

Papanikolaou, V. K., & Kappos, A. J. (2007). Confinement-sensitive plasticity constitutive model for concrete in triaxial compression. *International Journal of Solids and Structures, 44*(21), 7021–7048.

Potts, D. M., Cui, W. J., & Zdravković, L. (2021). A coupled thm finite element formulation for unsaturated soils and a strategy for its nonlinear solution. *Computers and Geotechnics, 136*, 104221.

Roscoe, K. H., & Burland, J. B. (1968). *On the generalized stress-strain behaviour of wet clay: Engineering plasticity* (pp. 535–609). Cambridge University Press.

Roscoe, K. H., Schofield, A., & Thurairajah, A. (1963). Yielding of clays in states wetter than critical. *Geotechnique, 13*(3), 211–240.

Roscoe, K. H., Schofield, A., & Wroth, A. P. (1958). On the yielding of soils. *Geotechnique, 8*(1), 22–53.

Sánchez, P. J., Huespe, A. E., Oliver, J., Diaz, G., & Sonzogni, V. E. (2012). A macroscopic damage-plastic constitutive law for modeling quasi-brittle fracture and ductile behavior of concrete. *International Journal for Numerical & Analytical Methods in Geomechanics, 36*(5), 546–573.

Sarkar, S., Singh, I. V., & Mishra, B. K. (2022). A localizing gradient plasticity model for ductile fracture. *Computer Methods in Applied Mechanics and Engineering, 388*, 114205.

Scalet, G., & Auricchio, F. (2018). Computational methods for elastoplasticity: An overview of conventional and less-conventional approaches. *Archives of Computational Methods in Engineering, 25*(3), 545–589.

Scherzinger, W. M. (2017). A return mapping algorithm for isotropic and anisotropic plasticity models using a line search method. *Computer Methods in Applied Mechanics and Engineering, 317*, 526–553.

Schmidt-Baldassari, M. (2003). Numerical concepts for rate-independent single crystal plasticity. *Computer Methods in Applied Mechanics and Engineering, 192*, 1261–1280.

Shen, W. Q., Liu, S. Y., Xu, W. Y., & Shao, J. F. (2022). An elastoplastic damage constitutive model for rock-like materials with a fractional plastic flow rule. *International Journal of Rock Mechanics and Mining Sciences, 156*, 105140.

Shterenlikht, A., & Alexander, N. A. (2012). Levenberg-Marquardt vs Powell's dogleg method for Gurson-Tvergaard-Needleman plasticity model. *Computer Methods in Applied Mechanics and Engineering, 237–240*, 1–9.

Simo, J. C., & Hughes, T. J. (2006). *Computational inelasticity*. Springer Science & Business Media, New York.

Simo, J. C., & Ortiz, M. (1985). A unified approach to finite deformation elastoplastic analysis based on the use of hyperelastic constitutive equations. *Computer Methods in Applied Mechanics and Engineering, 49*(2), 221–245.

Starman, B., Halilovič, M., Vrh, M., & Atok, B. (2014). Consistent tangent operator for cutting-plane algorithm of elasto-plasticity. *Computer Methods in Applied Mechanics and Engineering, 272*, 214–232.

Strömberg, L., & Ristinmaa, M. (1996). Fe-formulation of a nonlocal plasticity theory. *Computer Methods in Applied Mechanics and Engineering, 136*(1–2), 127–144.

Sumelka, W. (2014). A note on non-associated Drucker-Prager plastic flow in terms of fractional calculus. *Journal of Theoretical and Applied Mechanics, 52*(2), 571–574.

Sun, Y. F., Gao, Y. F., & Zhu, Q. Z. (2018). Fractional order plasticity modelling of state-dependent behaviour of granular soils without using plastic potential. *International Journal of Plasticity, 102*, 53–69.

Sun, Y. F., & Xiao, Y. (2017). Fractional order plasticity model for granular soils subjected to monotonic triaxial compression. *International Journal of Solids and Structures, 118–119*, 224–234.

Wang, G. S., Lu, D. C., Du, X. L., Zhou, X., & Cao, S. T. (2018). A true 3D frictional hardening elastoplastic constitutive model of concrete based on a united hardening/softening function. *Journal of the Mechanics and Physics of Solids, 119*, 250–273.

Wang, W., Datcheva, M., Schanz, T., & Kolditz, O. (2006). A sub-stepping approach for elastoplasticity with rotational hardening. *Computational Mechanics, 37*(3), 266–278.

Wang, Z. L., Dafalias, Y. F., Li, X. S., & Makdisi, F. I. (2002). State pressure index for modeling sand behavior. *Journal of Geotechnical and Geoenvironmental Engineering, 128*(6), 511–519.

Wu, J. Y., Li, J., & Faria, R. (2006). An energy release rate-based plastic-damage model for concrete. *International Journal of Solids and Structures, 43*(3–4), 583–612.

Xiao, Y., Zhang, X. M., Wang, C. G., Cui, H., & Liu, H. L. (2023). Breakage-dependent fractional plasticity model for sands. *International Journal of Geomechanics, 23*(3), 4022299.

Yao, Y. P., Hou, W., & Zhou, A. N. (2009). Uh model: Three-dimensional unified hardening model for overconsolidated clays. *Geotechnique, 59*(5), 451–469.

Yao, Y. P., Liu, L., Luo, T., Tian, Y., & Zhang, J. M. (2019). Unified hardening (UH) model for clays and sands. *Computers and Geotechnics, 110*, 326–343.

Yao, Y. P., Sun, D. A., & Matsuoka, H. (2008). A unified constitutive model for both clay and sand with hardening parameter independent on stress path. *Computers and Geotechnics, 35*(2), 210–222.

Yao, Y. P., & Zhou, A. N. (2013). Non-isothermal unified hardening model: A thermo-elasto-plastic model for clays. *Géotechnique, 63*(15), 1328–1345.

Yoon, S. Y., Lee, S. Y., & Barlat, F. (2020). Numerical integration algorithm of updated homogeneous anisotropic hardening model through finite element framework. *Computer Methods in Applied Mechanics and Engineering, 372*, 113449.

Yu, T., Teng, J. G., Wong, Y. L., & Dong, S. L. (2010). Finite element modeling of confined concrete-I: Drucker–prager type plasticity model. *Engineering Structures, 32*(3), 665–679.

Zhao, X. G., & Cai, M. (2010). Influence of plastic shear strain and confinement-dependent rock dilation on rock failure and displacement near an excavation boundary. *International Journal of Rock Mechanics and Mining Sciences, 47*(5), 723–738.

Zheng, B. T., & Teng, J. G. (2022). A plasticity constitutive model for concrete under multiaxial compression. *Engineering Structures, 251*, 113435.

Zheng, H., Zhang, T., & Wang, Q. S. (2020). The mixed complementarity problem arising from non-associative plasticity with non-smooth yield surfaces. *Computer Methods in Applied Mechanics and Engineering, 361*, 112756.

Zhou, F. X., Wang, L. Y., & Liu, H. B. (2021). A fractional elasto-viscoplastic model for describing creep behavior of soft soil. *Acta Geotechnica, 16*, 67–76.

Zhou, X., Lu, D. C., Du, X. L., Wang, G. S., & Meng, F. P. (2020). A 3D non-orthogonal plastic damage model for concrete. *Computer Methods in Applied Mechanics and Engineering, 360*, 112716.

Zhou, X., Lu, D. C., Su, C. C., Gao, Z. W., & Du, X. L. (2022a). An unconstrained stress updating algorithm with the line search method for elastoplastic soil models. *Computers and Geotechnics, 143*, 104592.

Zhou, X., Lu, D. C., Zhang, Y. N., Du, X. L., & Rabczuk, T. (2022b). An open-source unconstrained stress updating algorithm for the modified cam-clay model. *Computer Methods in Applied Mechanics and Engineering, 390*, 114356.

Chapter 2
Multi-physical and Non-orthogonal Geotechnical Plasticity Theory

The plastic strain of the material is irreversible and nonlinear. The material produces new plastic strain when loading. The total plastic strain is the result superimposed by plastic strain increments and depends on the loading history of the material. It is difficult to directly establish the relationship between total plastic strain and stress state. Therefore, the stress–strain relationship in the plastic deformation stage is generally expressed in the form of the increment:

$$d\boldsymbol{\sigma} = \mathbb{D} : (d\boldsymbol{\varepsilon} - d\boldsymbol{\varepsilon}^{\mathrm{p}}) \tag{2.1}$$

where $d\boldsymbol{\sigma}$ is the stress increment. $d\boldsymbol{\varepsilon}$ and $d\boldsymbol{\varepsilon}^{\mathrm{p}}$ represent the total strain increment and irreversible plastic strain increment. The ":" denotes the double dot product of tensor operation. $\mathbb{D}$ is the elastic stiffness matrix.

$$\mathbb{D} = \frac{E}{(1+\nu)(1-2\nu)}\begin{bmatrix} 1-\nu & \nu & \nu & 0 & 0 & 0 \\ \nu & 1-\nu & \nu & 0 & 0 & 0 \\ \nu & \nu & 1-\nu & 0 & 0 & 0 \\ 0 & 0 & 0 & \frac{1-2\nu}{2} & 0 & 0 \\ 0 & 0 & 0 & 0 & \frac{1-2\nu}{2} & 0 \\ 0 & 0 & 0 & 0 & 0 & \frac{1-2\nu}{2} \end{bmatrix} \tag{2.2}$$

where E and ν are Young's modulus and Poisson's ratio, respectively. The plastic strain increment $d\varepsilon^{\mathrm{p}}$ can be expressed as the product of magnitude and direction.

$$d\varepsilon^{\mathrm{p}} = d\phi\mathbf{r} \tag{2.3}$$

where $d\phi$ is a nonnegative plastic multiplier, representing the magnitude of plastic strain increment. $\mathbf{r}$ denotes the direction of plastic strain increment, which represents the proportional relationship among the components of plastic strain increment. In the calculation of elastoplasticity models, it is typically to calculate the stress increment

D. Lu et al., *Multi-physical and Non-orthogonal Geotechnical Plasticity Modeling and Algorithms*, https://doi.org/10.1007/978-981-96-2369-3_2

based on the given strain increment, or vice versa. It can be seen from Eqs. (2.1) and (2.3) that the task here is to determine the magnitude and direction of plastic strain increment.

In the geotechnical plasticity theory, there are two limitations for current modeling methods in determining the plastic strain increment. One is that the environmental factors affecting the mechanical properties of materials are not considered as constitutive variables. The second is that an additional plastic potential function is required to determine the direction of plastic strain increment. The two limitations restrict the development of the plasticity model for geomaterials in unsteady environments. In this chapter, the authors propose a multi-physical and non-orthogonal plasticity modeling method to establish the incremental stress–strain-environment relationship in unsteady environments. When determining the magnitude of plastic strain increment, the environmental factors are used as the constitutive variables to describe the plastic strain behavior in unsteady environments. When determining the direction of plastic strain increment, a non-orthogonal flow rule without plastic potential function is proposed to obtain the non-orthogonal gradient of yield function. In addition, based on the geometric relationship between yield surface and generalized load increment, which is composed of mechanical load increment and environmental load increment, a loading/unloading criterion in unsteady environments is established in the strain-environment space, and it is applicable to both hardening and softening stages. It is worth noting that environmental factors not only affect the plastic deformation properties of materials but also their elastic deformation properties. In this book, our research interest mainly focuses on the former.

2.1 Magnitude of Plastic Strain Increment

The plastic multiplier $d\phi$ can be obtained by the consistency condition together with the yield function and the hardening function. The yield function describes the material state when a plastic flow occurs, which distinguishes the boundary between elastic and plastic deformation. In a broad sense, the yield function is the function of the stress and environmental variables (e.g., temperature, saturation, time, strain rate, ionic concentration, etc.) that can affect the plastic deformation process of materials.

$$f\left[\boldsymbol{\sigma}, H(\boldsymbol{\varepsilon}^{\mathrm{p}}), \psi\right] = 0 \tag{2.4}$$

where H indicates the hardening function that is generally a function of plastic strain, recording the loading history of the material. ψ denotes one or more environmental factors and is also constitutive variables similar to stress or strain. In unsteady environments, some properties of the material will change with the environmental variables. For example, the concrete strengths enlarge with increasing strain rate (Kong et al., 2018; Shkolnik, 2008; Zeng et al., 2023) but decrease with rising temperature (Chang et al., 2006; Chen et al., 2015; He et al., 2021), where strain rate and temperature are two typical environmental variables. In plasticity modeling, some model parameters characterizing the material behavior are no longer constants but should be functions of environmental variables. Therefore, the environmental variables can be introduced into the yield functions from the perspective of material parameters:

$$f\left[\boldsymbol{\sigma}, H(\boldsymbol{\varepsilon}^{\mathrm{p}}), \mathbf{M}(\psi)\right] = 0 \tag{2.5}$$

where the vector **M** represents the material parameters that depend on the environmental variables. Equation (2.5) shows that environmental variables can be introduced into plasticity models by using material parameters as a medium. When deriving the constitutive relationship, we may still use Eq. (2.4), which will not affect the final result.

The consistency condition ensures the consistency of the stress state, the hardening function, and the environmental variable. As shown in Fig. 2.1, during plastic deformation, the yield surface follows the current state point, which is composed of stress and environment variable. In the plastic calculation, the consistency condition can be used to close the set of constitutive equations to further determine the plastic multiplier $\mathrm{d}\phi$. The consistency condition can be expressed as (Lu et al., 2020, 2023):

$$\mathrm{d}f = \frac{\partial f}{\partial \boldsymbol{\sigma}} : \mathrm{d}\boldsymbol{\sigma} + \frac{\partial f}{\partial H}\frac{\partial H}{\partial \boldsymbol{\varepsilon}^{\mathrm{p}}} : \mathrm{d}\boldsymbol{\varepsilon}^{\mathrm{p}} + \frac{\partial f}{\partial \mathbf{M}}\frac{\partial \mathbf{M}}{\partial \psi}\mathrm{d}\psi = 0 \tag{2.6}$$

where $(\partial \mathbf{M}/\partial \psi)\mathrm{d}\psi$ represents the change of material property caused by the environmental change. For convenience, Eq. (2.6) can also be written as:

$$\mathrm{d}f = \frac{\partial f}{\partial \boldsymbol{\sigma}} : \mathrm{d}\boldsymbol{\sigma} + \frac{\partial f}{\partial H}\frac{\partial H}{\partial \boldsymbol{\varepsilon}^{\mathrm{p}}} : \mathrm{d}\boldsymbol{\varepsilon}^{\mathrm{p}} + \frac{\partial f}{\partial \psi}\mathrm{d}\psi = 0 \tag{2.7}$$

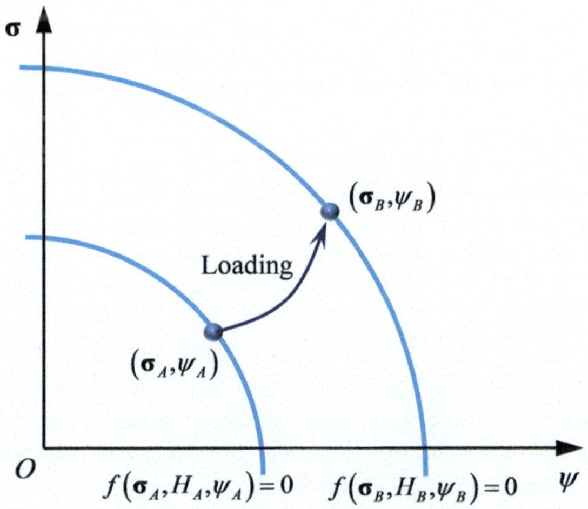

Fig. 2.1 Consistency condition

In Eq. (2.7), the status of environmental variables is equivalent to stress and plastic strain and is considered as the constitutive variables. Equation (2.7) is also known as the unsteady consistency condition. By submitting Eq. (2.3) into Eq. (2.7), we can obtain:

$$df = \frac{\partial f}{\partial \boldsymbol{\sigma}} : d\boldsymbol{\sigma} + d\phi \frac{\partial f}{\partial H} \frac{\partial H}{\partial \boldsymbol{\varepsilon}^{p}} : \mathbf{r} + \frac{\partial f}{\partial \psi} d\psi = 0 \tag{2.8}$$

Then, the expression of $d\phi$ under the given stress increment $d\boldsymbol{\sigma}$ and environmental variable increment $d\psi$ can be obtained as follows:

$$d\phi = \frac{1}{H_{p}^{\sigma}} \left(\frac{\partial f}{\partial \boldsymbol{\sigma}} : d\boldsymbol{\sigma} + \frac{\partial f}{\partial \psi} d\psi \right) \tag{2.9}$$

where H_{p}^{σ} is the plastic modulus. $H_{p}^{\sigma} > 0$ for the hardening stage and $H_{p}^{\sigma} < 0$ for the softening stage.

$$H_{p}^{\sigma} = -\frac{\partial f}{\partial H} \frac{\partial H}{\partial \boldsymbol{\varepsilon}^{p}} : \mathbf{r} \tag{2.10}$$

Similarly, under the condition of the given strain increment $d\boldsymbol{\varepsilon}$ and environmental variable increment $d\psi$, the expression of $d\phi$ can be derived by submitting Eqs. (2.1) and (2.3) into Eq. (2.7).

$$df = \frac{\partial f}{\partial \boldsymbol{\sigma}} : \mathbb{D} : (d\boldsymbol{\varepsilon} - d\phi \mathbf{r}) + d\phi \frac{\partial f}{\partial H} \frac{\partial H}{\partial \boldsymbol{\varepsilon}^{p}} : \mathbf{r} + \frac{\partial f}{\partial \psi} d\psi = 0 \tag{2.11}$$

After simple operations, we can obtain:

$$d\phi = \frac{1}{H_{p}^{\varepsilon}} \left(\frac{\partial f}{\partial \boldsymbol{\sigma}} : \mathbb{D} : d\boldsymbol{\varepsilon} + \frac{\partial f}{\partial \psi} d\psi \right) \tag{2.12}$$

where H_{p}^{ε} can be understood as the plastic modulus in the strain space. $H_{p}^{\varepsilon} > 0$ for both the hardening stage and the softening.

$$H_{p}^{\varepsilon} = \frac{\partial f}{\partial \boldsymbol{\sigma}} : \mathbb{D} : \mathbf{r} - \frac{\partial f}{\partial H} \frac{\partial H}{\partial \boldsymbol{\varepsilon}^{p}} : \mathbf{r} \tag{2.13}$$

Equations (2.9) and (2.12) denote that the plastic strain of the material is not only caused by changes in stress or strain but may also be caused by the change in the environmental variables. On the other hand, these two equations also imply that the loading/unloading criterion in unsteady environments needs to consider the influence of environmental variables.

In steady environments, the environmental variables are constant and always maintain their initial state ψ_0. The material parameters in the plasticity model remain unchanged during loading. The yield function defined by Eq. (2.4) will degenerate into

$$f\left[\boldsymbol{\sigma}, H(\boldsymbol{\varepsilon}^{\mathrm{p}}), \psi_0\right] = 0 \tag{2.14}$$

The consistency condition in steady environments can be obtained by performing total differentiation on Eq. (2.14) or setting $\mathrm{d}\psi = 0$ in Eq. (2.11)

$$\mathrm{d}f = \frac{\partial f}{\partial \boldsymbol{\sigma}} : \mathrm{d}\boldsymbol{\sigma} + \frac{\partial f}{\partial H}\frac{\partial H}{\partial \boldsymbol{\varepsilon}^{\mathrm{p}}} : \mathrm{d}\boldsymbol{\varepsilon}^{\mathrm{p}} = 0 \tag{2.15}$$

Similarly, by submitting Eqs. (2.1) and (2.3) into Eq. (2.15), we can obtain expressions of the plastic multiplier in steady environments:

$$\mathrm{d}\phi = \frac{1}{H_{\mathrm{p}}^{\sigma}}\frac{\partial f}{\partial \boldsymbol{\sigma}} : \mathrm{d}\boldsymbol{\sigma} \tag{2.16}$$

$$\mathrm{d}\phi = \frac{1}{H_{\mathrm{p}}^{\varepsilon}}\frac{\partial f}{\partial \boldsymbol{\sigma}} : \mathbb{D} : \mathrm{d}\boldsymbol{\varepsilon} \tag{2.17}$$

where the plastic flow direction $\mathbf{r}$ is still unknown and needs to be determined.

2.2 Direction of Plastic Strain Increment

The classical plasticity theory adopts the orthogonal flow rule, assuming that the plastic flow direction is orthogonal to the plastic potential surface and that the plastic potential function has the same form as the yield function. Consequently, the obtained plastic flow direction is orthogonal to the yield surface. This approach is well-suited for metallic materials. However, for geomaterials, researchers have observed from numerical and physical experiments that the direction of plastic flow is non-orthogonal to the yield surface. This leads to limitations in describing the plastic deformation behavior of geomaterials using the orthogonal flow rule due to the unknown form of the plastic potential function. In the following content, we will first verify the non-orthogonal phenomenon of geomaterials through numerical experiments. Subsequently, a non-orthogonal flow rule is proposed, which directly obtains the non-orthogonal gradient of the yield function as the plastic flow direction of materials through fractional derivatives, thereby avoiding the need to construct a plastic potential function.

2.2.1 Non-orthogonal Phenomena in Numerical Experiments

Numerical experiments can offer repeatable initial loading tests and thus are more cost-effective than laboratory tests. In this section, the Discrete Element Method (DEM) is employed to investigate the non-orthogonal phenomena of plastic flow direction as it can effectively replicate the material's cohesion-friction deformation mechanism of granular materials. Figure 2.2a presents a cubic numerical model composed of rigid convex polyhedral particles. Its edge length is 0.12 m, and the porosity ratio is 0.669.

The selection of particle shape draws inspiration from Mohamed's research results, and this shape can well reflect the mechanical properties of Toyoura sand (Mohamed et al., 2022). In addition, the method of stress probing is used to determine the plastic flow direction and the tangential direction of the yield surface at a certain point. First, the specimen is loaded to point A along an isotropic compression path, then loaded to point B along a conventional triaxial compression path, as shown in Fig. 2.2b. Then, a series of stress probes of the same magnitude but in different directions are applied to obtain the desired plastic strain response.

The details of the stress probing method are depicted in Fig. 2.3a. According to plasticity theory, if the current stress point is on the yield surface, stress probes directed outward from the yield surface will produce significant plastic strain, while those directed inward will cause minimal plastic deformation, negligible enough to be disregarded. Therefore, by analyzing the magnitude of plastic strain increments caused by stress probes with different directions, the tangent of the yield surface at the current point can be determined. On the other hand, existing studies have shown that the plastic strain increments obtained by stress probes will form a narrow ellipse, with its major axis direction serving as the direction of plastic flow, as illustrated in Fig. 2.3b.

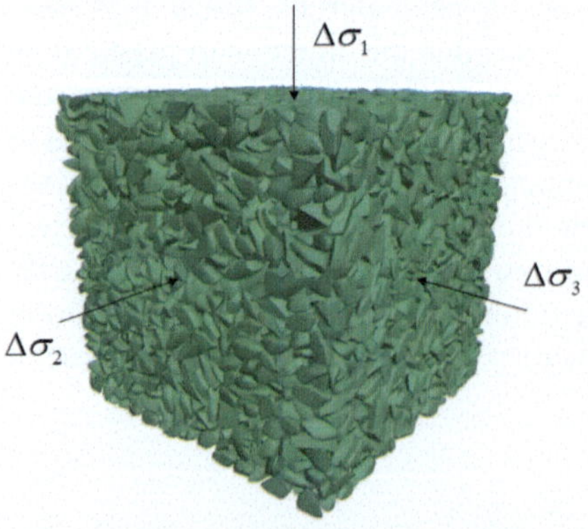

(a) Virtual specimen of Toyoura sand

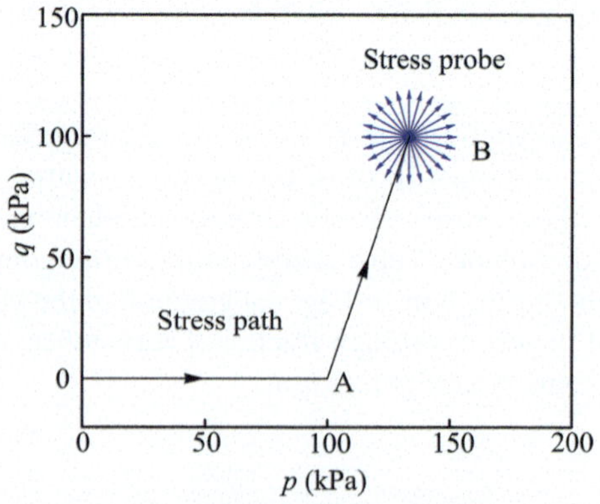

(b) Loading and detection paths

Fig. 2.2 Numerical test of Toyoura sand

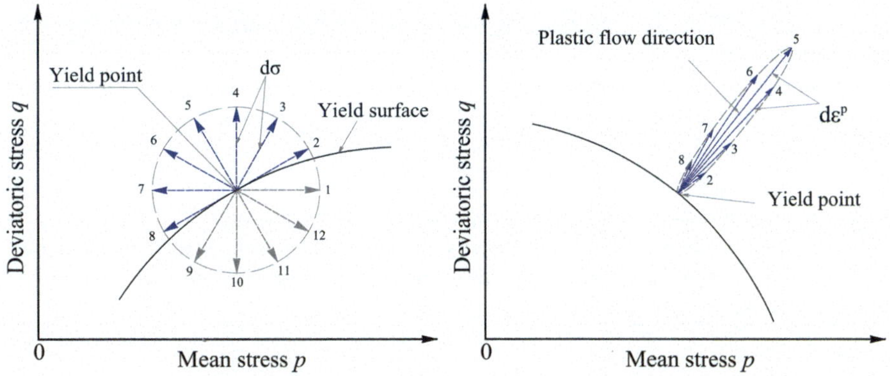

(a) Stress probe (b) Plastic strain increment

Fig. 2.3 Schematic diagrams of stress probe and plastic strain increment in the meridian plane

Figure 2.4a presents the total strain increment and elastic strain increment obtained from stress probing at point B, where the plastic strain increments can be calculated by subtracting the elastic part from the total strain increment. It's observed that the plastic strain increments under different probes concentrate in a narrow region passing through the origin, as shown in Fig. 2.4b, with the pink dashed line representing the direction of plastic flow at point B. The pink dashed line in the figure represents the plastic flow direction at point B.

Furthermore, the results from Fig. 2.4a are plotted in polar coordinates, as shown in Fig. 2.5, where the radial distance represents the magnitude of strain increment, and the polar angle represents the direction of the stress path of the stress probe. It can be seen that the plastic strain increments primarily concentrate in the loading

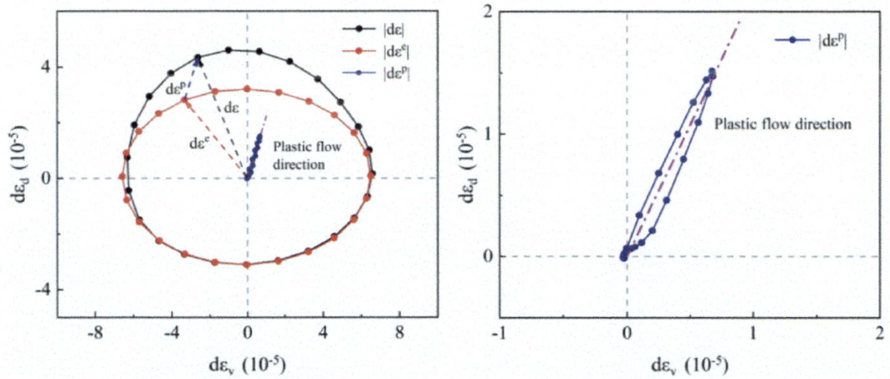

(a) Stress probe (b) Plastic strain increment

Fig. 2.4 Strain increment at point B

path range of 25° to 205°. Hence, the normal to the yield surface at point B can be determined to be at 115°.

The direction of plastic strain increment and the orthogonal gradient direction of the yield surface are plotted on the meridian plane, as shown in Fig. 2.6. The stress probing results at different stress points all indicate that the plastic flow direction deviates from the outer normal direction of the yield surface, namely the non-orthogonal phenomena of plastic strain behavior. Moreover, the degree of non-orthogonality is stress-state-dependent, adding complexity to the construction of the plastic potential function. A new flow rule that can conveniently describe this non-orthogonal phenomenon is eagerly desired.

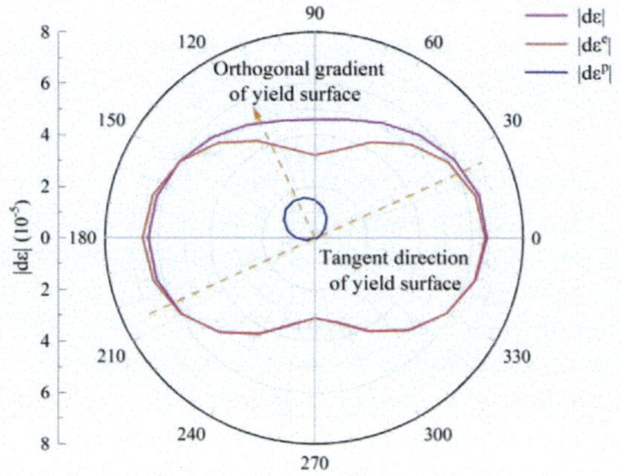

Fig. 2.5 Strain increments in the polar coordinate system

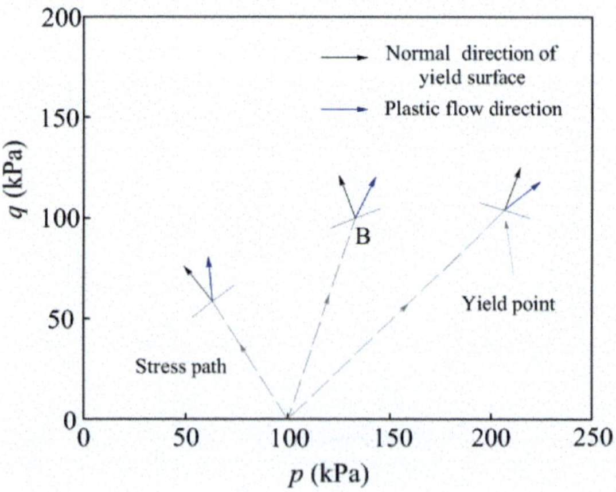

Fig. 2.6 Plastic flow direction and orthogonal yield surface direction measured by DEM

2.2.2 Non-orthogonal Plastic Flow Rule

For the non-orthogonal plastic flow rule, the plastic flow direction is determined by the non-orthogonal gradient of the existing yield function with the aid of a fractional derivative. The non-orthogonal flow rule can be expressed as follows:

$$\mathbf{r} = \left(\frac{\partial^{\mu_1} f}{\partial S_1^{\mu_1}}, \frac{\partial^{\mu_2} f}{\partial S_2^{\mu_2}}, \frac{\partial^{\mu_3} f}{\partial S_3^{\mu_3}}\right) : \frac{\partial \mathbf{S}}{\partial \boldsymbol{\sigma}} \tag{2.18}$$

where $\mathbf{S} = (S_1, S_2, S_3)$ is a set of coordinate bases consisting of stress invariants. $\partial^{\mu_1} f / \partial S_1^{\mu_1}$, $\partial^{\mu_2} f / \partial S_2^{\mu_2}$, and $\partial^{\mu_3} f / \partial S_3^{\mu_3}$ are the fractional derivatives of yield function f with respect to stress invariants S_1, S_2, and S_3, respectively. $\left(\partial^{\mu_1} f / \partial S_1^{\mu_1}, \partial^{\mu_2} f / \partial S_2^{\mu_2}, \partial^{\mu_3} f / \partial S_3^{\mu_3}\right)$ is the fractional gradient of the yield surface in $\mathbf{S}$ coordinate space, which is generally non-orthogonal to the yield surface. Although hundreds of definitions of fractional derivatives have been reported (De Oliveira & Tenreiro Machado, 2014; Teodoro et al., 2019), the most commonly used definitions for plasticity modeling are Riemann-Liouville (RL) fractional derivatives and Caputo fractional derivatives (Kilbas et al., 2006). The basic properties of these two fractional derivatives can be found in Appendix A. The geometric interpretation of the non-orthogonal flow rule is shown in Fig. 2.7a and b from the local and global views.

Compared with the orthogonal flow rule, the non-orthogonal flow rule avoids the necessity of constructing plastic potential function but the ability to adjust the plastic flow direction is remained. The proportional relationship among fractional

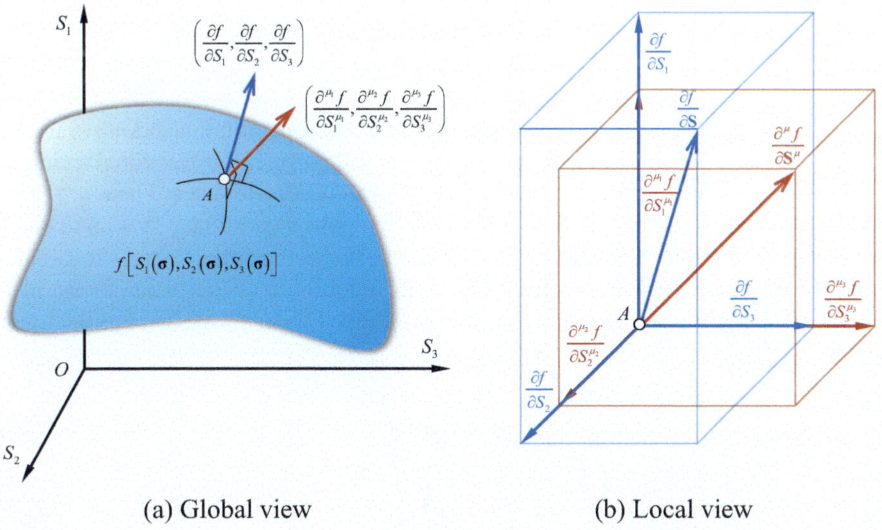

Fig. 2.7 Non-orthogonal flow rule

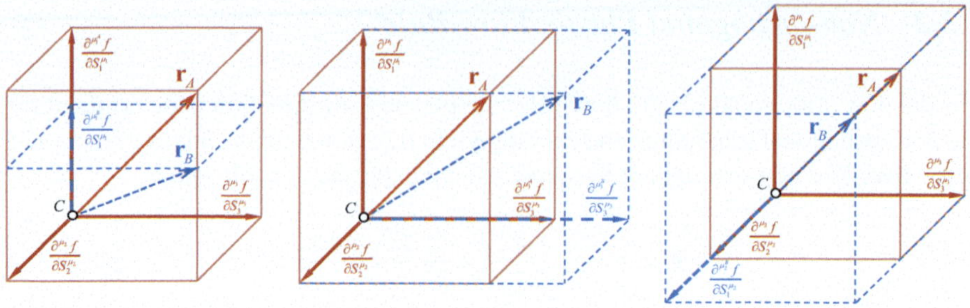

Fig. 2.8 Influence of fractional order on the plastic flow direction

gradient components (i.e., $\partial^{\mu_1} f / \partial S_1^{\mu_1}$, $\partial^{\mu_2} f / \partial S_2^{\mu_2}$, and $\partial^{\mu_3} f / \partial S_3^{\mu_3}$) is related to the fractional orders (μ_1, μ_2, and μ_3). The vector direction $\partial^{\mu} f / \partial \mathbf{S}^{\mu}$ composed of these components can be adjusted properly by changing fractional orders. When the values of μ_1, μ_2 and μ_3 are determined individually based on the physical meaning indicated by $\partial^{\mu_1} f / \partial S_1^{\mu_1}$, $\partial^{\mu_2} f / \partial S_2^{\mu_2}$, and $\partial^{\mu_3} f / \partial S_3^{\mu_3}$, the non-orthogonal flow rule may be more powerful in describing the direction of plastic strain increment. For example, as shown in Fig. 2.8, when μ_2 and μ_3 remain unchanged, adjusting μ_1 can change the value of the component $\partial^{\mu_1} f / \partial S_1^{\mu_1}$, while the values of $\partial^{\mu_2} f / \partial S_2^{\mu_2}$ and $\partial^{\mu_3} f / \partial S_3^{\mu_3}$ remain unchanged. It is worth noting that theoretically, only two of the three fractional orders are independent of each other because, in the stress invariant space, only two independent parameters are needed to adjust the proportional relationship of the three plastic strain increment components.

In the case of $\mu_1 = \mu_2 = \mu_3 = \mu$, Eq. (2.18) will degenerate into the following form.

$$\mathbf{r} = \frac{\partial^{\mu} f}{\partial \mathbf{S}^{\mu}} : \frac{\partial \mathbf{S}}{\partial \boldsymbol{\sigma}} \tag{2.19}$$

where $\partial \mathbf{S}/\partial \boldsymbol{\sigma}$ is the covariant transformation in tensor operation, also known as the coordinate transformation. In the non-orthogonal flow rule, the fractional gradient $\partial^{\mu} f / \partial \mathbf{S}^{\mu}$ of the yield surface is first determined in $\mathbf{S}$ coordinate space. Then, $\partial^{\mu} f / \partial \mathbf{S}^{\mu}$ is transformed by $\partial \mathbf{S}/\partial \boldsymbol{\sigma}$ from $\mathbf{S}$ coordinate system into $\boldsymbol{\sigma}$ coordinate system to characterize the direction of plastic strain increment. For geomaterials, stress invariants (p, q, θ) are often used to construct the yield function. p is the mean stress, also known as the hydrostatic pressure, q is the generalized shear stress, and θ is the Lode angle.

$$p = \frac{1}{3}\boldsymbol{\sigma} : \mathbf{1} \tag{2.20}$$

$$q = \sqrt{\frac{3}{2}} \|\mathbf{s}\| \tag{2.21}$$

$$\theta = \frac{1}{3}\arccos\left(\frac{3\sqrt{3}J_3}{2J_2^{3/2}}\right) \tag{2.22}$$

where $\mathbf{1}$ is the 2nd-order unit tensor. $\mathbf{s} = \boldsymbol{\sigma} - p\mathbf{1}$ is the deviatoric stress tensor. $J_2 = \mathbf{s} : \mathbf{s}/2$ and $J_3 = \det(\mathbf{s})$ are the second and third stress invariants of deviatoric stress, respectively. In this situation, based on the coaxial assumption, $\partial^\mu f \big/ \partial p^\mu$, $\partial^\mu f \big/ \partial q^\mu$, and $\partial^\mu f \big/ \partial \theta^\mu$ represent the directions of plastic volume strain increment $d\varepsilon_v^p$, equivalent plastic shear strain increment $d\varepsilon_d^p$, and the plastic strain increment $d\varepsilon_\theta^p$ caused by the Lode angle change. Equation (2.19) can be rewritten as follows:

$$\mathbf{r} = \frac{\partial^\mu f}{\partial p^\mu}\frac{\partial p}{\partial \boldsymbol{\sigma}} + \frac{\partial^\mu f}{\partial q^\mu}\frac{\partial q}{\partial \boldsymbol{\sigma}} + \frac{\partial^\mu f}{\partial \theta^\mu}\frac{\partial \theta}{\partial \boldsymbol{\sigma}} \tag{2.23}$$

The fractional order μ is the model parameter, and its physical meaning depends on the form of the yield function and the properties of the material. When μ equals to 1, $\partial^\mu f \big/ \partial \mathbf{S}^\mu$ is orthogonal to the yield surface and the non-orthogonal flow rule degenerates into the orthogonal flow rule.

2.3 Multi-physical Loading/Unloading Criterion

The mechanical behaviors of materials are vastly different in the elastic and plastic states. The yield surface determines the boundary of the current elastic domain. The state point inside the yield surface is the elastic state, and the state point on the yield surface is the plastic state. For steady environments, the state point can be represented by the stress point or the strain point, which depends on whether the yield surface is located in the stress space or the strain space. Accordingly, for unsteady environments, the state point is composed of stress and environment variables, or strain and environment variables. In mathematics, the elastic and plastic states can be defined as follows:

$$\begin{cases} f < 0, & \text{Elastic state} \\ f = 0, & \text{Plastic state} \end{cases} \tag{2.24}$$

Generally, the loading means that the material undergoes a plastic deformation process under the current load increment. From a geometric perspective, it manifests as the state point moving towards the outside of the current yield surface, as shown in Fig. 2.9. In the hardening stage, the plastic deformation will also cause the shape of the yield surface to change and expand outward so that the new state point will always remain on the subsequent yield surface. If the state point moves towards the inside of the yield surface under the current load increment, this process is called unloading. At this time, only elastic deformation occurs, and the shape and size of the yield surface remain unchanged. Of course, the state point may also move along

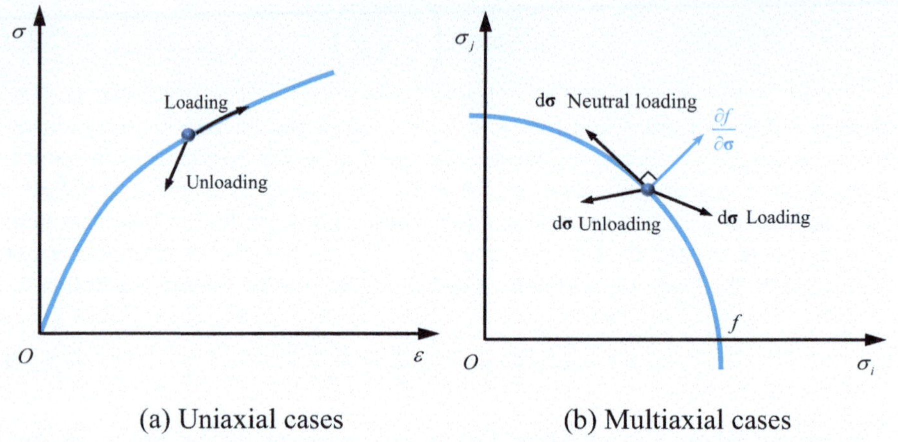

Fig. 2.9 Loading/unloading estimations in the stress space for steady environments

the current yield surface. This process is called neutral loading, and the deformation generated is also elastic.

Without considering the influence of environmental variables, the above judgment process can be expressed by the following formula:

$$\begin{cases} \dfrac{\partial f}{\partial \boldsymbol{\sigma}} : \mathrm{d}\boldsymbol{\sigma} > 0 \quad \text{Loading} \\ \dfrac{\partial f}{\partial \boldsymbol{\sigma}} : \mathrm{d}\boldsymbol{\sigma} = 0 \quad \text{Neutral loading} \\ \dfrac{\partial f}{\partial \boldsymbol{\sigma}} : \mathrm{d}\boldsymbol{\sigma} < 0 \quad \text{Unloading} \end{cases} \tag{2.25}$$

Equation (2.25) is the loading/unloading criterion for strengthening materials in steady-state environments, which can also be obtained through the Drucker postulate. In finite element analysis, the known load increment is the strain increment, while Eq. (2.25) requires a known stress increment. In addition, for most geomaterials, two phenomena, strain hardening and strain softening, will occur successively throughout the whole loading process. In the softening stage, the yield surface in the stress space shrinks inward, which limits the application of Eq. (2.25). Therefore, for geomaterials, the loading/unloading criterion is generally constructed in the strain space, where the yield surface constantly expands outwards during the whole loading process. The yield function expressed by the strain is written as follows:

$$F(\boldsymbol{\varepsilon}, H) = 0 \tag{2.26}$$

In the strain space, the load vector is characterized by the strain increment, as shown in Fig. 2.10. When the included angle between the strain increment $\mathrm{d}\boldsymbol{\varepsilon}$ and the external normal direction $\partial F / \partial \boldsymbol{\varepsilon}$ of the yield surface is acute, this indicates that

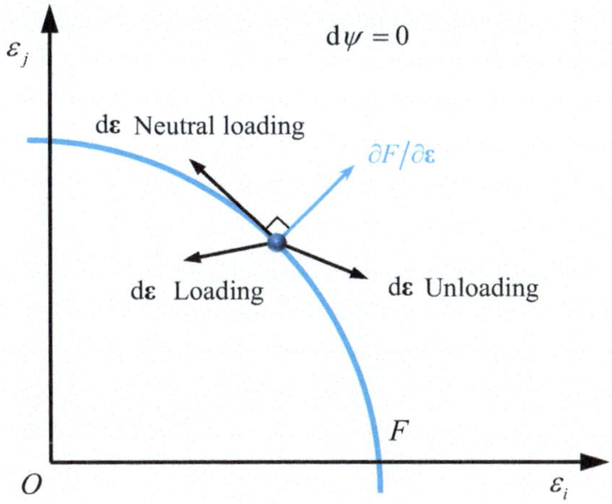

Fig. 2.10 Loading/unloading criterion under strain space in steady environments

the strain state of the material tends to move out of the yield surface and the current load step is loading. The material will produce plastic deformation under the action of strain increment $d\boldsymbol{\varepsilon}$. If the included angle is the obtuse angle, this indicates that the strain state of the material will move towards the interior of the yield surface and the current load step is unloading. The right angle represents the neutral loading. The strain state of the material will moves along the current yield surface. The loading/unloading criterion in the strain space can be expressed as follows:

$$\begin{cases} \dfrac{\partial F}{\partial \boldsymbol{\varepsilon}} : d\boldsymbol{\varepsilon} > 0 \quad \text{Loading} \\ \dfrac{\partial F}{\partial \boldsymbol{\varepsilon}} : d\boldsymbol{\varepsilon} = 0 \quad \text{Neutral loading} \\ \dfrac{\partial F}{\partial \boldsymbol{\varepsilon}} : d\boldsymbol{\varepsilon} < 0 \quad \text{Unloading} \end{cases} \tag{2.27}$$

Equation (2.27) is suitable for both hardening and softening stages. However, the yield function of plastic models is typically formulated in the stress form. For convenience, the loading/unloading criterion defined by Eq. (2.27) is converted into that expressed by stress. Utilizing Hooke's law, the relationship between Eqs. (2.26) and (2.4) is given as follows:

$$F(\boldsymbol{\varepsilon}, H) = f\left[\mathbb{D} : (\boldsymbol{\varepsilon} - \boldsymbol{\varepsilon}^{\mathrm{p}}), H\right] = f(\boldsymbol{\sigma}, H) \tag{2.28}$$

The partial differential of Eq. (2.28) with respect to $\boldsymbol{\varepsilon}$ will produce:

$$\frac{\partial F}{\partial \boldsymbol{\varepsilon}} = \frac{\partial f}{\partial \boldsymbol{\sigma}} : \frac{\partial \boldsymbol{\sigma}}{\partial \boldsymbol{\varepsilon}} = \frac{\partial f}{\partial \boldsymbol{\sigma}} : \mathbb{D} \tag{2.29}$$

Then, for geomaterials in steady environments, the loading/unloading criterion can be reformulated as:

$$\frac{\partial F}{\partial \boldsymbol{\varepsilon}} : \mathrm{d}\boldsymbol{\varepsilon} = \begin{cases} \frac{\partial f}{\partial \boldsymbol{\sigma}} : \mathbb{D} : \mathrm{d}\boldsymbol{\varepsilon} > 0 & \text{Loading} \\ \frac{\partial f}{\partial \boldsymbol{\sigma}} : \mathbb{D} : \mathrm{d}\boldsymbol{\varepsilon} = 0 & \text{Neutral loading} \\ \frac{\partial f}{\partial \boldsymbol{\sigma}} : \mathbb{D} : \mathrm{d}\boldsymbol{\varepsilon} < 0 & \text{Unloading} \end{cases} \tag{2.30}$$

In what follows, we discuss further the loading/unloading criterion for geomaterials under unsteady environments. The yield function defined by Eq. (2.26) can be rewritten as follows:

$$F(\boldsymbol{\varepsilon}, H, \psi) = 0 \tag{2.31}$$

In unsteady environments, the loading process of the material can be driven not only by the strain increment but also by the change of environment variables. The multi-physical loading/unloading criterion needs to be established in ($\boldsymbol{\varepsilon}$, ψ) coordinate space, as shown in Fig. 2.11. $(\partial F/\partial \boldsymbol{\varepsilon}, \partial F/\partial \psi)$ denotes the normal direction of the yield surface. The load vector $(\mathrm{d}\boldsymbol{\varepsilon}, \mathrm{d}\psi)$ comprises the strain increment $\mathrm{d}\boldsymbol{\varepsilon}$ and the environmental increment $\mathrm{d}\psi$. The scenarios that the included angle of $(\mathrm{d}\boldsymbol{\varepsilon}, \mathrm{d}\psi)$ and $(\partial F/\partial \boldsymbol{\varepsilon}, \partial F/\partial \psi)$ are less than 90°, equal to 90°, and more than 90° represent the loading, neutral loading, and unloading, respectively. By analogy with the derivation of Eqs. (2.28) to (2.30), we can obtain the expression of the multi-physical loading/unloading criterion under the unsteady environments.

$$\left(\frac{\partial F}{\partial \boldsymbol{\varepsilon}}, \frac{\partial F}{\partial \psi}\right)(\mathrm{d}\boldsymbol{\varepsilon}, \mathrm{d}\psi)^{\mathrm{T}} = \begin{cases} \left(\frac{\partial f}{\partial \boldsymbol{\sigma}}\mathbb{D}, \frac{\partial f}{\partial \psi}\right)(\mathrm{d}\boldsymbol{\varepsilon}, \mathrm{d}\psi)^{\mathrm{T}} > 0 & \text{Loading} \\ \left(\frac{\partial f}{\partial \boldsymbol{\sigma}}\mathbb{D}, \frac{\partial f}{\partial \psi}\right)(\mathrm{d}\boldsymbol{\varepsilon}, \mathrm{d}\psi)^{\mathrm{T}} = 0 & \text{Neutral loading} \\ \left(\frac{\partial f}{\partial \boldsymbol{\sigma}}\mathbb{D}, \frac{\partial f}{\partial \psi}\right)(\mathrm{d}\boldsymbol{\varepsilon}, \mathrm{d}\psi)^{\mathrm{T}} < 0 & \text{Unloading} \end{cases} \tag{2.32}$$

The loading/unloading criterion can also be established directly by determining the sign of the plastic multiplier from an algebraic perspective. For a given load increment $(\mathrm{d}\boldsymbol{\varepsilon}, \mathrm{d}\psi)$, if the calculated result of the plastic multiplier is greater than 0, i.e., $\mathrm{d}\phi > 0$, this means that the material undergoes plastic deformation under the action of load increment. The deformation process is considered as the loading case. If the calculated result of the plastic multiplier is less than 0, i.e., $\mathrm{d}\phi < 0$, this result will be discarded and replaced by $\mathrm{d}\phi = 0$ due to the non-negativity of $\mathrm{d}\phi$. The deformation process is considered as the unloading case and the elastic constitutive equation will be used. The result of $\mathrm{d}\phi = 0$ represents neutral loading with only elastic deformation.

$$\begin{cases} \mathrm{d}\phi > 0 & \text{Loading} \\ \mathrm{d}\phi = 0 & \text{Neutral loading} \\ \mathrm{d}\phi < 0 \Rightarrow \mathrm{d}\phi = 0 & \text{Unloading} \end{cases} \tag{2.33}$$

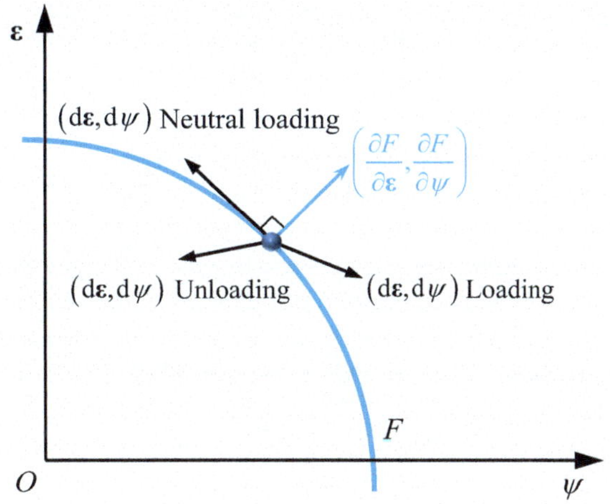

Fig. 2.11 Multi-physical loading/unloading criterion under strain-environment space in unsteady environments

Comparing Eqs. (2.12) and (2.32), we can note that the numerator of the plastic multiplier expression is the discriminative indicator of the loading/unloading criterion. Considering that $H_{\mathrm{p}}^{\varepsilon}$ is always greater than 0 in the hardening and softening stages, we can conclude that the loading/unloading judgment based on the sign of the plastic multiplier is completely consistent with the one based on the geometric relationship between yield surface and generalized load increment.

2.4 Constitutive Equations of Multi-physical and Non-orthogonal Plasticity Theory

In the multi-physical and non-orthogonal geotechnical plasticity theory, the environmental variables that affect the properties of the material are used as constitutive variables to describe the plastic deformation mechanism of materials in unsteady environments. The incremental stress–strain constitutive relationship ($\mathrm{d}\boldsymbol{\sigma} - \mathrm{d}\boldsymbol{\varepsilon}$) in the classical elastoplastic theory has been developed into a stress–strain-environment incremental constitutive relationship ($\mathrm{d}\boldsymbol{\sigma} - \mathrm{d}\boldsymbol{\varepsilon} - \mathrm{d}\psi$). Submitting Eq. (2.12) and (2.3) into (2.1), we can obtain:

$$\mathrm{d}\boldsymbol{\sigma} = \mathbb{D}^{\mathrm{ep}} : \mathrm{d}\boldsymbol{\varepsilon} + \mathbf{D}^{\psi}\mathrm{d}\psi \tag{2.34}$$

where $\mathbb{D}^{\mathrm{ep}}$ is the elastoplastic stiffness matrix. $\mathbf{D}^{\psi}$ is a second-order tensor used to describe stress changes caused by environmental load increment.

$$\mathbb{D}^{\mathrm{ep}} = \mathbb{D} - \frac{(\mathbb{D} : \mathbf{r}) \otimes \left(\frac{\partial f}{\partial \boldsymbol{\sigma}} : \mathbb{D}\right)}{\frac{\partial f}{\partial \boldsymbol{\sigma}} : \mathbb{D} : \mathbf{r} - \frac{\partial f}{\partial H}\frac{\partial H}{\partial \boldsymbol{\varepsilon}^{\mathrm{p}}} : \mathbf{r}} \tag{2.35}$$

$$\mathbf{D}^{\psi} = -\frac{(\mathbb{D} : \mathbf{r}) \otimes \frac{\partial f}{\partial \psi}}{\frac{\partial f}{\partial \boldsymbol{\sigma}} : \mathbb{D} : \mathbf{r} - \frac{\partial f}{\partial H} \frac{\partial H}{\partial \boldsymbol{\varepsilon}^{p}} : \mathbf{r}} \tag{2.36}$$

where the "⊗" denotes the tensorial product.

Equation (2.34) is the constitutive equations of multi-physical and non-orthogonal geotechnical plasticity theory in the loading case, which can capture the mechanical response caused by environmental changes even under constant stress or strain conditions, such as creep and relaxation, thermal stress and thermal strain, and so on. In steady environments, the constitutive relationship in the form of $\mathrm{d}\boldsymbol{\sigma} - \mathrm{d}\boldsymbol{\varepsilon} - \mathrm{d}\psi$ will degenerate into the one of $\mathrm{d}\boldsymbol{\sigma} - \mathrm{d}\boldsymbol{\varepsilon}$ when the environmental variables are constant.

$$\mathrm{d}\boldsymbol{\sigma} = \mathbb{D}^{\mathrm{ep}} : \mathrm{d}\boldsymbol{\varepsilon} \tag{2.37}$$

In the unloading case, Eq. (2.34) will degenerate into the classical elastic constitutive relationship as follows:

$$\mathrm{d}\boldsymbol{\sigma} = \mathbb{D} : \mathrm{d}\boldsymbol{\varepsilon} \tag{2.38}$$

2.5 Determination Method for Model Parameters

Reasonable parameter calibration is a prerequisite for applying constitutive models for analyzing engineering problems. Model parameters with a clear physical meaning can usually be calibrated by performing specific experiments. When experimental conditions are not available, the parameters need to be determined by empirical estimation. Empirical estimation is highly influenced by subjective factors, which can affect the applicability of the model. With the development of artificial intelligence technology, more and more intelligent optimization algorithms are used to determine the parameters of the constitutive models (Obrzud et al., 2009; Zhang et al., 2013; Zhao et al., 2015).

The reasonableness of the parameter being calibrated needs to be checked by experiment curves. That is, the experiment conditions are set in advance, and the degree of agreement between the stress–strain curve and the experiment data is checked by adding loads step by step. The error between the calculated curve and the experiment data is expressed through the error function shown in Eq. (2.39).

$$\text{Error} = \frac{1}{N}\left(\sum_{i=1}^{N} \left| U_{\text{exp}}^{i} - U_{\text{pre}}^{i} \right|\right) \tag{2.39}$$

where N is the number of data points of the experiment, U_{exp}^{i} is the experiment value of measurement point i and U_{pre}^{i} is the simulated value of measurement point i.

The incremental method is frequently used in the numerical calculation of soil constitutive models. The strain difference between neighboring measurement points is taken as the strain increment and the environmental difference is taken as the environmental increment. Stress increment is obtained by bringing the strain increment and environmental increment into Eq. (2.34). The σ can be obtained as shown in equation Eq. (2.40).

$$\sigma = \sum_{i=1}^{N} \Delta\sigma_i \tag{2.40}$$

In 1962, John Holland (1962) first proposed the idea that population evolution simulates adaptation to new systems and introduced population, fitness values, selection, crossover, and variation. The genetic algorithm (GA) is first proposed by Holland's student Bagley in 1967 (Bagley, 1967). Compared with traditional heuristic optimization algorithms (hill-climbing, simulated annealing), the main features of GA are the population search strategy and simple genetic operators. The error between the experimental results and the numerical prediction are operated as a fitness value, which reduces the dependence of the heuristic algorithm on human–computer interaction during the search process. As shown in Fig. 2.12, GA is a typical iterative process with the main operators of selection, crossover, and mutation in the following steps: (1) Randomly generate multiple model parameters and set the current algebra $k = 1$; (2) Determine whether the current algebra reaches the maximum algebra $k_{\max}$. If yes, output the result. If not then go to step 3; (3) Selection, crossover, and mutation of current populations; (4) The parameters are inputted into the constitutive model to calculate; (5) Calculation error with experiment data, set $k = k + 1$, and go to step 2.

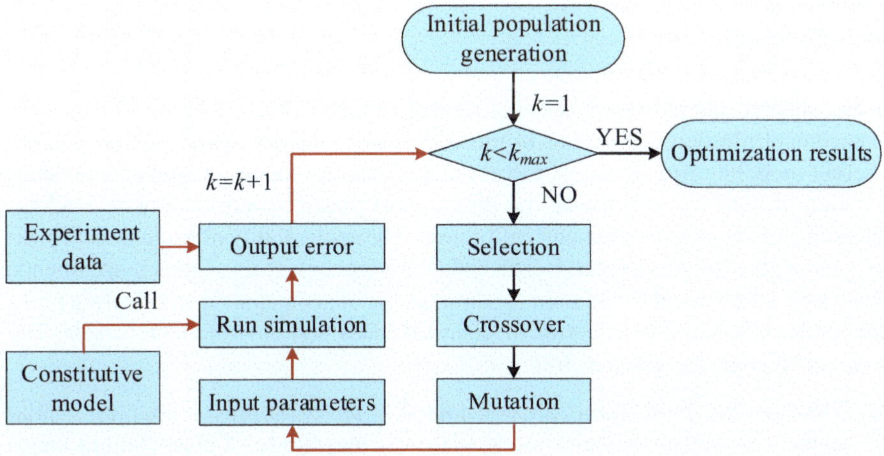

Fig. 2.12 Basic flow of genetic algorithm

The parameters of the constitutive model are considered genes, and the error with respect to the experimental results is used as a fitness function in GA. Each individual in the population represents a selection solution for the parameters of the constitutive model. The mechanism for tournament selection is that two individuals are randomly selected, with the lesser-adapted individual being chosen to enter the crossover pool. For a population $A = (\mathbf{M}_1, \mathbf{M}_2, \ldots, \mathbf{M}_n)$, the fitness value of the individual $\mathbf{M}_i \in A$ is $fit(\mathbf{M}_i)$. Individuals $\mathbf{M}_i$ and $\mathbf{M}_j$ are randomly selected, if $fit(\mathbf{M}_i) > fit(\mathbf{M}_j)$, $\mathbf{M}_i$ is entered into the crossover pool, else $\mathbf{M}_j$ is selected. Single-point crossover is often used in crossover operations. Two individuals are randomly selected from the crossover pool, denoted as:

$$\mathbf{M}_1 = M_{11}, M_{12}, \ldots, M_{1i_1}, M_{1i_2}, \ldots, M_{1n} \tag{2.41}$$

$$\mathbf{M}_2 = M_{21}, M_{22}, \ldots, M_{2i_1}, M_{2i_2}, \ldots, M_{2n} \tag{2.42}$$

Randomly select a cross bit $k_{\text{bit}} \in (1,2, \ldots, n-1)$, assuming $k_{\text{bit}} = i_1$. Then, the genes in the right part of the position are swapped to produce two new individuals:

$$\mathbf{M'}_1 = M_{11}, M_{12}, \ldots, M_{1i_1}, M_{1i_2}, \ldots, M_{1n} \tag{2.43}$$

$$\mathbf{M}'_2 = M_{21}, M_{22}, \ldots, M_{2i_1}, M_{1i_2}, \ldots, M_{1n} \tag{2.44}$$

Mutation is a phenomenon that mimics the mutation of genes on chromosomes in nature. Set the mutation probability $R < 1$, then randomly select a gene and randomly give an $r \in [0,1]$. If $r < R$, reinitialize the gene.

2.6 Summary

In this chapter, the authors propose a rigorous multi-physical and non-orthogonal geotechnical plasticity modeling in unsteady environments, which solves the two inherent modeling limitations of the classical plasticity theory, including the determination of both the magnitude and direction of plastic strain increment, loading/unloading criterion, and parameter calibration. The applicability of geotechnical plasticity theory has been expanded from steady environments to unsteady environments, which lays a theoretical foundation for the analysis of geotechnical engineering problems under the coupling effect of mechanical loads and environmental loads. The main conclusions are given as follows:

(1) The environmental factors are considered as the constitutive variables, similar to those of stress or strain, to calculate the magnitude of plastic strain increment, and thus the plastic deformation caused by environmental and mechanical loads can be reasonably captured. The multi-physical loading/unloading

criterion in unsteady environments is established from both geometric and algebraic perspectives, i.e., the geometric relationship between yield surface and generalized load increment in the strain-environment space as well as the sign of plastic multiplier. Based on the proposed modeling method, the incremental stress–strain relationship has been developed into the incremental stress–strain-environment relationship, which provides an effective theoretical tool for the mechanical analysis of material behavior and engineering structural performance in unsteady environments.

(2) Compared with the orthogonal flow rule, the proposed non-orthogonal flow rule can directly obtain the non-orthogonal gradient of yield function by the fractional derivatives, without the need for constructing the additional plastic potential function. The classical orthogonal plasticity theory has been developed into a non-orthogonal plasticity theory. On the one hand, the non-orthogonal flow rule can be used to improve the performance of existing plasticity models due to its powerful ability to adjust the non-orthogonal direction of the yield surface. On the other hand, the non-orthogonal flow rule effectively reduces the difficulty of plasticity modeling, especially for geomaterials and new materials with limited test data.

References

Bagley, J. D. (1967). *The behavior of adaptive systems which employ genetic and correlation algorithms*. University of Michigan.

Chang, Y. F., Chen, Y. H., Sheu, M. S., & Yao, G. C. (2006). Residual stress–strain relationship for concrete after exposure to high temperatures. *Cement and Concrete Research, 36*(10), 1999–2005.

Chen, L., Fang, Q., Jiang, X. Q., Ruan, Z., & Hong, J. (2015). Combined effects of high temperature and high strain rate on normal weight concrete. *International Journal of Impact Engineering, 86*, 40–56.

He, Z. J., Ma, Y. N., Wang, Z. W., Zhang, X. J., Sheng, Z. X., Ding, M. J., & Fu, C. (2021). Triaxial strength and deformation characteristics and its constitutive model of high-strength concrete before and after high temperatures. *Structures, 30*, 1127–1138.

Holland, J. H. (1962). Outline for a logical theory of adaptive systems. *Journal of the ACM, 9*(3), 297–314.

Kilbas, A. A., Srivastava, H. M., & Trujillo, J. J. (2006). *Theory and applications of fractional differential equations*. Elsevier.

Kong, X. Z., Fang, Q., Chen, L., & Wu, H. (2018). A new material model for concrete subjected to intense dynamic loadings. *International Journal of Impact Engineering, 120*, 60–78.

Lu, D. C., Zhou, X., Du, X. L., & Wang, G. S. (2020). 3D dynamic elastoplastic constitutive model of concrete within the framework of rate-dependent consistency condition. *Journal of Engineering Mechanics, 146*(11), 4020124.

Lu, D. C., Meng, F. P., Zhou, X., Zhuo, Y. H., Gao, Z. W., & Du, X. L. (2023). A dynamic elastoplastic model of concrete based on a modeling method with environmental factors as constitutive variables. *Journal of Engineering Mechanics, 149*(12), 04023102.

Mohamed, T., Duriez, J., Veylon, G., & Peyras, L. (2022). DEM models using direct and indirect shape descriptions for toyoura sand along monotonous loading paths. *Computers and Geotechnics, 142*, 104551.

Obrzud, R. F., Vulliet, L., & Truty, A. (2009). A combined neural network/gradient-based approach for the identification of constitutive model parameters using self-boring pressuremeter tests. *International Journal for Numerical and Analytical Methods in Geomechanics, 33*(6), 817–849.

De Oliveira, E. C., & Tenreiro Machado, J. A. (2014) A review of definitions for fractional derivatives and integral. *Mathematical Problems in Engineering.*

Shkolnik, I. E. (2008). Influence of high strain rates on stress–strain relationship, strength and elastic modulus of concrete. *Cement and Concrete Composites, 30*(10), 1000–1012.

Teodoro, G. S., Machado, J. T., & De Oliveira, E. C. (2019). A review of definitions of fractional derivatives and other operators. *Journal of Computational Physics, 388*, 195–208.

Zeng, Y. Q., Xu, L. H., Chi, Y., Yu, M., & Huang, L. (2023). Compressive behavior of circular GFRP tube-confined UHPC-filled steel-encased stub columns. *Composite Structures, 309*, 116730.

Zhang, Y. L., Gallipoli, D., & Augarde, C. (2013). Parameter identification for elasto-plastic modelling of unsaturated soils from pressuremeter tests by parallel modified particle swarm optimization. *Computers and Geotechnics, 48*, 293–303.

Zhao, B. D., Zhang, L. L., Jeng, D. S., Wang, J. H., & Chen, J. J. (2015). Inverse analysis of deep excavation using differential evolution algorithm. *International Journal for Numerical and Analytical Methods in Geomechanics, 39*(2), 115–134.

Chapter 3
Non-orthogonal Elastoplastic Model for Soil in Steady Environments

The steady environment condition is a prerequisite for traditional elastoplastic constitutive models. However, traditional elastoplastic constitutive models often oversimplify the elastoplastic constitutive model framework by assuming orthogonality, disregarding the complex nature of soil deformation. Recognizing the limitations of the orthogonal elastoplastic constitutive modeling of soil, recent research has paved the way for the development of the non-orthogonal elastoplastic constitutive models. By considering the effects of non-orthogonality, engineers and researchers can gain deeper insights into the mechanical properties and deformation response of saturated soil.

This chapter present a framework of non-orthogoanl elastoplastic constitutive model specifically designed for soil under steady environmental conditions, where temperature, water content, and other environment variables are all constant. The concept of characteristic stress is provided first, which aims to capture of the 3D failure conditions of soil, and gives a deep insight for comprehensively understand the deformation behaviours of soil. By incorporating the non-orthogonal plastic flow rule (Lu et al., 2019a, 2019b) into the characteristic stress space, an approach to establishing a 3D NOEP constitutive model for soil is proposed (Liang et al., 2020; Lu et al., 2019a). We explore its applications in clays and sands under stready environment conditions, by which the strength and deformation behaviour of soils under 3D stress conditions are directly described. Through a comprehensive analysis of experimental data and model predictions, this study provides an in-depth understanding of model's capabilities and limitations, thereby shedding light on its potential for capturing mechanical behaviours of soils under steady environment conditions.

D. Lu et al., *Multi-physical and Non-orthogonal Geotechnical Plasticity Modeling and Algorithms*, https://doi.org/10.1007/978-981-96-2369-3_3

3.1 Characteristic Stress Method

The concept of characteristic stress was originally proposed to establish a simplified and unified strength criterion (Lu et al., 2017). It has been further developed as a novel approach for establishing three-dimensional elastoplastic models for soils (Lu et al., 2019a; Ma et al., 2017). The expression of the characteristic stress tensor can be written as:

$$\mathbf{c} = p_r \left[\left(\frac{\sigma_1}{p_r}\right)^{\beta} \mathbf{n}_i^{(1)} \otimes \mathbf{n}_j^{(1)} + \left(\frac{\sigma_2}{p_r}\right)^{\beta} \mathbf{n}_i^{(2)} \otimes \mathbf{n}_j^{(2)} + \left(\frac{\sigma_3}{p_r}\right)^{\beta} \mathbf{n}_i^{(3)} \otimes \mathbf{n}_j^{(3)} \right] \quad (3.1)$$

where $\mathbf{c}$ is the characteristic stress tensor whose principal directions coincide with those of ordinary stress tensor $\boldsymbol{\sigma}$. And the expression of characteristic principal stress $c_k = p_\text{r}(\sigma_k/p_\text{r})^\beta$ can be obtained based on Eq. (3.1), where σ_k ($k = 1, 2, 3$) indicates the principal stress. p_r is the reference stress for the dimensionless transformation, and $p_\text{r} = 1$ kPa is commonly used. β is the characteristic stress index, which is a non-dimensional material parameter that ranges from 0 to 1. $\mathbf{n}_i^{(k)}$ in Eq. (3.1) is the corresponding directional cosine of the direction vector $\boldsymbol{e}^{(k)}$ of the principal stress, that is, the principal stress vector can be expressed as $c_k\boldsymbol{e}^{(k)}$. Therefore, the spectral decomposition of characteristic stress tensor can be illustrated by the schematic diagram shown in Fig. 3.1.

Based on the friction rule on the octahedral plane under the action of $\mathbf{c}$, a nonlinear unified strength criterion (Lu et al., 2017), known as the β criterion, has been developed, which can be expressed as:

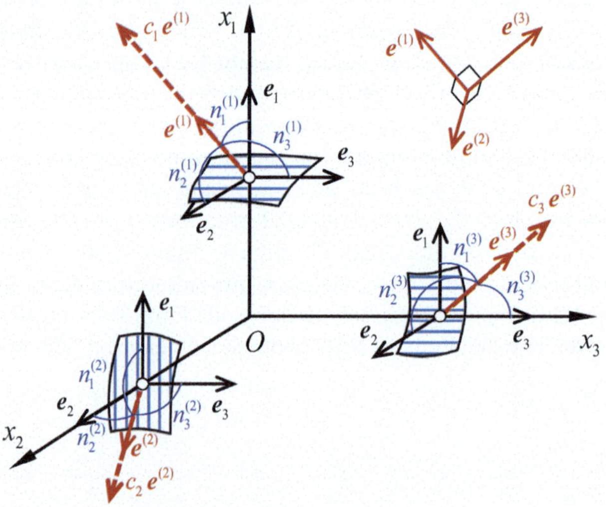

Fig. 3.1 Spectral decomposition of characteristic stress tensor

$$\frac{c_{\mathrm{s}}}{c_{\mathrm{n}}} = F_{\mathrm{f}} \tag{3.2}$$

where F_{f} is the characteristic failure stress ratio. The specific expressions of characteristic stress variables and parameters are listed in Table 3.1, which are compared with the ordinary stress.

The concept of characteristic stress can be utilized as a failure condition for geomaterials, making it a suitable choice in describing deformation behaviours under three-dimensional stress conditions. This enables the developed elastoplastic model in the **c** stress space be born with the potential to capture three-dimensional mechanical properties of soils. Therefore, this chapter endeavors to combine the characteristic stress with the non-orthogonal plastic flow rule to develop a framework for the 3D-NOEP model.

Based on the elastoplastic theory, the total strain $\boldsymbol{\varepsilon}$ and its increment $\mathbf{d}\boldsymbol{\varepsilon}$ can be decomposed into the reversible elastic strain component and the irreversible plastic strain component:

$$\boldsymbol{\varepsilon} = \boldsymbol{\varepsilon}^{\mathrm{e}} + \boldsymbol{\varepsilon}^{\mathrm{p}}, \quad \mathbf{d}\boldsymbol{\varepsilon} = \mathbf{d}\boldsymbol{\varepsilon}^{\mathrm{e}} + \mathbf{d}\boldsymbol{\varepsilon}^{\mathrm{p}} \tag{3.3}$$

where $\boldsymbol{\varepsilon}^{\mathrm{e}}$ and $\boldsymbol{\varepsilon}^{\mathrm{p}}$ are the elastic strain and plastic strain, respectively. And $\mathbf{d}\boldsymbol{\varepsilon}^{\mathrm{e}}$ and $\mathbf{d}\boldsymbol{\varepsilon}^{\mathrm{p}}$ are their increments.

For the plastic strain increment, the non-orthogonal plastic flow rule (Lu et al., 2019a, 2019b) is adopted and can be expressed as:

$$\mathbf{d}\boldsymbol{\varepsilon}^{\mathrm{p}} = \Lambda \mathbf{m} = \Lambda \frac{\partial^{\mu} f}{\partial \mathbf{S}^{\mu}} : \frac{\partial \mathbf{S}}{\partial \boldsymbol{\sigma}} \tag{3.4}$$

Table 3.1 Comparison between the ordinary stress and the characteristic stress

	Ordinary stress	Characteristic stress
Stress tensor	$\boldsymbol{\sigma}$	$\mathbf{c}$
Principal stress	$\sigma_i, \quad i = 1, 2, 3$	$c_i = p_r\left(\frac{\sigma_i}{p_r}\right)^{\beta}, \quad i = 1, 2, 3$
Mean stress	$p = \boldsymbol{\sigma}:\mathbf{1}/3$	$c_n = \mathbf{c}:\mathbf{1}/3$
General shear stress	$q = \sqrt{\frac{3}{2}(\boldsymbol{\sigma} - p\mathbf{1}):(\boldsymbol{\sigma} - p\mathbf{1})}$	$c_{\mathrm{s}} = \sqrt{\frac{3}{2}(\mathbf{c} - p\mathbf{1}):(\mathbf{c} - p\mathbf{1})}$
Stress Lode angle	$\theta = \arctan\left[\frac{\sqrt{3}(\sigma_2 - \sigma_3)}{2\sigma_1 - \sigma_2 - \sigma_3}\right]$	$\theta_{\beta} = \arctan\left[\frac{\sqrt{3}(c_2 - c_3)}{2c_1 - c_2 - c_3}\right]$
Stress ratio	$\eta = q/p$	$\chi = c_{\mathrm{s}}/c_{\mathrm{n}}$
Phase transition stress ratio	$M = \frac{6 \sin \varphi}{3 - \sin \varphi}$	$F = 3\frac{(1+\sin \varphi)^{\beta} - (1-\sin \varphi)^{\beta}}{(1+\sin \varphi)^{\beta} + 2(1-\sin \varphi)^{\beta}}$
Failure stress ratio	$M_{\mathrm{f}} = \frac{6 \sin \varphi_{\mathrm{f}}}{3 - \sin \varphi_{\mathrm{f}}}$	$F_{\mathrm{f}} = 3\frac{(1+\sin \varphi_{\mathrm{f}})^{\beta} - (1-\sin \varphi_{\mathrm{f}})^{\beta}}{(1+\sin \varphi_{\mathrm{f}})^{\beta} + 2(1-\sin \varphi_{\mathrm{f}})^{\beta}}$

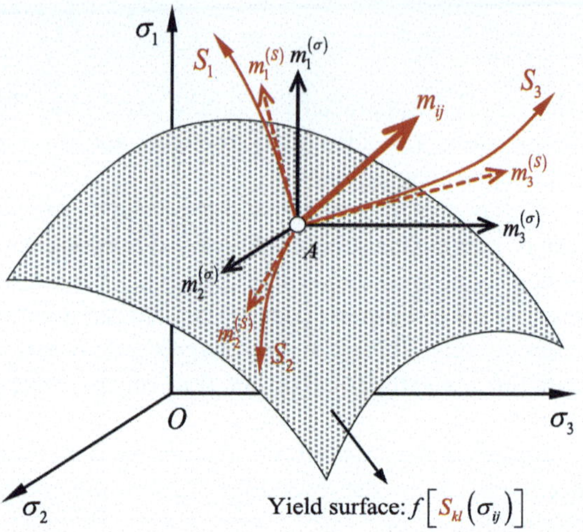

Fig. 3.2 Schematic diagram of the non-orthogonal plastic flow rule

where Λ is the plastic multiplier, and $\mathbf{m} = m_{ij}\mathbf{n}_i \otimes \mathbf{n}_j$ is the plastic flow direction. $\mathbf{S} = S_{kl}\mathbf{n}_k \otimes \mathbf{n}_l$ $(k, l = 1, 2, 3)$ is stress invariant that constructs the yield function and is associated with $\boldsymbol{\sigma}$. The non-orthogonal plastic flow rule is illustrated in Fig. 3.2: first determine the non-orthogonal gradient $\partial^\mu f/\partial\mathbf{S}^\mu$ of the yield function by the fractional derivative in the $\mathbf{S}$ coordinate system, and then transform the non-orthogonal gradient in the $\mathbf{S}$ coordinate system to the $\boldsymbol{\sigma}$ coordinate system by the coordinate transformation $\partial\mathbf{S}/\partial\boldsymbol{\sigma}$. The non-orthogonal plastic flow rule in tensor form is the kernel for determining the general expression of plastic strain increment, which is essential for the model implementation. The plastic flow direction $\mathbf{m}$ is non-orthogonal to the yield surface, which can also degenerate into the normal direction when $\mu = 1$.

The general form of the yield function in the $\mathbf{c}$ space can thus be expressed as:

$$f = f\left[\mathbf{c}, H\left(\boldsymbol{\varepsilon}^{\mathrm{p}}\right)\right] = 0 \tag{3.5}$$

where H is the hardening parameter indicating the hardening degree of soil due to the occurrence of plastic strain $\boldsymbol{\varepsilon}^{\mathrm{p}}$. In view of the superior abilities of $\mathbf{c}$ in reflecting 3D properties of soil only through two characteristic stress invariants c_{n} and c_{s} (Lu et al., 2019a), the yield function can be written as:

$$f = f\left[c_{\mathrm{n}}, c_{\mathrm{s}}, H\left(\boldsymbol{\varepsilon}^{\mathrm{p}}\right)\right] = 0 \tag{3.6}$$

where c_{n} and c_{s} are the characteristic mean stress and the characteristic shear stress, respectively. Their expressions can be found in Table 3.1.

By applying the non-orthogonal plastic flow rule in Eq. (3.4) to the yield function in Eq. (3.6), the plastic flow direction $\mathbf{m}$ in tensor form can be derived in the $\mathbf{c}$ space:

$$\mathbf{m} = \frac{\partial^{\mu} f}{\partial c_{\mathrm{n}}^{\mu}} \frac{\partial c_{\mathrm{n}}}{\partial \mathbf{c}} + \frac{\partial^{\mu} f}{\partial c_{\mathrm{s}}^{\mu}} \frac{\partial c_{\mathrm{s}}}{\partial \mathbf{c}} \tag{3.7}$$

The plastic multiplier Λ can be determined by the consistency condition, which is the total differential of the yield function $f = 0$ in the $\mathbf{c}$ space:

$$df = \frac{\partial f}{\partial \boldsymbol{\sigma}} d\boldsymbol{\sigma} + \frac{\partial f}{\partial H} \frac{\partial H}{\partial \boldsymbol{\varepsilon}^{\mathrm{p}}} : d\boldsymbol{\varepsilon}^{\mathrm{p}} = 0 \tag{3.8}$$

where $\boldsymbol{\varepsilon}^{\mathrm{p}}$ is the plastic strain used for constructing the hardening parameter H, and its increment can be expressed as $\boldsymbol{\varepsilon}^{\mathrm{p}} = \Lambda \mathbf{m}$ based on Eq. (3.4). By substituting $\boldsymbol{\varepsilon}^{\mathrm{p}} = \Lambda \mathbf{m}$ into Eq. (3.8), the multiplier Λ can thus be expressed as:

$$\Lambda = -\frac{\frac{\partial f}{\partial \mathbf{c}} d\mathbf{c}}{\frac{\partial f}{\partial H} \frac{\partial H}{\partial \boldsymbol{\varepsilon}^{\mathrm{p}}} : \mathbf{m}} \tag{3.9}$$

By substituting the plastic flow direction $\mathbf{m}$ in Eq. (3.7) and the plastic multiplier Λ in Eq. (3.9) into Eq. (3.4), the plastic strain increment in the tensor form can be written as:

$$d\boldsymbol{\varepsilon}^{\mathrm{p}} = \Lambda \mathbf{m} = -\frac{\frac{\partial f}{\partial \mathbf{c}} d\mathbf{c}}{\frac{\partial f}{\partial H} \frac{\partial H}{\partial \boldsymbol{\varepsilon}^{\mathrm{p}}} : \mathbf{m}} \left(\frac{\partial^{\mu} f}{\partial c_{\mathrm{n}}^{\mu}} \frac{\partial c_{\mathrm{n}}}{\partial \mathbf{c}} + \frac{\partial^{\mu} f}{\partial c_{\mathrm{s}}^{\mu}} \frac{\partial c_{\mathrm{s}}}{\partial \mathbf{c}} \right) \tag{3.10}$$

For the elastic strain increment in Eq. (3.3), we can make use of the elastic theory, e.g. Hooke's law:

$$d\boldsymbol{\varepsilon}^{\mathrm{e}} = \frac{1+\nu}{E} d\boldsymbol{\sigma} - \frac{\nu}{E} (d\boldsymbol{\sigma} : \mathbf{1}) \otimes \mathbf{1} \tag{3.11}$$

where ν is the Poisson's ratio, and E is the Young's modulus. The expression of E can be obtained from the e-ln$p$ relationship of the swelling line under isotropic compression conditions:

$$E = 3(1 - 2\nu) \frac{1 + e_0}{\kappa} p \tag{3.12}$$

where κ is the swelling index, and e_0 is the initial void ratio at $p = p_0$. In order to express the framework of the NOEP model in the $\mathbf{c}$ space, the elastic principal strain can also be formulated in the same space based on Eqs. (3.11) and (3.1), as follows:

$$\begin{aligned}
d\varepsilon_1^e &= \frac{1}{\beta E}c_1^{\frac{1}{\beta}-1}dc_1 - \frac{\nu}{\beta E}\left(c_2^{\frac{1}{\beta}-1}dc_2 + c_3^{\frac{1}{\beta}-1}dc_3\right)\\
d\varepsilon_2^e &= \frac{1}{\beta E}c_2^{\frac{1}{\beta}-1}dc_2 - \frac{\nu}{\beta E}\left(c_3^{\frac{1}{\beta}-1}dc_3 + c_1^{\frac{1}{\beta}-1}dc_1\right)\\
d\varepsilon_3^e &= \frac{1}{\beta E}c_3^{\frac{1}{\beta}-1}dc_3 - \frac{\nu}{\beta E}\left(c_1^{\frac{1}{\beta}-1}dc_1 + c_2^{\frac{1}{\beta}-1}dc_2\right)
\end{aligned} \tag{3.13}$$

The Young's elastic modulus E can also be expressed as the function of the mean characteristic stress, as follows:

$$E = 3(1 - 2\nu)\frac{1 + e_0}{\kappa}c_n^{\frac{1}{\beta}} \tag{3.14}$$

Finally, by substituting Eqs. (3.11) and (3.10) into Eq. (3.3), the total strain increment is thus determined within the framework of the 3D non-orthogonal elastoplastic model.

3.2 Non-orthogonal Elastoplastic Model for Clay

Three essential parts for developing a NOEP model, including the yield function, the non-orthogonal plastic flow rule and the hardening parameter, are all established in the characteristic stress space. Among them, the combination of the non-orthogonal plastic flow rule and the characteristic stress is used to capture the plastic deformation properties of soils under true 3D stress conditions.

3.2.1 Yield Function

For the sake of simplicity, the yield function established in the characteristic stress space takes the form of an ellipsoid. It has the same form and convexity with the yield function of the modified Cam-clay (MCC) model (Roscoe & Burland, 1968). The yield function in this paper can be expressed as:

$$f = c_s^2 + N^2\left(c_n^2 - c_{nx}c_n\right) = 0 \tag{3.15}$$

where N is the ratio of the vertical axis and the horizontal axis of the elliptic yield curve as shown in Fig. 3.3, which is the geometrical role of M in the yield function of the MCC model. The failure stress ratio F is the physical role of M in the MCC model for describing the critical state of soil. N and F equal to M when $\beta = 1$. The equivalent consolidation pressure c_{nx} is the intersection coordinate of the yield curve with the abscissa and can be expressed as:

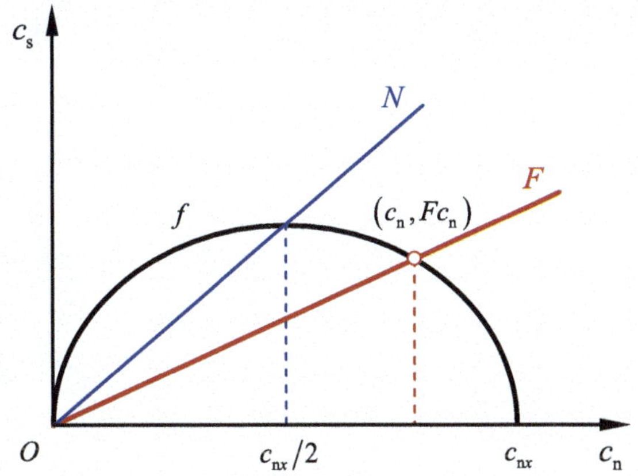

Fig. 3.3 Yield surface in the meridian plane

$$c_{nx} = \frac{c_s^2 + N^2 c_n^2}{N^2 c_n} \tag{3.16}$$

Only two stress invariants c_n and c_s are used to construct the yield function. Thus, the yield surface in the **c** space shown in Fig. 3.4a is an axially symmetric ellipsoid with respect to the c_n-axis. However, the corresponding yield surface in the $\boldsymbol{\sigma}$ space is an irregular, smooth and convex three-axis symmetric surface when $0 < \beta < 1$ as shown in Fig. 3.4b.

The yield curves in the meridian plane and the deviatoric plane of the $\boldsymbol{\sigma}$ space are shown in Fig. 3.5. All the yield curves pass through the critical state point $(p_0, M \cdot p_0)$. For a certain value of β $(0 < \beta < 1)$, yield curve is asymmetric with respect to the p-axis in the meridian plane as shown in Fig. 3.5a and is a curved triangle in the deviatoric plane as shown in Fig. 3.5b. The value of β reflects effects of Lode angle θ or the intermediate principal stress coefficient b on soil behaviour. The yield curves of $\beta = 0.1$ and $\beta = 0.01$ are almost coincident, thus $\beta = 0.1$ can be taken as the approximate lower bound in this paper for saving the computational cost. As q increases to the critical state along the stress path of constant p_0 under the monotonic loading, the yield curve of $\beta = 0.1$ in the meridian plane gradually evolves and expands, as shown in Fig. 3.6a. The yield curve in the deviatoric plane changes from an approximate circle to a curved triangle, as shown in Fig. 3.6b, which reflects the stress-induced anisotropy of geomaterials.

3.2.2 Direction of Plastic Strain Increment

The non-orthogonal plastic flow rule and the characteristic stress are independent in definition but complementary in property description: one for adjusting the plastic strain increment direction, and the other for describing the soil behaviour under true 3D stress conditions in a simple and unified way. By introducing the new proposed

Fig. 3.4 Yield surface

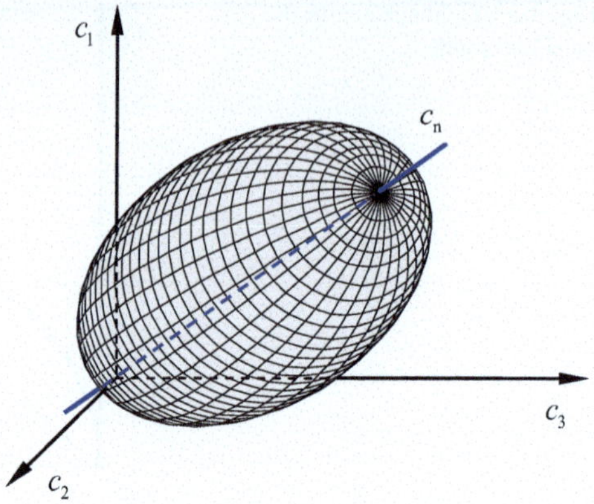

(a) In the $\mathbf{c}$ space;

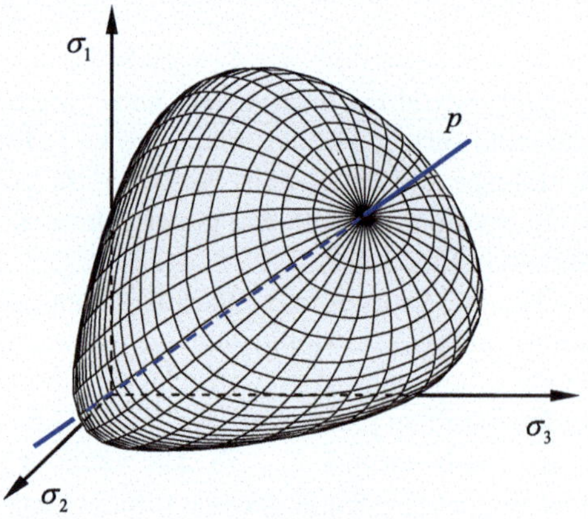

(b) In the $\boldsymbol{\sigma}$ space

non-orthogonal plastic flow rule into the characteristic stress space, a novel sight to describe the soil behaviour is proposed. The non-orthogonal plastic flow rule in the characteristic stress space is thus established by referring Eq. (3.4) as follows:

$$\mathrm{d}\boldsymbol{\varepsilon}^{\mathrm{p}} = \Lambda\left(\frac{\partial^{\mu} f}{\partial c_{\mathrm{n}}^{\mu}}\frac{\partial c_{\mathrm{n}}}{\partial \mathbf{c}} + \frac{\partial^{\mu} f}{\partial c_{\mathrm{s}}^{\mu}}\frac{\partial c_{\mathrm{s}}}{\partial \mathbf{c}}\right) \tag{3.17}$$

where the fractional partial derivative of the yield function with respect to c_{n} and c_{s} are determined as follows:

Fig. 3.5 Yield curves in the **σ** space with the decrease of β

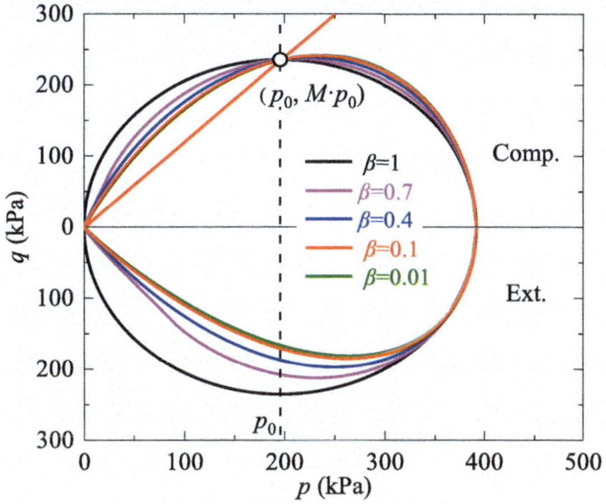

(a) In the meridian plane

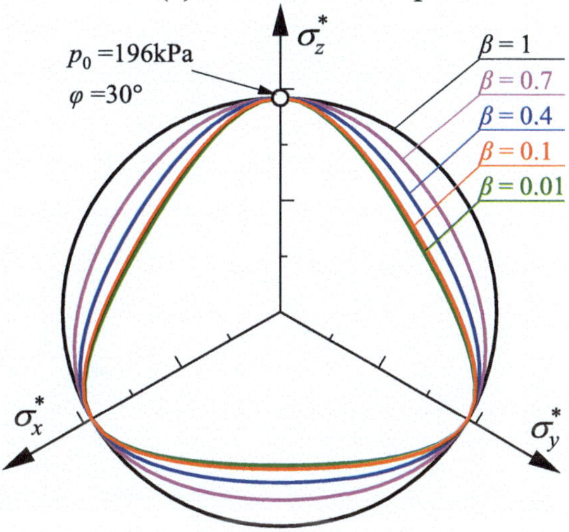

(b) In the deviatoric plane

$$
\begin{cases}
\dfrac{\partial^{\mu} f}{\partial c_{\mathrm{n}}^{\mu}} = \dfrac{2N^2 c_{\mathrm{n}}^{2-\mu}}{\Gamma(3-\mu)} - \dfrac{N^2 c_{\mathrm{n}x} c_{\mathrm{n}}^{1-\mu}}{\Gamma(2-\mu)} \\
\dfrac{\partial^{\mu} f}{\partial c_{\mathrm{s}}^{\mu}} = \dfrac{2c_{\mathrm{s}}^{2-\mu}}{\Gamma(3-\mu)}
\end{cases} \tag{3.18}
$$

Introducing $c_{\mathrm{n}x}$ expressed by Eq. (3.16) and the recursive property of gamma function, i.e., $\Gamma(x+1) = x\cdot\Gamma(x)$ into Eq. (3.18) yields:

Fig. 3.6 Yield curve in the **σ** space with the increase of q at a constant p

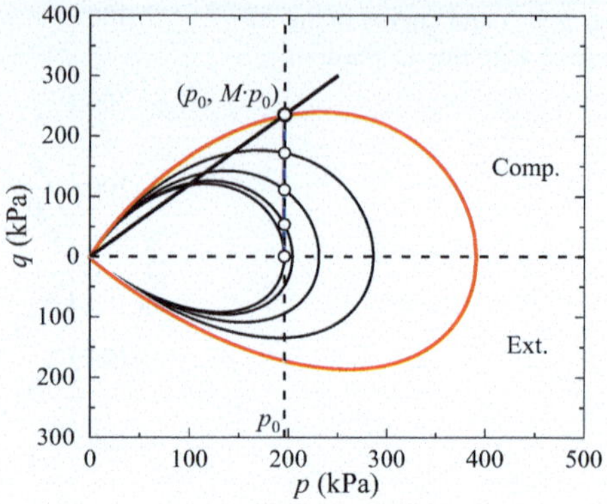

(a) In the meridian plane

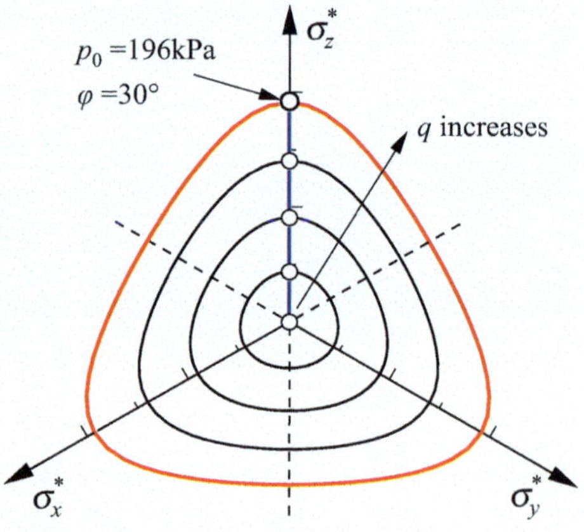

(b) In the deviatoric plane

$$\begin{cases} \dfrac{\partial^\mu f}{\partial c_{\mathrm{n}}^\mu} = \dfrac{\mu N^2 c_{\mathrm{n}}^2 - (2-\mu) c_{\mathrm{s}}^2}{\Gamma(3-\mu) c_{\mathrm{n}}^\mu} \\ \dfrac{\partial^\mu f}{\partial c_{\mathrm{s}}^\mu} = \dfrac{2 c_{\mathrm{s}}^{2-\mu}}{\Gamma(3-\mu)} \end{cases} \tag{3.19}$$

The covariant transformation coefficient, i.e., the partial differential of c_{n} and c_{s} with respect to **c**, can be derived as follows:

$$\begin{cases} \dfrac{\partial c_{\mathrm{n}}}{\partial \mathbf{c}} = \dfrac{1}{3}\mathbf{1} \\ \dfrac{\partial c_{\mathrm{s}}}{\partial \mathbf{c}} = \dfrac{3(\mathbf{c} - c_{\mathrm{n}}\mathbf{1})}{2 c_{\mathrm{s}}} \end{cases} \tag{3.20}$$

By substituting Eqs. (3.19) and (3.20) into Eq. (2.14), the plastic strain increment can be expressed as:

$$\mathrm{d}\boldsymbol{\varepsilon}^{\mathrm{p}} = \Lambda\left[\frac{\mu N^2 c_{\mathrm{n}}^2 - (2-\mu)c_{\mathrm{s}}^2}{\Gamma(3-\mu)c_{\mathrm{n}}^{\mu}}\frac{1}{3}\mathbf{1} + \frac{2c_{\mathrm{s}}^{2-\mu}}{\Gamma(3-\mu)}\frac{3(\mathbf{c}-c_{\mathrm{n}}\mathbf{1})}{2c_{\mathrm{s}}}\right] \tag{3.21}$$

The ratio of the plastic volumetric strain increment $\mathrm{d}\varepsilon_{\mathrm{v}}^{\mathrm{p}}$ to the plastic deviatoric strain increment $\mathrm{d}\varepsilon_{\mathrm{s}}^{\mathrm{p}}$ (i.e., the stress-dilatancy ratio $d = \mathrm{d}\varepsilon_{\mathrm{v}}^{\mathrm{p}}/\mathrm{d}\varepsilon_{\mathrm{s}}^{\mathrm{p}}$) can thus be obtained from Eq. (2.14) together with Eq. (3.19). The stress-dilatancy ratio d is the reflection of the plastic strain increment direction, which can be expressed as follows.

$$d = \frac{\mathrm{d}\varepsilon_{\mathrm{v}}^{\mathrm{p}}}{\mathrm{d}\varepsilon_{\mathrm{s}}^{\mathrm{p}}} = \frac{\Lambda\frac{\partial^{\mu} f}{\partial c_{\mathrm{n}}^{\mu}}}{\Lambda\frac{\partial^{\mu} f}{\partial c_{\mathrm{s}}^{\mu}}} = \frac{\mu N^2 - (2-\mu)\chi^2}{2\chi^{2-\mu}} \tag{3.22}$$

The condition of $\varphi_{\mathrm{c}} = 30°$ and $\beta = 0.4$ is taken as a special case for demonstration, and $N = 0.769$ and $\mu = 0.537$ can be determined by parameter determination methods. When the soil is loaded from the isotropic compression point A_0 to the point A_1 along the conventional triaxial compression path A_0A_{f}, the yield curve and the plastic strain increment directions are shown in Fig. 3.7. As shown in Fig. 3.7a, the yield curve in the meridian plane through the current stress point A_1 is indicated by the black ellipsoidal curve. The plastic strain increment directions on the yield curve for $\mu = 0.537$ are indicated by the red solid arrow. The normal gradient directions when $\mu = 1$ are orthogonal to the yield curve, and is indicated by the blue dash arrow. Since the expression of the yield function is independent of θ, the yield curve in the deviatoric plane is a circle as shown in Fig. 3.7b. Therefore, the plastic strain increment direction is perpendicular to the yield circle in the deviatoric plane and independent of μ. The overview of the yield surface and the plastic strain increment directions in the **c** space are shown in Fig. 3.8.

For the condition of $\mu = 0.537$, the variation rules of the plastic strain increments with increasing χ are shown in Fig. 3.9. When $\chi = 0$, the plastic strain increment direction is horizontal to the right and is independent of μ, which means that plastic deviatoric strain does not occur under the isotropic compression conditions, i.e., $\mathrm{d}\varepsilon_{\mathrm{s}}^{\mathrm{p}} = 0$. With the increase of χ, the plastic strain increment directions gradually change from horizontal to vertical, and $\mathrm{d}\varepsilon_{\mathrm{v}}^{\mathrm{p}} > 0$. When χ reaches F, the stress state reaches the critical state, but not the peak of the yield curve. The plastic strain increment direction is vertically upward, which means that plastic volumetric strain does not occur, i.e., $\mathrm{d}\varepsilon_{\mathrm{v}}^{\mathrm{p}} = 0$. It should be noticed that for a fixed value of χ, the plastic strain increment directions on different yield curves are the same, as shown in Fig. 3.9.

The stress-dilatancy relationship actually reflects the variation rules of the plastic strain increment direction. The stress-dilatancy relationship d–χ at different values of β is shown in Fig. 3.10. $\varphi_{\mathrm{c}} = 30°$ is used for demonstration. The model parameters N and μ correspond to $\beta = 0.1$, 0.4, 0.7 and 1.0 are listed in Table 3.2. Once the

Fig. 3.7 Plastic strain increment direction in the **c** space at the current loading point

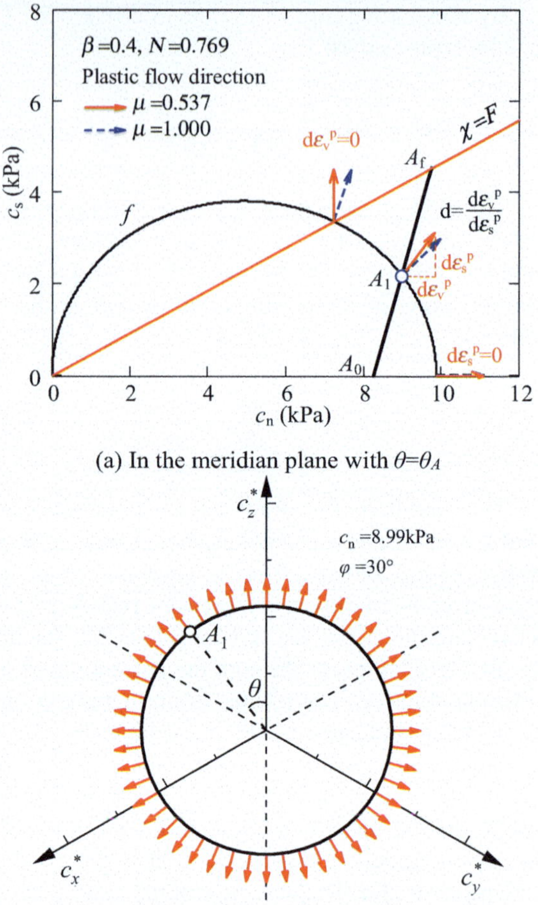

(a) In the meridian plane with $\theta=\theta_A$

(b) In the deviatoric plane with $p=p_A$

material parameters φ_c and β are obtained, the one-to-one model parameters N and μ can be determined by parameter determination methods. For $\beta = 1$, N and F equal to M, μ degenerates to 1 and the d–χ relationship is the same with the stress-dilatancy relationship of MCC model. With the decrease of β, the determined values of N and μ decrease accordingly as listed in Table 3.2, and the d-χ relationship curves decline with the decreasing β. Different levels of stress-dilatancy under compression and extension conditions can be reflected as shown in Fig. 3.10b, which benefits from the characteristic stress. The difference degree of the stress-dilatancy under triaxial compression and extension conditions enlarges with the decreasing β. Figure 3.11 shows the stress-dilatancy curves in the characteristic stress space and the general stress space for $\beta = 0.1$. Stress-dilatancy curves at $b = 0, 0.25, 0.5, 0.75$ and 1 in the characteristic stress space are coincident, as shown in Fig. 3.11a. These coincident

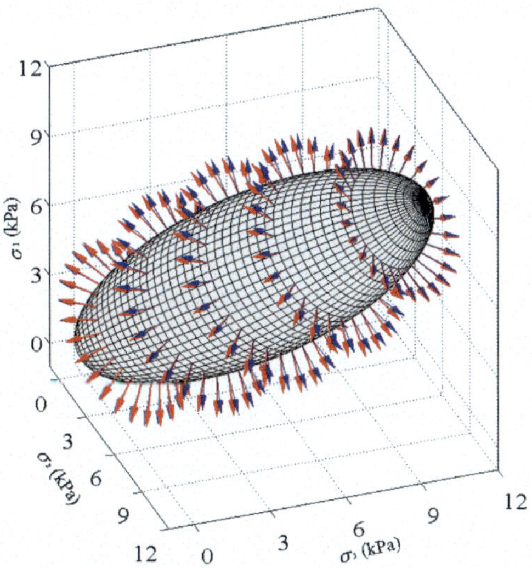

Fig. 3.8 Plastic strain increment direction in the **c** space

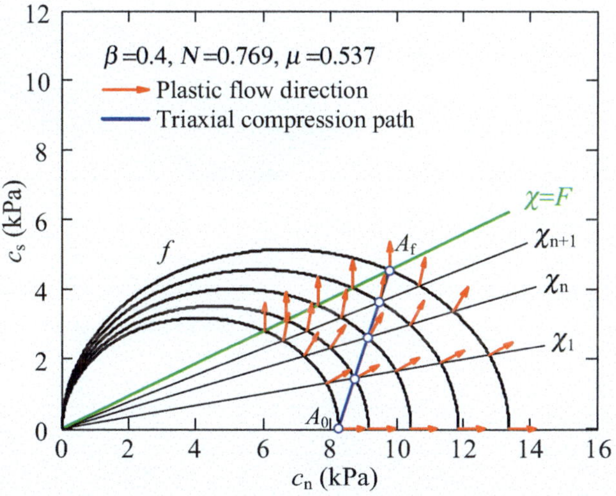

Fig. 3.9 Plastic strain increment direction during the loading process

curves in the **c** space correspond to a series of stress-dilatancy curves in the **σ** space, as shown in Fig. 3.11b.

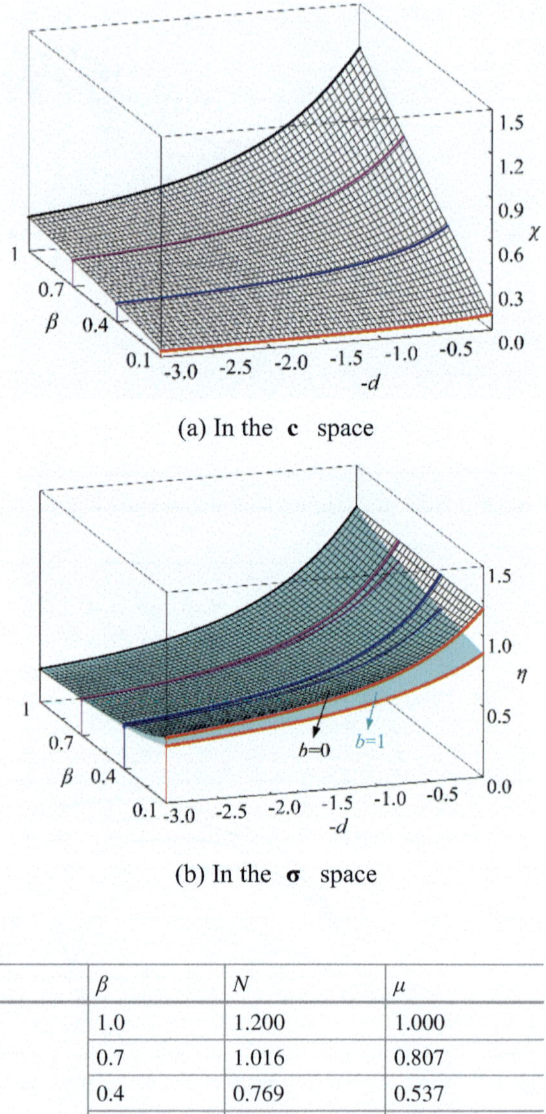

Fig. 3.10 Stress-dilatancy relationship for different β-values

Table 3.2 Parameters in the characteristic stress spaces

φ_c	β	N	μ
30°	1.0	1.200	1.000
	0.7	1.016	0.807
	0.4	0.769	0.537
	0.1	0.381	0.158

Fig. 3.11 Stress-dilatancy curves when $\beta = 0.1$

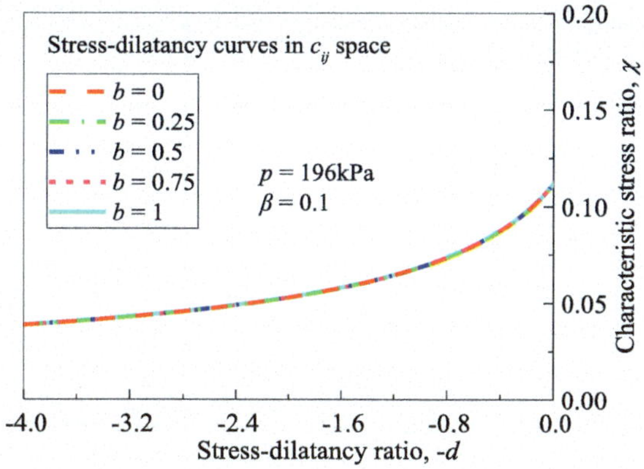

(a) In the **c** space

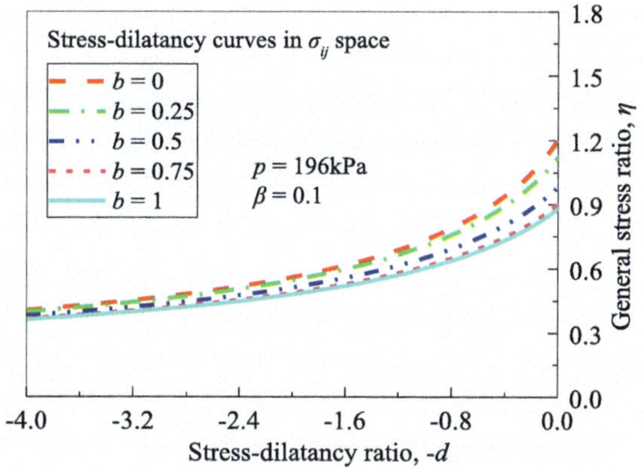

(b) In the $\boldsymbol{\sigma}$ space

3.2.3 *Magnitude of Plastic Strain Increment*

Hardening parameter can be used to describe the hardening/softening rule as well as the evolution of the yield surface. Some valuable studies on hardening parameter have been done to describe the strain hardening and softening properties (Wang et al., 2018; Yao et al., 2009), the stress-dilatancy and state-dependency (Li & Dafalias, 2000) and the particle breakage effect (Xiao et al., 2017). Similar with the MCC model, the plastic volumetric strain is chosen as the hardening parameter in this paper, which can be derived from the linear relationship between e and ln$p$ for normally consolidated clay under isotropic compression conditions ($\sigma_1 = \sigma_2 = \sigma_3$). As shown in Fig. 3.12, the relationship between the void ratio e and ln$c_{\rm n}$ can be easily obtained from the linear relationship of e – ln$p$ under the isotropic compression conditions. Corresponding to the compression index λ and the swelling index κ in the general

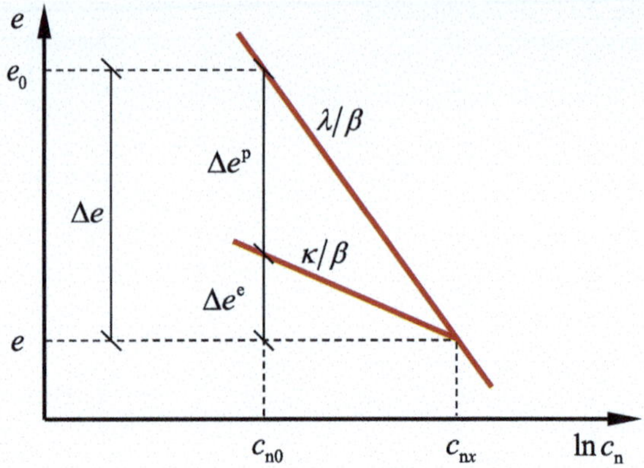

Fig. 3.12 e-lnc_n relationship in the **c** space.

stress space, the compression index and the swelling index in the characteristic stress space become λ/β and κ/β.

The way to derive the relationship between the plastic volumetric strain ε_v^p and the equivalent consolidation pressure c_{nx} in the **c** space is listed in Table 3.3. The equivalent consolidation pressure can be written as follows:

$$c_{nx} = c_{n0}\exp\left(\frac{\varepsilon_v^p}{\rho_p}\right) \tag{3.23}$$

where $\rho_p = (\lambda - \kappa)/\beta/(1 + e_0)$. e_0 is the initial void ratio. c_{n0} is the initial mean characteristic stress.

The consistency condition, i.e., the total differential of the yield function Eq. (3.15), can be expressed as follows:

$$df = \frac{\partial f}{\partial c_n}dc_n + \frac{\partial f}{\partial c_s}dc_s + \frac{\partial f}{\partial c_{nx}}\frac{\partial c_{nx}}{\partial \varepsilon_v^p}d\varepsilon_v^p = 0 \tag{3.24}$$

Table 3.3 Rule of the strain hardening for clay in the characteristic stress space

	Characteristic stress space
Initial mean characteristic stress	$c_{n0} = p_0^\beta$
Void ratio e	$\Delta e = e - e_0 = -\frac{\lambda}{\beta}\ln\frac{c_{nx}}{c_{n0}}$
volumetric strain ε_v	$\varepsilon_v = \frac{-\Delta e}{1+e_0} = \frac{\lambda}{\beta(1+e_0)}\ln\frac{c_{nx}}{c_{n0}}$
Elastic volumetric strain $\varepsilon_v{}^e$	$\varepsilon_v^e = \frac{-\Delta e^e}{1+e_0} = \frac{\kappa}{\beta(1+e_0)}\ln\frac{c_{nx}}{c_{n0}}$
Plastic volumetric strain $\varepsilon_v{}^p$	$\varepsilon_v^p = \varepsilon_v - \varepsilon_v^e = \frac{\lambda-\kappa}{\beta(1+e_0)}\ln\frac{c_{nx}}{c_{n0}}$
Equivalent consolidation pressure c_{nx}	$c_{nx} = c_{n0}\exp\left(\beta\frac{1+e_0}{\lambda-\kappa}\varepsilon_v^p\right) = c_{n0}\exp\left(\frac{\varepsilon_v^p}{\rho_p}\right)$

where $\partial f / \partial c_{\mathrm{n}} = N^2(2c_{\mathrm{n}} - c_{\mathrm{n}x})$, $\partial f / \partial c_{\mathrm{s}} = 2c_{\mathrm{s}}$, $\partial f / \partial c_{\mathrm{n}x} = -N^2 c_{\mathrm{n}}$. The derivative of $c_{\mathrm{n}x}$ with respect to $\varepsilon_{\mathrm{v}}^{\mathrm{p}}$ can derived from Eq. (3.23):

$$\frac{\partial c_{\mathrm{n}x}}{\partial \varepsilon_{\mathrm{v}}^{\mathrm{p}}} = \frac{c_{\mathrm{n}x}}{\rho_{\mathrm{p}}} \tag{3.25}$$

By substituting $\mathrm{d}\varepsilon_{\mathrm{v}}^{\mathrm{p}}$ derived from Eq. (3.21) together with Eq. (3.25) into Eq. (3.24), the plastic multiplier Λ can be derived as follows:

$$\Lambda = \rho_{\mathrm{p}} \frac{\Gamma(3-\mu)c_{\mathrm{n}}^{\mu}\left[N^2(2c_{\mathrm{n}} - c_{\mathrm{n}x})\mathrm{d}c_{\mathrm{n}} + 2c_{\mathrm{s}}\mathrm{d}c_{\mathrm{s}}\right]}{N^2 c_{\mathrm{n}} c_{\mathrm{n}x}\left[\mu N^2 c_{\mathrm{n}}^2 - (2-\mu)c_{\mathrm{s}}^2\right]} \tag{3.26}$$

3.2.4 *Stress–Strain Relationship*

The plastic strain increment is derived by substituting the plastic multiplier Λ in Eq. (3.26) together with $c_{\mathrm{n}x}$ in Eq. (3.16) into Eq. (3.21).

$$\mathrm{d}\boldsymbol{\varepsilon}^{\mathrm{p}} = \rho_{\mathrm{p}}\left[\frac{\mathbf{1}}{3} + \frac{3c_{\mathrm{n}}^{\mu}c_{\mathrm{s}}^{1-\mu}(\mathbf{c} - c_{\mathrm{n}}\mathbf{1})}{\mu N^2 c_{\mathrm{n}}^2 - (2-\mu)c_{\mathrm{s}}^2}\right]\frac{\left(N^2 c_{\mathrm{n}}^2 - c_{\mathrm{s}}^2\right)\mathrm{d}c_{\mathrm{n}} + 2c_{\mathrm{n}}c_{\mathrm{s}}\mathrm{d}c_{\mathrm{s}}}{c_{\mathrm{n}}\left(N^2 c_{\mathrm{n}}^2 + c_{\mathrm{s}}^2\right)} \tag{3.27}$$

The stress–strain relationship in Eq. (3.27) is totally expressed by the characteristic stresses. It can also degenerate into the MCC model when $\beta = 1$. The parameters used in Eq. (3.27) can all be determined by the traditional triaxial tests.

3.2.5 *Parameter Calibration and Model Validation*

(1) ***Parameter Calibration***

There are only five independent material parameters included in the proposed non-orthogonal elastoplastic model, including λ, κ, v, φ_{c} and β, each of which has clear physical significance. Among them, λ, κ, v and φ_{c} are the same as the material parameters included in the MCC model. Only one material parameter β is added, which can be determined by combining φ_{c} for $b = 0$ with an additional internal friction angle φ_b for $b \neq 0$. According to the critical stress ratio F in the $\mathbf{c}$ space, which is independent of the b-value, the equation $F|_{b=0} = \mathrm{F}|_b$ can be used to determine β.

$$\frac{\sqrt{2}\left(R_c^{\beta}-1\right)}{R_c^{\beta}+2}=\frac{\sqrt{\left[R_b^{\beta}-(bR_b-b+1)^{\beta}\right]^2+\left[(bR_b-b+1)^{\beta}-1\right]^2+\left(R_b^{\beta}-1\right)^2}}{(bR_b-b+1)^{\beta}+R_b^{\beta}+1} \tag{3.28}$$

where $R_c = (1 + \sin\varphi_c)/(1 - \sin\varphi_c)$ and $R_b = (1 + \sin\varphi_b)/(1 - \sin\varphi_b)$. Especially, β can be determined with the internal friction angle φ_e for $b = 1$ by the following equation.

$$\frac{(1+\sin\varphi_c)^{\beta}-(1-\sin\varphi_c)^{\beta}}{(1+\sin\varphi_c)^{\beta}+2(1-\sin\varphi_c)^{\beta}}=\frac{(1+\sin\varphi_e)^{\beta}-(1-\sin\varphi_e)^{\beta}}{2(1+\sin\varphi_e)^{\beta}+(1-\sin\varphi_e)^{\beta}} \tag{3.29}$$

Thus, β actually reflects the proportional relation of φ_c and φ_e.

In addition, two model parameters, i.e., N and μ, are include in the proposed model, which can be determined by material parameters without the need for additional tests. The determined methods of the model parameters are given as follows:

Under the triaxial compression condition with p = constant, $\varepsilon_v^e = 0$. Therefore, $\varepsilon_v = \varepsilon_v^p$. When the soil is loaded to the critical state from A to B shown in Fig. 3.13, the variation of the void ratio equals to the vertical distance between the NCL and the CSL, which can be expressed as $\Delta\Pi = (\lambda-\kappa)\cdot\ln\xi$ (Liang et al., 2019; Yu, 1998). Thus, ε_v^p can be calculated by the following equation:

$$\varepsilon_v^p=\frac{\Delta\Pi}{1+e_0}=\frac{(\lambda-\kappa)\cdot\ln\xi}{1+e_0} \tag{3.30}$$

Therefore, c_{nx} can be derived by substituting Eq. (3.30) into Eq. (3.23):

$$c_{nx}=c_{n0}\exp\left(\varepsilon_v^p/\rho_p\right)=\xi^{\beta}c_{n0} \tag{3.31}$$

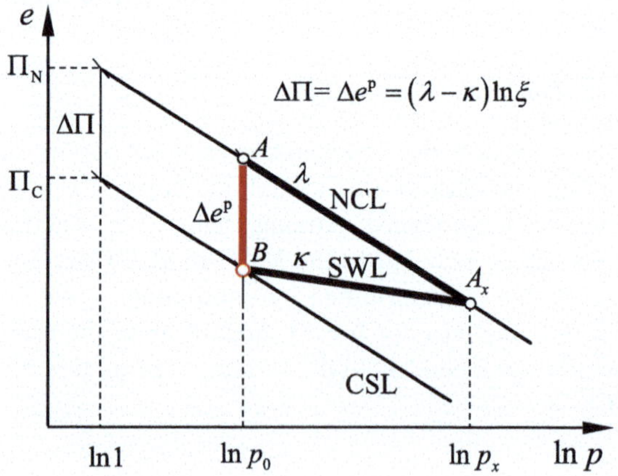

Fig. 3.13 Vertical distance between the NCL and the CSL

The critical principal stress $\sigma_{1z} = p_0(1 + 2M/3)$ and $\sigma_{3z} = p_0(1 - M/3)$. The critical mean characteristic stress $c_{nz} = ({\sigma_{1z}}^{\beta} + 2{\sigma_{3z}}^{\beta})/3$ is thus obtained. By substituting the critical stress state $(c_{nz}, c_{sz}) = (c_{nz}, F \cdot c_{nz})$ into Eq. (3.16), the equivalent consolidation pressure is:

$$c_{nx} = c_{nz}\left(F^2/N^2 + 1\right) \tag{3.32}$$

Combining Eqs. (3.31) with (3.32), one can get N as follows:

$$N = \sqrt{\frac{F^2 c_{nz}}{\xi^{\beta} c_{n0} - c_{nz}}} = F\sqrt{\frac{\left[(1+2M/3)^{\beta} + 2(1-M/3)^{\beta}\right]}{3\cdot\xi^{\beta} - \left[(1+2M/3)^{\beta} + 2(1-M/3)^{\beta}\right]}} \tag{3.33}$$

where F can be calculated by φ_c and β, and M is the function of φ_c. Therefore, N is actually determined only by φ_c and β.

When the critical state condition is reached, χ reaches F, and d in Eq. (3.22) equals 0. The fractional order μ can thus be determined by Eq. (3.22) as follows:

$$\mu = \frac{2F^2}{N^2 + F^2} \tag{3.34}$$

The fractional order μ is also determined by two material parameters, φ_c and β.

(2) ***Validation***

The existing elastoplastic models based on fractional derivative for soils cannot describe the deformation properties of soils under the true 3D stress conditions. The capability of the proposed non-orthogonal elastoplastic model in capturing the deformation properties of soil is thus examined by predicting test results and comparing with the predictions of the MCC model. The true triaxial test results of the new Fujinomori (NF) clay under the drained conditions (Chowdhury & Nakai, 1998) and the Grundite clay under the undrained conditions (Lade & Musante, 1978) are used in this paper.

(a) ***Fujinomori clay***

Drained true triaxial tests were performed on the normally consolidated NF clay at a constant mean stress $p = 196$ kPa. The values of the intermediate principal stress coefficient b were set as 0, 0.268, 0.5 and 0.732, corresponding to different stress Lode angles $\theta = 0°$, 15°, 30°, 45°, respectively. The values of material parameters for the NF clay used in the models are $\varphi_c = 33.7°$, $\lambda/(1 + e_0) = 0.0508$, $\kappa/(1 + e_0) = 0.0112$ and $\nu = 0.3$. The characteristic material parameter $\beta = 0.1$ is used for predictions. Accordingly, the model parameters $N = 0.4248$ and $\mu = 0.1654$ at the condition that $\xi = 2$ can be determined by Eqs. (3.33) and (3.34).

The experimental data from the literature (Chowdhury & Nakai, 1998) are arranged in terms of the relation among principal stress ratio σ_1/σ_3, volumetric strain

ε_v and principal strains (ε_1, ε_2 and ε_3), as indicated by the open circles in Fig. 3.14. These true triaxial test data at different b-values are predicted by the proposed 3D non-orthogonal elastoplastic model and the MCC model, which are indicated by the red solid line and the black dash line, respectively. Both the proposed model and the MCC model can well predict the stress–strain behaviour under triaxial compressions conditions, i.e., $b = 0$, as shown in Fig. 3.14a. However, for $b = 0.268$, 0.5 and 0.732, the experimental stress–strain behaviour is predicted by the proposed non-orthogonal elastoplastic model better than the MCC model, as shown in Fig. 3.14b–d. That is, the proposed non-orthogonal elastoplastic model can well capture the strength property and the deformation behaviour of the normally consolidated NF clay under the drained triaxial conditions.

The relationship between the strains under true 3D stress conditions is actually a reflection of the deformation flow direction of soil. Test results with different b-values are arranged in terms of the relation between ε_2 and ε_1 and the relation between ε_3 and ε_1, as shown by dots in Fig. 3.15. Test results shown in Fig. 3.15a indicate that ε_2 is negative, i.e., extensive, when $b = 0$ and $b = 0.268$, and is positive, i.e., compressive, when $b = 0.5$ and $b = 0.732$. Figure 3.15b shows that ε_3 is extensive in

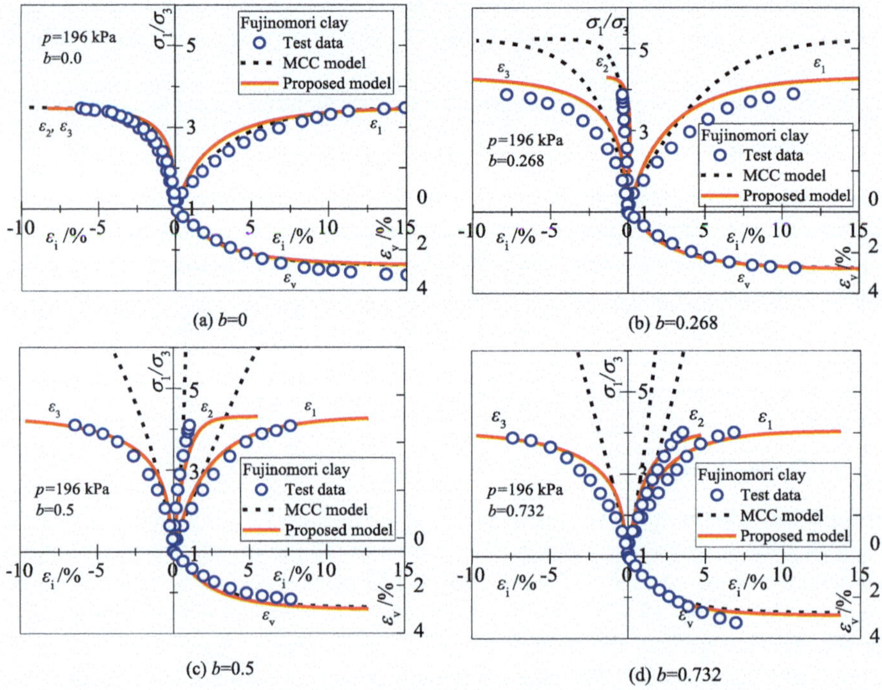

Fig. 3.14 Comparison between test results of Fujinomori clay and predicted results under true 3D stress conditions at $p = 196$ kPa with certain b-value

all the cases, and the extent of extensive gradually decreases for higher b-values. The experimental relation between ε_2 and ε_1 and between ε_3 and ε_1 for a constant b-value, as well as the variation rule for different b-values, are all reasonably captured by the proposed non-orthogonal elastoplastic model better than the MCC model.

(b) ***Grundite clay***

Undrained true triaxial experimental study was performed to study the influence of b on the stress–strain behaviour, pore pressure and strength properties of normally consolidated Grundite clay (Lade & Musante, 1978). Cubical specimens of the remolded Grundite clay were tested at a consolidation pressure of $\sigma_c = 147$ kPa and all three principal stresses are independently controlled. b was maintained constant in each test, and $b = 0.00$, 0.21, 0.40, 0.70 and 0.95, separately. Soil parameters for predictions are from the literature (Yao et al., 2004) and listed as: $\varphi_c = 28.2°$, $\lambda/(1$

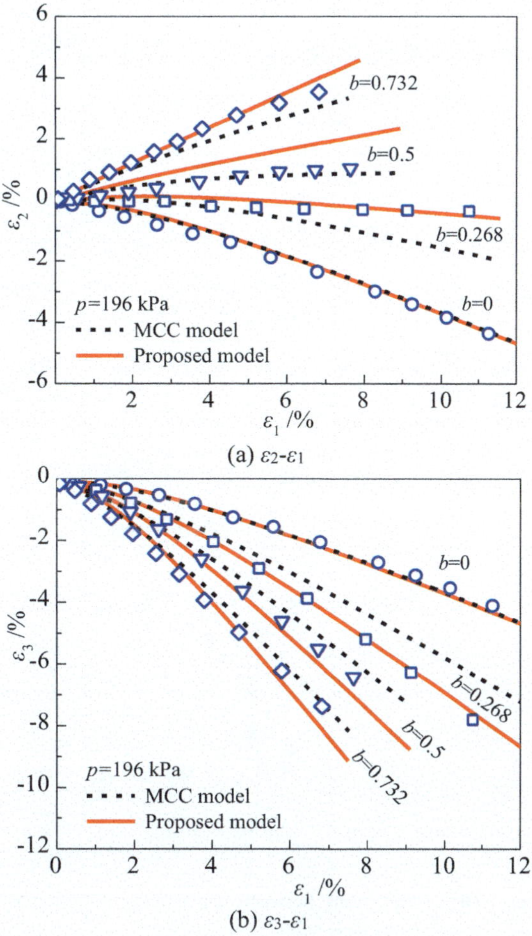

Fig. 3.15 Comparisons between test results of Fujinomori clay and predicted results between principal stains

$+ e_0) = 0.0846$, $\kappa/(1 + e_0) = 0.0169$ and $\nu = 0.27$. $\varphi_e = 31.18°$ from the test with $b = 0.95$ is used to determine $\beta = 0.303$. Correspondingly, the model parameters $N = 0.6289$ and $\mu = 0.4222$ at the condition that $\xi = 2$ can be determined from Eqs. (3.33) and (3.34).

Comparison between test data and predicted lines are depicted in Fig. 3.16, which are arranged in terms of relations among the principal stress ratio σ_1/σ_3, the pore pressure ratio u/σ_c and the principal strain ε_1, separately. When $b = 0$, the predicted

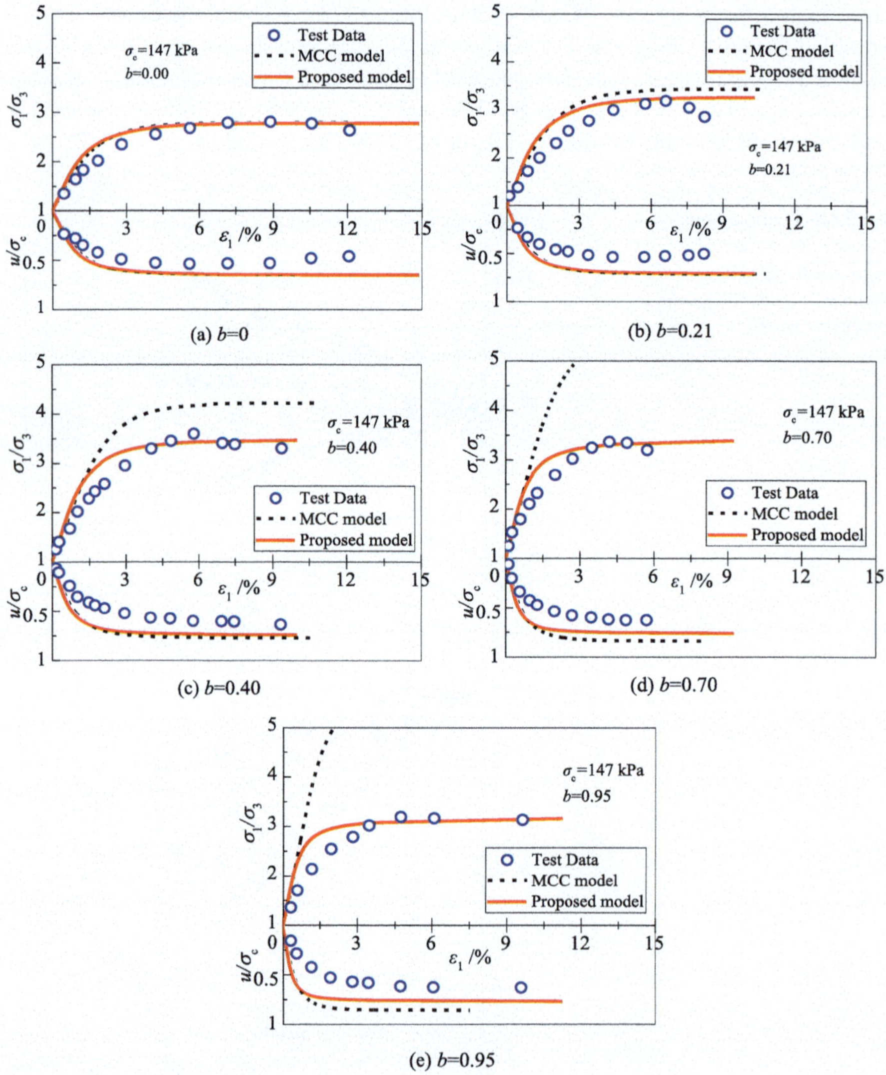

Fig. 3.16 Comparisons between test results of Grundite clay and predicted results with $\sigma_c = 147$ kPa and a certain b-value

lines of the proposed non-orthogonal elastoplastic model are almost the same with the predicted lines of the MCC model, and they both capture the experimental stress–strain behaviour well, as shown in Fig. 3.16a. The shear strengths of the Grundite clay with $b = 0.21, 0.40, 0.70$ and 0.95 are well captured by the proposed model, but obviously overpredicted by the MCC model, as shown in Fig. 3.16b–d. In addition, the proposed non-orthogonal elastoplastic model better predicts the relationship between u/σ_c and ε_1 for $b \neq 0$ than the MCC model.

The test data between principal strains under undrained conditions are compared with the predicted lines of the proposed non-orthogonal elastoplastic model and the MCC model. Test results are represented by dots. The predictions of the proposed model and the MCC model are represented by the red solid lines and the black dash

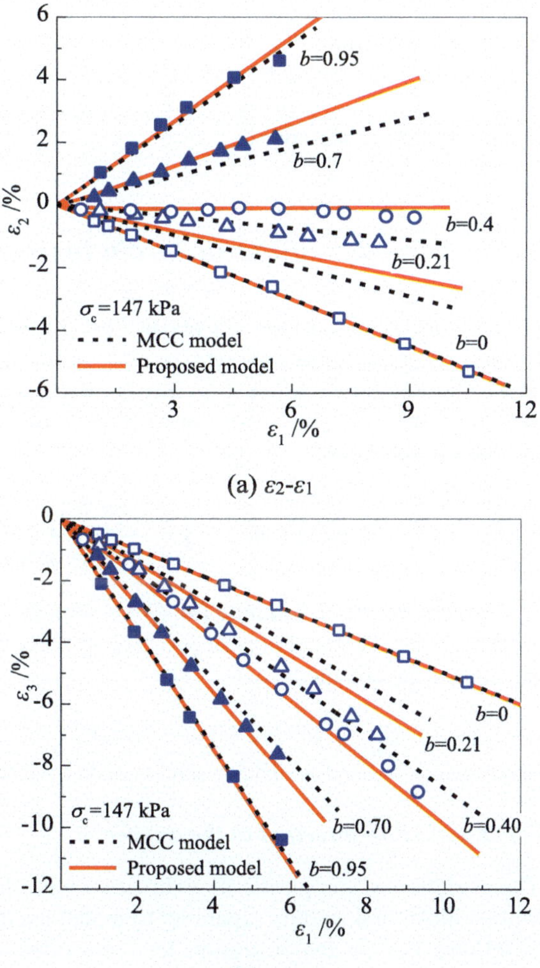

(a) ε_2-ε_1

(b) ε_3-ε_1

Fig. 3.17 Comparisons between test results of Grundite clay and predicted results between principal stains lines, respectively. The test results and the predictions of ε_2 and ε_3 are plotted versus ε_1 for different b-values, as shown in Fig. 3.17. As shown in Fig. 3.17a, ε_2 changes from extension to compression as b increases. ε_3 is under extension for all b-values, and the extent of extension gradually decreases with a larger b-value, as shown in Fig. 3.17b. By comparison, the ability of the proposed model to capture the relationship between the principal strains under true 3D stress conditions is demonstrated to be better than the MCC model.

3.3 Non-orthogonal Elastoplastic Model for Sand with Dilatancy

The non-orthogonal flow rule is not only suitable for clays but also holds potential for describing the strength properties and deformation behaviors of sands. Here, the same yield function, expressed by Eq. (3.15), is employed, which plays the key role in determining Λ and **m**.

3.3.1 Direction of Plastic Strain Increment

By applying the non-orthogonal plastic flow rule in Eq. (3.4) to the yield function, and further incorporating the expression of p_x in Eq. (3.16) and the recursive properties of the Gamma function $\Gamma(x+1) = x\cdot\Gamma(x)$, the non-orthogonal gradient, denotes as $\boldsymbol{m} = (m_\mathrm{n}, m_\mathrm{s})$, remains consistent with the expression given in Eq. (3.19), and can also be expressed as:

$$\mathbf{m} = (m_\mathrm{n}, m_\mathrm{s}) = \left(\frac{\mu N^2 c_\mathrm{n}^2 + (\mu - 2)c_\mathrm{s}^2}{\Gamma(3-\mu)c_\mathrm{n}^\mu}, \frac{2c_\mathrm{s}^{2-\mu}}{\Gamma(3-\mu)} \right) \tag{3.35}$$

Furthermore, by combining Eq. (3.35) with the coordinate transformation coefficient $\partial c_n/\partial \mathbf{c} = \mathbf{1}/3$ and $\partial c_s/\partial \mathbf{c} = 3/2(\mathbf{c} - c_n\mathbf{1})/c_s$, the plastic flow direction tensor **m** can thus be obtained:

$$\mathbf{m} = \frac{\mu N^2 c_\mathrm{n}^2 + (\mu - 2)c_\mathrm{s}^2}{\Gamma(3-\mu)c_\mathrm{n}^\mu} \frac{1}{3}\mathbf{1} + \frac{2c_\mathrm{s}^{2-\mu}}{\Gamma(3-\mu)} \frac{3(\mathbf{c} - c_\mathrm{n}\mathbf{1})}{2c_\mathrm{s}} \tag{3.36}$$

(1) ***Plastic flow direction in the* c *space***

The evolution of the plastic flow direction with the expansion of the yield surface will be illustrated, and its capability to model the dilatancy of sand will be explored as follows. For the demonstration, let us consider the following material parameters: $M = 0.95$, $M_\mathrm{f} = 1.65$ and $\beta = 0.4$, which are typical for dense sands. Therefore, the characteristic phase transition stress ratio $F = 0.366$ and the characteristic failure stress ratio $F_{\mathrm{f}} = 0.664$ are correspondingly obtained based on expressions listed in Table 3.1, and $N = 0.62$ and $\mu = 0.52$ can be determined by the parameter's determination method. Typical stress paths under true triaxial conditions under $p =$ 196 kPa at different stress Lode angles θ ($\theta = 0°$, 15°, 30°, 45° and 60°) are used for illustrating the variation rule of the plastic flow direction.

The simple expression of the yield function makes the yield surface in a simple ellipsoidal shape in the **c** space, and the ellipsoidal yield surface expands only as the size changes rather than the shape changes, as shown in Fig. 3.18. The elliptic yield curve in the meridian plane and the circular yield curve in the deviatoric plane are shown in Fig. 3.19a and b, respectively. With the expansion of the ellipsoidal yield surface, the plastic flow directions for $\mu = 0.52$ and $\mu = 1.00$ correspondingly change, as shown in Figs. 3.18 and 3.19. The plastic flow direction that is non-orthogonal to the yield surface when $\mu = 0.52$ is indicated by the red arrow, and the plastic flow direction that is orthogonal to the yield surface when $\mu = 1.00$ is indicated by the black arrow. As the shear stress increases in the meridian plane, the non-orthogonal plastic flow direction shown in Fig. 3.19a changes from horizontal to the right at $\chi = 0$ to vertically upward at $\chi = F$ for describing the negative dilatancy, and finally turns to the direction pointing to the left until χ reaches F_{f} for capturing the positive dilatancy. As for the plastic flow direction in the deviatoric plane, it is independent of the fractional order μ, since no variable associated with the stress Lode angle contains in the yield function. That is, the plastic flow directions in the deviatoric plane for $\mu = 0.52$ coincide with the plastic flow directions for $\mu = 1$. Here, the coincident plastic flow directions are represented by the red arrows, as shown in Fig. 3.19b. The plastic flow direction in the deviatoric plane is the radial direction along the current characteristic stress Lode angle θ_β defined in Table 3.1. The θ_β-values on the stress paths under the condition that $\theta = 0°$ and 60° are constant, while θ_β-values on the characteristic stress paths at $\theta = 15°$, 30° and 45° are changeable and increase

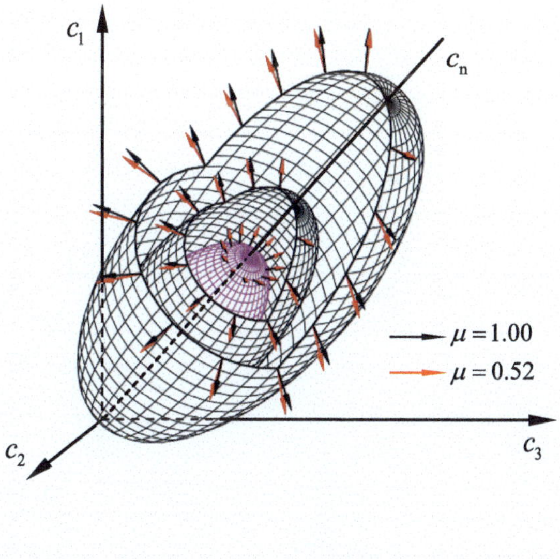

Fig. 3.18 Evolution of the non-orthogonal plastic flow direction with the expanding yield surface in the **c** space

with increasing shear stress. Therefore, the plastic flow directions along the radial direction change with θ_β-values accordingly, as shown in Fig. 3.19b.

(2) *Plastic flow direction in the* $\boldsymbol{\sigma}$ *space*

The expansion of the ellipsoidal yield surface in the **c** space corresponds to the evolution of the yield surface in the $\boldsymbol{\sigma}$ space, as shown in Fig. 3.20. The plastic flow direction determined in the **c** space is applied to the corresponding stress state point on the yield surface in the $\boldsymbol{\sigma}$ space. Corresponding to the expansion of the elliptic yield curve shown in Fig. 3.19a, the evolution of the unsymmetrical yield curves in the meridian plane of the $\boldsymbol{\sigma}$ space is shown in Fig. 3.21a. As the shear stress increases

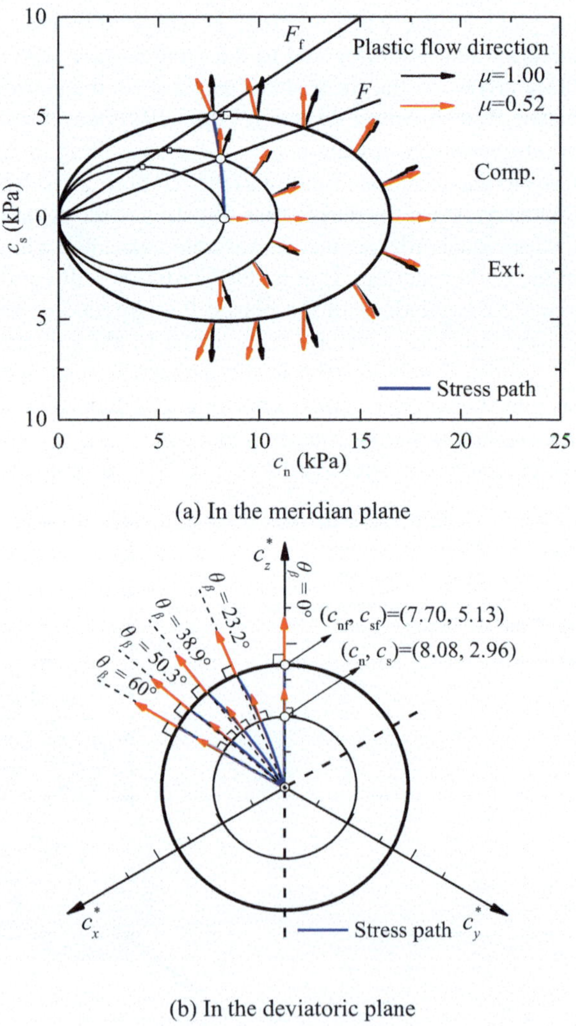

Fig. 3.19 Yield curve and non-orthogonal plastic flow direction in the **c** space

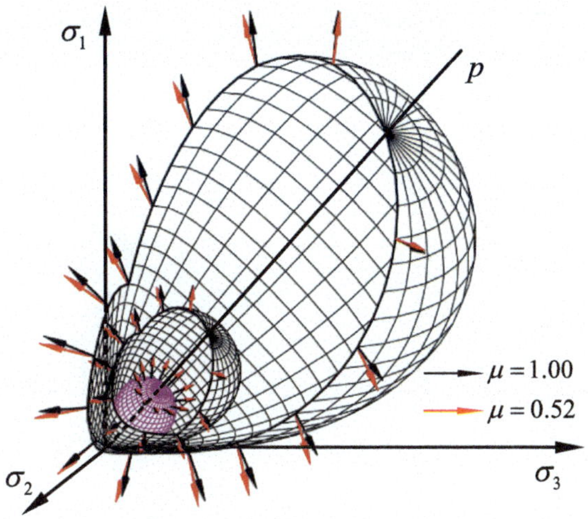

Fig. 3.20 Evolution of the yield surface and the non-orthogonal plastic flow direction in the $\boldsymbol{\sigma}$ space

along the stress path under triaxial compression conditions with a constant p, the plastic flow direction in Fig. 3.21a changes in the same way as that in Fig. 3.19a. It changes from horizontal at stress ratio $\eta = 0$ to vertically upward at $\eta = M$ and finally turns to the direction pointing to the left until η reaches M_f for capturing the dilatancy behaviour from negative to positive. In the deviatoric plane of the $\boldsymbol{\sigma}$ space, the yield curve tends to change from an approximate circle to a curved triangle with the increase of η, and the plastic flow directions at different θ-values correspondingly change, as shown in Fig. 3.21b. The plastic flow directions at $\theta = 0°$ and $60°$ coincide with the radial directions, while the plastic flow directions at $\theta = 15°$, $30°$ and $45°$ deviate leftward from the radial direction, and the degree of deviation increases with the increase of η. This means the stress-induced anisotropy associated with the yield behaviour and plastic flow direction is reflected.

The above analysis illustrates that the non-orthogonal plastic flow rule and characteristic stress are complementary in capturing soil behaviours under 3D stress conditions (Lu et al., 2019a). The non-orthogonal plastic flow rule makes the plastic flow direction suitable for capturing the dilatancy of sand from negative to positive, while the characteristic stress extends the ability to capture dilatancy to 3D stress conditions. The combination of these two concepts gives rise to a new approach to capture the plastic flow direction for the sand under 3D stress conditions.

3.3.2 Magnitude of Plastic Strain Increment

Hardening parameter is an internal variable reflecting the hardening behaviours during the loading process. It is closely related to the determination of the plastic multiplier Λ in Eq. (3.9). In order to incorporate the dilatancy of sand from negative to positive into a non-orthogonal model, the hardening parameter should be formulated

Fig. 3.21 Yield curve and non-orthogonal plastic flow direction in the **σ** space

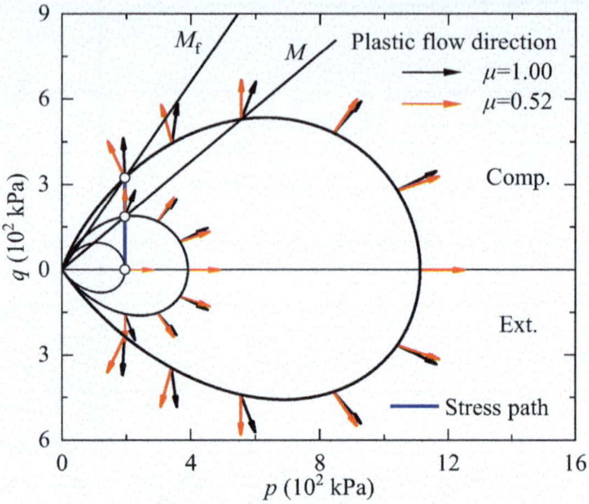

(a) In the meridian plane

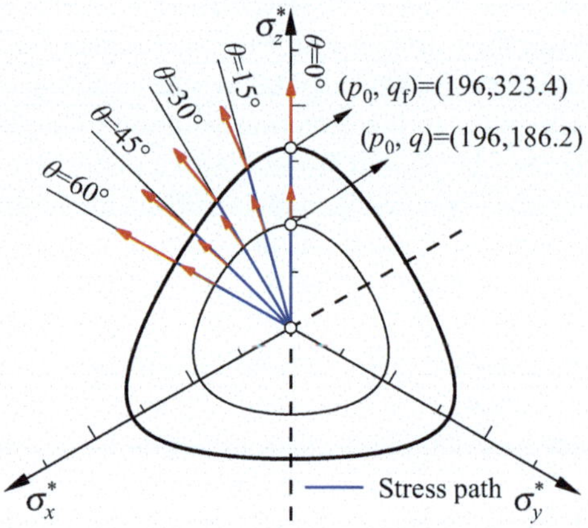

(b) In the deviatoric plane

to be coordinated with the plastic flow direction. During the formulation process, the hardening parameter should also be independent of stress path (Nakai, 1989; Yao et al., 2008). Obviously, the plastic volumetric strain as a hardening parameter is effective for normally consolidated clay. However, it is not suitable for describing initial negative and subsequent positive dilatancy of sand. By referring to the method proposed in the literature (Yao et al., 2008), a stress-path-dependency factor Ω is introduced and combined with the volumetric strain to construct the hardening parameter H, which aims to make H stress-path-independent and capable of capturing the dilatancy of sand.

$$H = \int \Omega \mathrm{d}\varepsilon_{\mathrm{v}}^{\mathrm{p}} \tag{3.37}$$

For normally consolidated clay, $\varepsilon_{\mathrm{v}}^{\mathrm{p}}$ is used as the hardening parameter, so $\Omega = 1$ can be obtained. The relationship $\varepsilon_{\mathrm{v}}^{\mathrm{p}} = l_{\mathrm{p}} \ln c_{\mathrm{n}x} / c_{\mathrm{n}0}$ can be derived based on the linear relationship between e and $\ln(c_{\mathrm{n}})$ under isotropic compression conditions (Lu et al., 2019a), where $l_{\mathrm{p}} = (\lambda - \kappa)/[\beta \cdot (1 + e_0)]$, and λ is the compression index. Therefore, the equivalent characteristic consolidation pressure can be expressed as $c_{\mathrm{n}x} = c_{\mathrm{n}0} \exp(\varepsilon_{\mathrm{v}}^{\mathrm{p}} / l_{\mathrm{p}})$. For sand, H in Eq. (3.37) is adopted to replace $\varepsilon_{\mathrm{v}}^{\mathrm{p}}$, $c_{\mathrm{n}x}$ can thus be expressed as:

$$c_{\mathrm{n}x} = c_{\mathrm{n}0} \exp\left(\frac{H}{l_{\mathrm{p}}}\right) = c_{\mathrm{n}0} \exp\left[\frac{1}{l_{\mathrm{p}}} \int \Omega \mathrm{d}\varepsilon_{\mathrm{v}}^{\mathrm{p}}\right] \tag{3.38}$$

The partial derivative of the yield function f with respect to $\varepsilon_{\mathrm{v}}^{\mathrm{p}}$ is required for determining Λ in Eq. (3.9) and can be obtained based on Eqs. (3.15) and (3.38):

$$\frac{\partial f}{\partial \varepsilon_{\mathrm{v}}^{\mathrm{p}}} = \frac{\partial f}{\partial c_{\mathrm{n}x}} \frac{\partial c_{\mathrm{n}x}}{\partial \varepsilon_{\mathrm{v}}^{\mathrm{p}}} = -\frac{\Omega}{l_{\mathrm{p}}} N^2 c_{\mathrm{n}} c_{\mathrm{n}x} \tag{3.39}$$

In addition, the following differential equation is also needed and can be derived from Eq. (3.15):

$$\frac{\partial f}{\partial c_{ij}} \mathrm{d}c_{ij} = N^2 (2c_{\mathrm{n}} - c_{\mathrm{n}x}) \mathrm{d}c_{\mathrm{n}} + 2c_{\mathrm{s}} \mathrm{d}c_{\mathrm{s}} \tag{3.40}$$

By substituting Eqs. (3.39) and (3.40) together with Eq. (3.16) into Eq. (3.9), the specific expression of Λ is obtained:

$$\Lambda = \frac{l_{\mathrm{p}}}{\Omega} \frac{(N^2 c_{\mathrm{n}}^2 - c_{\mathrm{s}}^2) \mathrm{d}c_{\mathrm{n}} + 2c_{\mathrm{s}} c_{\mathrm{n}} \mathrm{d}c_{\mathrm{s}}}{c_{\mathrm{n}} (c_{\mathrm{s}}^2 + N^2 c_{\mathrm{n}}^2) m_{\mathrm{n}}} \tag{3.41}$$

By substituting Λ in Eq. (3.41) and $(m_{\mathrm{n}}, m_{\mathrm{s}})$ in Eq. (3.35) into Eq. (3.10), together with the stress ratio $\chi = c_{\mathrm{n}}/c_{\mathrm{s}}$, the plastic volumetric strain and the plastic shear strain can be expressed as:

$$\begin{cases} \mathrm{d}\varepsilon_{\mathrm{v}}^{\mathrm{p}} = \Lambda m_{\mathrm{n}} = \dfrac{l_{\mathrm{p}}}{\Omega} \dfrac{(N^2 - \chi^2) \mathrm{d}c_{\mathrm{n}} + 2\chi \mathrm{d}c_{\mathrm{s}}}{c_{\mathrm{n}} (\chi^2 + N^2)} \\ \mathrm{d}\varepsilon_{\mathrm{s}}^{\mathrm{p}} = \Lambda m_{\mathrm{s}} = \dfrac{l_{\mathrm{p}}}{\Omega} \dfrac{(N^2 - \chi^2) \mathrm{d}c_{\mathrm{n}} + 2\chi \mathrm{d}c_{\mathrm{s}}}{c_{\mathrm{n}} (\chi^2 + N^2)} \dfrac{2\chi^{2-\mu}}{\mu N^2 - (2 - \mu)\chi^2} \end{cases} \tag{3.42}$$

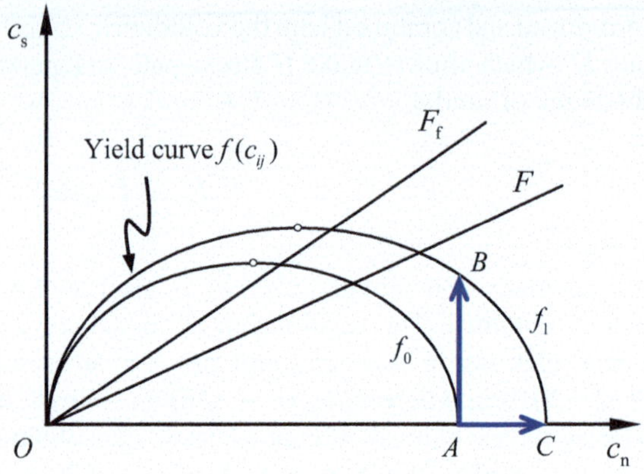

Fig. 3.22 Stress paths for deriving the hardening parameter H

The factor Ω is required for the explicit expression of the plastic strain increment. By referring the literature (Yao et al., 2008), the two stress paths shown in Fig. 3.22, include the constant characteristic mean stress path AB and the isotropic compression stress path AC, are used to derive the stress-path-dependent factor Ω and an explicit expression of H.

(1) ***Constant characteristic mean stress path AB*** **($\mathbf{d}c_n = 0$)**

Along the stress path AB, $\mathrm{d}c_\mathrm{n} = 0$ and c_n is a constant. Therefore, $\mathrm{d}\varepsilon_\mathrm{s}^\mathrm{p}$ in Eq. (3.42) can be simplified as:

$$\mathrm{d}\varepsilon_\mathrm{s}^\mathrm{p} = \frac{1}{\Omega}\frac{4l_\mathrm{p}\chi^{3-\mu}\mathrm{d}\chi}{(\chi^2 + N^2)\left[\mu N^2 - (2-\mu)\chi^2\right]} \tag{3.43}$$

For the normally consolidated clay, $\Omega = 1$. $\mathrm{d}\varepsilon_\mathrm{s}^\mathrm{p}$ is expressed as:

$$\mathrm{d}\varepsilon_\mathrm{s}^\mathrm{p} = \frac{4l_\mathrm{p}\chi^{3-\mu}\mathrm{d}\chi}{(N^2 + \chi^2)\left[\mu N^2 - (2-\mu)\chi^2\right]} \tag{3.44}$$

Similar to Eq. (3.44), the expression of $\mathrm{d}\varepsilon_\mathrm{s}^\mathrm{p}$ for sand can be assumed as:

$$\mathrm{d}\varepsilon_\mathrm{s}^\mathrm{p} = \rho\frac{4l_\mathrm{p}\chi^{3-\mu}\mathrm{d}\chi}{(N_\mathrm{f}^2 + \chi^2)\left[\mu N_\mathrm{f}^2 - (2-\mu)\chi^2\right]} \tag{3.45}$$

Here, ρ is an undetermined coefficient that reflects the ratio of the magnitude of $\mathrm{d}\varepsilon_\mathrm{s}^\mathrm{p}$ between clay and sand along the stress path AB. N in Eq. (3.44) is changed to N_f that related to the failure stress ratio F_f of sand. In fact, F_f is not constant in describing the shearing process with some complex behaviours, but for the sake of simplicity, it is assumed to be a constant. When sand reaches its failure state at $\chi = F_\mathrm{f}$, the expression of $d\varepsilon_s^p$ in Eq. (3.45) approaches infinity, which is similar to Eq. (3.44) for clay. Thus, $N_f = F_f\sqrt{2/\mu - 1}$ can be obtained.

Furthermore, by combining Eq. (3.43) with Eq. (3.45), the stress-path-dependent factor Ω is derived and can be expressed as:

$$\Omega = \frac{1}{\rho}\frac{(2-\mu)\left(F_f^2 - \chi^2\right)\left[(2-\mu)F_f^2 + \mu\chi^2\right]}{\mu\left(N^2 + \chi^2\right)\left[\mu N^2 - (2-\mu)\chi^2\right]} \tag{3.46}$$

Substituting Eq. (3.46) into Eq. (3.37) yields:

$$H = \int \frac{1}{\rho}\frac{(2-\mu)\left(F_f^2 - \chi^2\right)\left[(2-\mu)F_f^2 + \mu\chi^2\right]}{\mu\left(N^2 + \chi^2\right)\left[\mu N^2 - (2-\mu)\chi^2\right]}d\varepsilon_v^p \tag{3.47}$$

(2) ***Isotropic compression stress path AC*** **($dc_s = 0$)**

Under isotropic compression conditions ($dc_s = 0$), χ is equal to 0. Equation (3.47) can thus be simplified as follows:

$$H = \int \frac{1}{\rho}\frac{(2-\mu)^2 F_f^4}{\mu^2 N^4}d\varepsilon_v^p \tag{3.48}$$

Along stress path *AC*, only the plastic volumetric strain ε_v^p occurs for sand. That is, $H = \int d\varepsilon_v^p$ can be used to depict the hardening rule of sand under isotropic compression conditions. The expression of ρ can thus be derived by combining $H = \int d\varepsilon_v^p$ with Eq. (3.48).

$$\frac{1}{\rho} = \frac{\mu^2 N^4}{(2-\mu)^2 F_f^4} \tag{3.49}$$

By substituting Eq. (3.49) into Eq. (3.46), the stress-path-dependent factor Ω can be expressed as:

$$\Omega = \frac{\mu N^4}{(2-\mu)F_f^4}\frac{\left(F_f^2 - \chi^2\right)\left[(2-\mu)F_f^2 + \mu\chi^2\right]}{\left(N^2 + \chi^2\right)\left[\mu N^2 - (2-\mu)\chi^2\right]} \tag{3.50}$$

Finally, the new hardening parameter is obtained by substituting Eq. (3.50) into Eq. (3.37) and can be expressed as follows:

$$H = \int \frac{\mu N^4}{(2-\mu)F_f^4}\frac{\left(F_f^2 - \chi^2\right)\left[(2-\mu)F_f^2 + \mu\chi^2\right]}{\left(N^2 + \chi^2\right)\left[\mu N^2 - (2-\mu)\chi^2\right]}d\varepsilon_v^p \tag{3.51}$$

With the proposed hardening parameter *H*, the negative and positive dilatancy behaviours of sand can be described. The negative and positive dilatancy can be explained and indicated by the following equation derived from Eq. (3.51). Moreover, the following conclusions listed in Table 3.4 can be drawn from Eq. (3.52) based on the non-negative dH.

$$\mathrm{d}\varepsilon_{\mathrm{v}}^{\mathrm{p}} = \frac{(2-\mu)F_{\mathrm{f}}^4}{\mu N^4}\frac{(N^2+\chi^2)\left[\mu N^2-(2-\mu)\chi^2\right]}{(F_{\mathrm{f}}^2-\chi^2)\left[(2-\mu)F_{\mathrm{f}}^2+\mu\chi^2\right]}\mathrm{d}H \tag{3.52}$$

By adopting H with the properties indicated above, the shear responses transiting from contraction to dilation are captured. When $M_{\mathrm{f}} = M$, $F_{\mathrm{f}} = F$ can be obtained, H will degenerate to $\varepsilon_{\mathrm{v}}^{\mathrm{p}}$, only negative dilatancy will be described. For the condition that $\beta = 1$, $\mathbf{c}$ becomes $\boldsymbol{\sigma}$, $F = M$ and $F_{\mathrm{f}} = M_{\mathrm{f}}$ are obtained. When $N = M$, $\mu = 1$ can be derived, so H can degenerate to the hardening parameter proposed in the literature (Yao et al., 2008).

(3) ***Stress-path-independency of the hardening parameter***

The stress-path-independence of the hardening parameter H can be interpreted by the change of H along different paths, as shown in Figs. 3.23 and 3.24. The yield curve passes through the stress state point $A_0(c_{\mathrm{n}x0}, 0)$ which is the only intersection point of the yield surface f_0 and the c_{n}-axis. According to Eq. (3.38), the unique intersection point corresponds to a unique value of H_0. That is, the yield surface f_0 is actually an equivalent surface of H_0. As the loading continues along different paths to points B_1, C_1 and D_1 that are all on the yield surface f_1 as shown in Fig. 3.23a, the yield surface gradually expands, and the intersection moves along the c_{n}-axis to the point A_1 $(c_{\mathrm{n}x1}, 0)$. Accordingly, the value of the hardening parameter changes to H_1, which is equivalent for any stress state on f_1. Therefore, the difference $\Delta H = H_1 – H_0$ can be calculated and is independent on the stress path. This is exactly the basic properties that hardening parameters need to satisfy. The corresponding yield surface and stress paths in the $\boldsymbol{\sigma}$ space are shown in Fig. 3.23b. Figure 3.24 shows that, as the sand is loaded along different stress paths from f_0 to f_1, the value of the H changes from H_0 to H_1. The results of equal variation in H shown in Fig. 3.24 indicates that H is stress-path-independence.

With the newly derived stress-path-dependent factor Ω, the explicit expression of the plastic multiplier Λ can be obtained based on Eq. (3.41):

Table 3.4 Variation rule of $\mathrm{d}\varepsilon_{\mathrm{v}}^{\mathrm{p}}$ with the increase of stress ratio η

Stress state	Characteristic stress state	dH	$\mathrm{d}\varepsilon_{\mathrm{v}}^{\mathrm{p}}$	Current condition
$\eta = 0$	$\chi = 0$	$\mathrm{d}H = \mathrm{d}\varepsilon_{\mathrm{v}}^{\mathrm{p}}$	$\mathrm{d}\varepsilon_{\mathrm{v}}^{\mathrm{p}}>0$	Isotropic compression
$0 < \eta < M$	$0 < \chi < F$	$\mathrm{d}H > 0$	$\mathrm{d}\varepsilon_{\mathrm{v}}^{\mathrm{p}}>0$	Negative dilatancy
$\eta = M$	$\chi = F$	$\mathrm{d}H > 0$	$\mathrm{d}\varepsilon_{\mathrm{v}}^{\mathrm{p}}=0$	Phase transition state
$M < \eta \le M_{\mathrm{f}}$	$F < \chi \le F_{\mathrm{f}}$	$\mathrm{d}H > 0$	$\mathrm{d}\varepsilon_{\mathrm{v}}^{\mathrm{p}}<0$	Positive dilatancy

Fig. 3.23 Stress paths for verification of the stress-path-independence of H

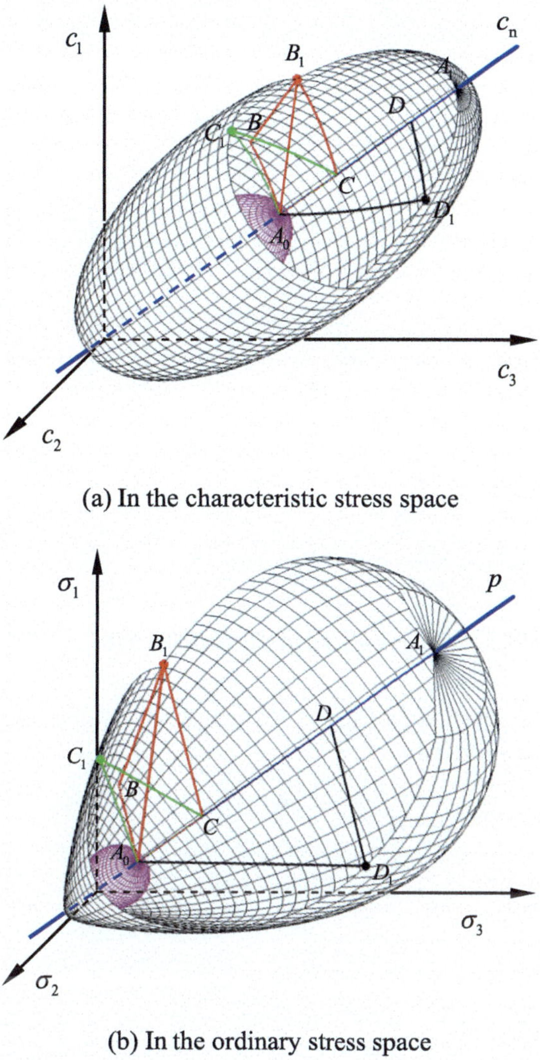

$$\Lambda = l_{\mathrm{p}} \frac{(2-\mu)F_{\mathrm{f}}^4}{\mu N^4} \frac{(N^2+\chi^2)\left[\mu N^2-(2-\mu)\chi^2\right]}{(F_{\mathrm{f}}^2-\chi^2)\left[(2-\mu)F_{\mathrm{f}}^2+\mu\chi^2\right]} \frac{(N^2c_{\mathrm{n}}^2-c_{\mathrm{s}}^2)\mathrm{d}c_{\mathrm{n}}+2c_{\mathrm{s}}c_{\mathrm{n}}\mathrm{d}c_{\mathrm{s}}}{c_{\mathrm{n}}(c_{\mathrm{s}}^2+N^2c_{\mathrm{n}}^2)m_{\mathrm{n}}} \tag{3.53}$$

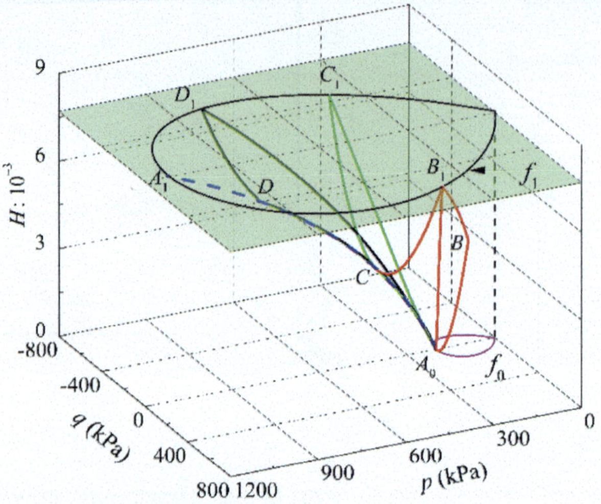

Fig. 3.24 Stress-path-independence of H

3.3.3 Stress–Strain Relationship

By substituting Eqs. (3.36) and (3.53) into Eq. (3.4), the plastic strain increment is thus derived:

$$\mathrm{d}\boldsymbol{\varepsilon}^{\mathrm{p}} = \frac{l_{\mathrm{p}}(2-\mu)F_{\mathrm{f}}^{4}\left[\mu N^{2}-(2-\mu)\chi^{2}\right]\left[\left(N^{2}-\chi^{2}\right)\mathrm{d}c_{\mathrm{n}}+2\chi\,\mathrm{d}c_{\mathrm{s}}\right]}{\mu N^{4}c_{\mathrm{n}}\left(F_{\mathrm{f}}^{2}-\chi^{2}\right)\left[(2-\mu)F_{\mathrm{f}}^{2}+\mu\chi^{2}\right]} \left[\frac{1}{3}\mathbf{1}+\frac{3\chi^{1-\mu}(\mathbf{c}/c_{\mathrm{n}}-\mathbf{1})}{\mu N^{2}-(2-\mu)\chi^{2}}\right] \tag{3.54}$$

As shown in Eq. (3.54), the expression of plastic strain increment is totally expressed by the characteristic stress. Benefits from the application of characteristic stress, the β criterion is used as the shear failure condition in the proposed model, which is similar to the use of the Drucker-Prager criterion in the MCC model. When $M_{\mathrm{f}} = M$, $F_{\mathrm{f}} = F$ can be obtained, and Eq. (3.54) can thus be simplified to determine the plastic strain increment for the loose sand with only negative dilatancy.

The total strain increment produced by the current stress increment can thus be obtained by Eq. (3.3), where the elastic strain increment for sand is expressed by Eq. (3.11) and the plastic strain increment is expressed by Eq. (3.54). The parameters involved in the non-orthogonal elastoplastic model can be easily determined, which will be discussed in the following section.

3.3.4 Parameter Calibration and Model Validation

(1) ***Parameter calibration***

There are seven independent material parameters in the proposed model, i.e., λ, κ, v, φ, φ_f, β and ξ. All material parameters have clear physical meaning and can be easily determined by conventional triaxial tests. The parameters λ, κ and v are the same with the parameters of the modified Cam-clay model. The parameters φ and φ_f are the internal friction angles of the sand under triaxial compression conditions at the phase transition state and the failure state, respectively. β is the characteristic stress index, which can be determined based on the expression of the characteristic failure stress ratio F_f listed in Table 3.1. Actually, F_f can be determined by combining the internal friction angle φ_{fb} at any intermediate principal stress coefficient b ($b = (\sigma_2 - \sigma_3)/(\sigma_1 - \sigma_3)$) with β based on the expression of the β criterion in Eq. (3.2). Note the value of F_f is independent of b, i.e., $F_f|_{b_1} = F_f|_{b_2}$, where the subscripts b_1 and b_2 are two different b-values. Based on this, the parameter β can be determined by φ_{b_1} and φ_{b_2}, which reflects the relative proportional relationship of shear strength at different b. In particular, φ_f at $b_1 = 0$ and φ_{fe} at $b_2 = 1$ are used to determine β-value based on the method represented by the literature (Lu et al., 2017). Parameter ξ is called the spacing ratio (Liang et al., 2019; Yu, 1998). It can also be used to describe the change of void ratio $\Delta\Pi$ of sand between the initial state and the phase transition state under triaxial compression conditions under $p =$ const. The change of void ratio has been derived and expressed as $\Delta\Pi = (\lambda - \kappa)\cdot\ln(p_x/p_{x0})$ (Liang et al., 2019), where p_{x0} and p_x are the equivalent mean principal stresses for the initial state and the phase transition state, respectively. Here, $p_x/p_{x0} = \xi$ is assumed, therefore, ξ can be determined by $\xi = \exp[\Delta\Pi/(\lambda - \kappa)]$, and $\xi = 2$ is a special case derived from the modified Cam-clay model.

In addition, N and μ are two model parameters, which can both be easily determined based on the above material parameters. The determination method of these two parameters is given below.

Since the equivalent mean stresses p_{x0} and p_x are both defined for isotropic compression conditions, the equivalent characteristic mean stresses can be expressed as $c_{nx0} = p_r\cdot(p_{x0}/p_r)^\beta$ and $c_{nx} = p_r\cdot(p_x/p_r)^\beta$. Therefore, based on $\xi = p_x/p_{x0}$ mentioned above, we can get $c_{nx} = \xi^\beta\cdot c_{nx0}$. On the other hand, substituting the phase transition stress state $(c_{nt}, c_{st}) = (c_{nt}, F\cdot c_{nt})$ into Eq. (3.16) yields:

$$c_{nx} = c_{nt}\left(\frac{F^2}{N^2} + 1\right) \tag{3.55}$$

By combining $c_{nx} = \xi^\beta\cdot c_{nx0}$ with Eq. (3.55), N can be derived as:

$$N = F\sqrt{\frac{c_{nt}}{\xi^\beta c_{nx0} - c_{nt}}} \tag{3.56}$$

where $c_{nt} = (c_{1t} + 2c_{3t})/3$, in which $c_{1t} = p_r \cdot (\sigma_{1t}/p_r)^\beta$ and $c_{3t} = p_r \cdot (\sigma_{3t}/p_r)^\beta$ can be obtained from Table 3.1. σ_{1t} and σ_{3t} are the major and minor principal stresses at the phase transition state, which can be calculated from $\sigma_{1t} = p_{x0}(1 + 2M/3)$ and $\sigma_{3t} = p_{x0}(1 - M/3)$, respectively. Therefore, parameter N can be determined by the following equation:

$$N = F\sqrt{\frac{(1+2M/3)^\beta + 2(1-M/3)^\beta}{3\xi^\beta - (1+2M/3)^\beta - 2(1-M/3)^\beta}} \tag{3.57}$$

The shape of the yield surface is known when N is determined. Therefore, the expression of plastic flow direction is obtained, and the stress-dilatancy ratio d can be correspondingly derived:

$$d = \frac{d\varepsilon_v^p}{d\varepsilon_s^p} = \frac{\mu N^2 - (2-\mu)\chi^2}{2\chi^{2-\mu}} \tag{3.58}$$

When the phase transition state reaches, χ reaches F, and d in Eq. (3.58) is equal to 0, then the fractional order μ can be determined by the following equation:

$$\mu = \frac{2F^2}{N^2 + F^2} \tag{3.59}$$

(2) ***Validation***

The performance of the proposed model can be evaluated and verified by comparing model predictions with test results of sand from literature. We consider the tests on saturated Toyoura sand under conventional triaxial compression ($\sigma_3 = p_0$, $p = p_0$ and $\sigma_1 = p_0$) and extension ($\sigma_3 = p_0$, $p = p_0$ and $\sigma_1 = p_0$) conditions and true triaxial conditions (θ = 15°, 30° and 45°) (Nakai, 1989). The mean diameter of sand is 0.2mm, the uniformity coefficient is 1.3 and the specimens were prepared in medium dense with an initial void ratio 0.68. The material parameters used for model prediction are listed in Table 3.5. The corresponding model parameters N = 0.439 and μ = 0.390 can thus be determined based on the material parameter's determination method.

Table 3.5 Material parameters used for model prediction

$\lambda/(1+e_0)$	$\kappa/(1+e_0)$	v	φ	φ_f	β	ξ
0.00403	0.00251	0.3	24.209°	40.475°	0.24	2.3

(a) ***Toyoura sand under conventional triaxial conditions***

Test data of saturated Toyoura sand under conventional triaxial compression ($\sigma_3 = p_0, p = p_0$ and $\sigma_1 = p_0$) and extension ($\sigma_3 = p_0, p = p_0$ and $\sigma_1 = p_0$) conditions are digitized from the literature (Yao et al., 2008) and indicated by open circles, as shown in Fig. 3.25. As shown in Fig. 3.25a and b, the negative dilatancy of sand is obvious, since p increases under triaxial compression and extension conditions when $\sigma_3 = p_0$. Less negative dilatancy can be found in Fig. 3.25c and d, since a constant p is applied. Moreover, due to the decrease of p under triaxial compression and extension conditions when $\sigma_1 = p_0$, the least negative dilatancy is found in Fig. 3.25e and f. The test results of Toyoura sand along six different stress paths are predicted, as shown in Fig. 3.25a–f. Due to the contributions of the non-orthogonal plastic flow direction and the newly constructed H, different degrees of dilatancy from negative to positive along six different stress paths are well predicted. Comparisons between the test results and the predictions show that the relationships of ε_1 versus σ_1/σ_3 and ε_3 versus σ_1/σ_3 are also well captured by the proposed model. On the other hand, the application of $\mathbf{c}$ makes the β criterion naturally used as the shear failure condition of sand. Therefore, the tendency that the shear strength of sand under extension conditions is greater than that under triaxial compression conditions can be well reflected.

(b) ***Toyoura sand under true triaxial conditions***

True triaxial drained tests under constant mean principal stress $p = 196$ kPa (Nakai, 1989) can be used to demonstrate the capability of the 3D elastoplastic models for sand. The values of θ are 15°, 30° and 45°, respectively. Triaxial tests under $p = 196$ when $\theta = 0°$ (triaxial compression) and 60° (triaxial extension) have been used to verify the proposed non-orthogonal model, as shown in Fig. 3.25c and d. The capability of the proposed model in capturing the behaviour of sand under 3D stress conditions can be verified by the predicting test results of sand, as shown in Fig. 3.26a–c. The comparisons are arrange in terms of the relations among the principal stress ratio σ_1/σ_3, volumetric strain ε_v, and principal strains ε_i ($i = 1, 2, 3$). In addition to well capture the volume change from negative to positive under 3D stress conditions, it is also critical for a 3D model for sand to reflect three-dimensional strength behaviour and the relative relationship between principal strains. For the strength property of sand under 3D stress conditions ($\theta = 0°$, 15°, 35°, 45° and 60°), the application of the β criterion makes the predictions in good agreement with the experiment results. Moreover, an appropriate description of the plastic flow direction in the deviatoric plane provides the proposed model with the ability to capture the relative relationship between different principal strains, as shown in Fig. 3.26a–c.

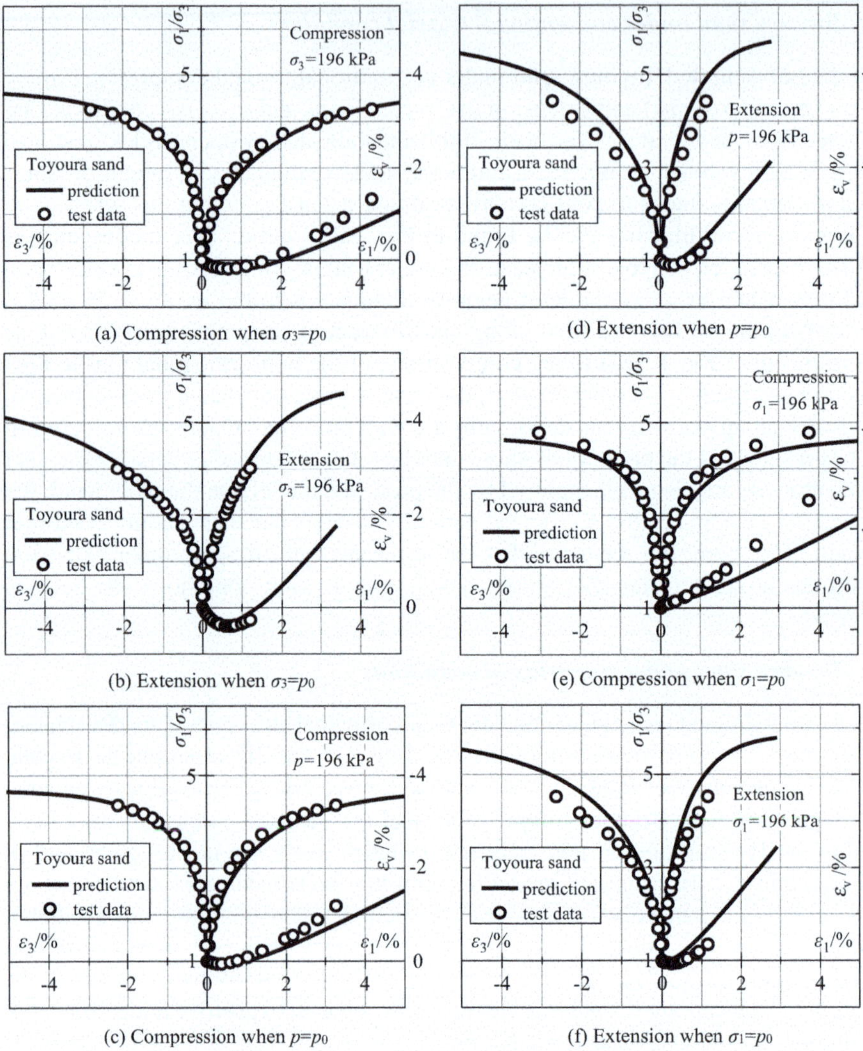

(a) Compression when $\sigma_3 = p_0$

(d) Extension when $p = p_0$

(b) Extension when $\sigma_3 = p_0$

(e) Compression when $\sigma_1 = p_0$

(c) Compression when $p = p_0$

(f) Extension when $\sigma_1 = p_0$

Fig. 3.25 Comparison between predicted and test results under conventional triaxial conditions

3.4 Conclusion

In this Chapter, a framework of the 3D non-orthogonal elastoplastic model was established by combining the non-orthogonal plastic flow rule and the characteristic stress, which makes the developed model be born with the potential to capture 3D mechanical behaviour of soil. The proposed model framework is then utilized to develop NOEP models for capturing the deformation behaviors of clays and sands.

Fig. 3.26 Comparison between predicted and test results under true traixial conditions when $p = p_0$

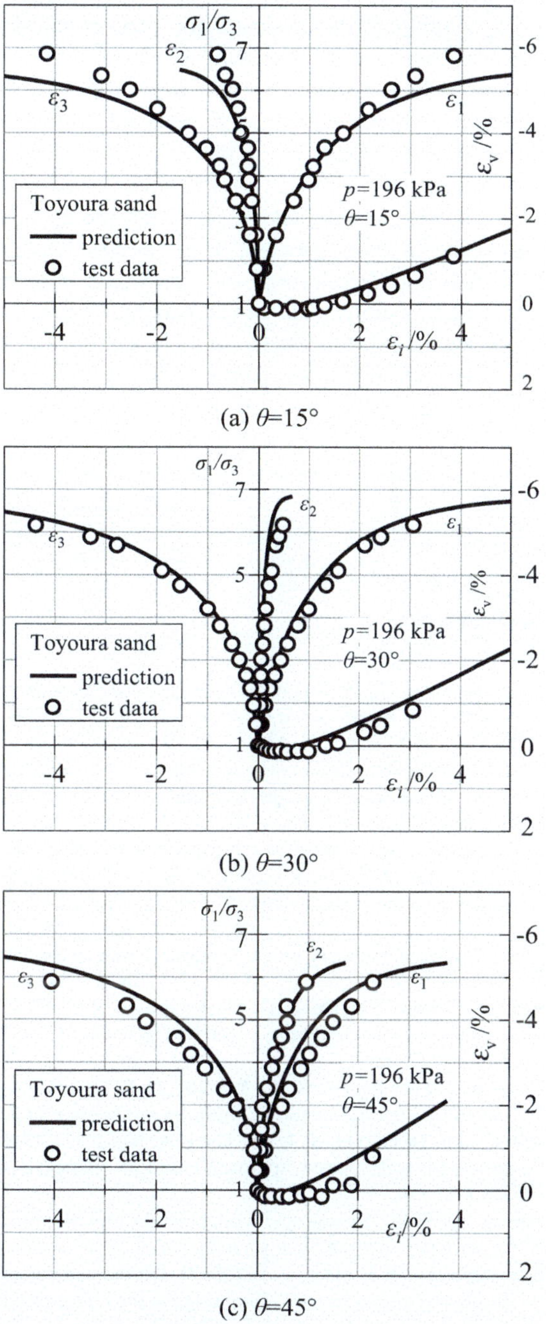

By combining the proposed non-orthogonal plastic flow rule with the ellipsoidal yield function, and further taking the plastic volumetric strain as the hardening parameter, a 3D NOEP model for clay was proposed. The proposed model has only one more material parameter, i.e. β, than the MCC model. It can well reflect the variation rule of the plastic strain increment direction under complex loading conditions and the properties of the critical state for soil. Furthermore, a hardening parameter is newly constructed for capturing the negative and positive dilatancy of sand during the shearing process, which is coordinated with the non-orthogonal plastic flow direction. In addition, the stress path independence of the hardening parameter is demonstrated. The developed NOEP model for sand with dilatancy properly describes the dilatancy behaviours of sand from initial negative to subsequent positive along different stress paths. Only seven material parameters with clear physical meanings are needed and can be easily determined by conventional triaxial tests. The performance of the proposed models was evaluated and compared with test results of clays and sands. The results indicate that the model effectively captures the strength and deformation behaviors of these materials under 3D stress conditions.

References

Chowdhury, E. Q., & Nakai, T. (1998). Consequences of the t_{ij}-concept and a new modeling approach. *Computers and Geotechnics, 23*(3), 131–164.

Lade, P. V., & Musante, H. M. (1978). Three-dimensional behavior of remolded clay. *Journal of the Geotechnical Engineering Division, 104*(2), 193–209.

Li, X. S., & Dafalias, Y. F. (2000). Dilatancy for cohesionless soils. *Géotechnique, 50*(4), 449–460.

Liang, J. Y., Lu, D. C., Zhou, X., Du, X. L., & Wu, W. (2019). Non-orthogonal elastoplastic constitutive model with the critical state for clay. *Computers and Geotechnics, 116*, 103200.

Liang, J. Y., Lu, D. C., Du, X. L., Wu, W., & Ma, C. (2020). Non-orthogonal elastoplastic constitutive model for sand with dilatancy. *Computers and Geotechnics, 118*, 103329.

Lu, D. C., Ma, C., Du, X. L., Jin, L., & Gong, Q. M. (2017). Development of a new nonlinear unified strength theory for geomaterials based on the characteristic stress concept. *International Journal of Geomechanics, ASCE, 17*(2), 04016058.

Lu, D. C., Liang, J. Y., Du, X. L., Ma, C., & Gao, Z. W. (2019a). Fractional elastoplastic constitutive model for soils based on a novel 3D fractional plastic flow rule. *Computers and Geotechnics, 105*, 277–290.

Lu, D. C., Zhou, X., Du, X. L., & Wang, G. S. (2019b). A 3D fractional elastoplastic constitutive model for concrete material. *International Journal of Solids and Structures, 165*, 160–175.

Ma, C., Lu, D. C., Du, X. L., & Zhou, A. N. (2017). Developing a 3D elastoplastic constitutive model for soils: A new approach based on characteristic stress. *Computers and Geotechnics, 86*, 129–140.

Nakai, T. (1989). An isotropic hardening elastoplastic model for sand considering the stress path dependency in three-dimensional stresses. *Soils and Foundations, 29*(1), 119–137.

Roscoe, K. H., & Burland, J. B. (1968). On the generalised stress-strain behaviour of 'wet' clay. In *Engineering plasticity*. Cambridge at the University Press, Cambridge(pp. 535–609).

Wang, G. S., Lu, D. C., Du, X. L., Zhou, X., & Cao, S. T. (2018). A true 3D frictional hardening elastoplastic constitutive model of concrete based on a unified hardening/softening function. *Journal of the Mechanics and Physics of Solids, 119*, 250–273.

Xiao, Y., Liu, H. L., Chen, Q. S., Ma, Q. F., Xiang, Y. Z., & Zheng, Y. R. (2017). Particle breakage and deformation of carbonate sands with wide range of densities during compression loading process. *Acta Geotechnica, 12*(5), 1177–1184.

Yao, Y. P., Lu, D. C., Zhou, A. N., & Zou, B. (2004). Generalized non-linear strength theory and transformed stress space. *Science in China Series E: Technological Sciences, 47*(6), 691–709.

Yao, Y. P., Sun, D. A., & Matsuoka, H. (2008). A unified constitutive model for both clay and sand with hardening parameter independent on stress path. *Computers and Geotechnics, 35*(2), 210–222.

Yao, Y. P., Hou, W., & Zhou, A. N. (2009). UH model: Three-dimensional unified hardening model for overconsolidated clays. *Géotechnique, 59*(5), 451–469.

Yu, H. S. (1998). Casm: A unified state parameter model for clay and sand. *International Journal for Numerical and Analytical Methods in Geomechanics, 22*(8), 621–653.

Chapter 4
Non-orthogonal Elastoplastic Model for Soil in Unsteady Water Environment

The state of water and air in the soil pores can be characterized through the water environment conditions in which the soil is located. Under the unsteady water environment, the water environment may change with the mechanical behaviour or other external factors. Variations in the water environment state cause the occurrence of deformation or the changes in the stress state. The constitutive model of unsaturated soil can be used to describe the mechanical behaviour of soil in the unsteady water environment. The constitutive model of unsaturated soil consists of two aspects: the mechanical constitutive relation and the hydraulic constitutive relation. For the determination of the mechanical constitutive relation, the basic constitutive variables are first selected to describe the stress state and the hydraulic state. Then, the isotropic compression behaviour under saturated and unsaturated conditions is determined, and the loading-collapse yield function is derived by taking the suitable hardening parameter. Finally, the mechanical constitutive relation under general stress conditions is obtained by combining the loading-collapse yield function with a constitutive model of saturated soil. The establishment of the hydraulic constitutive relation requires to capture the hydraulic hysteresis and the effect of the mechanical behaviour on the hydraulic behaviour.

This chapter presents a new modeling approach to describe the mechanical behaviour of soil in the unsteady water environment from the perspective of material parameters and stress state changing with the water environment. Based on a non-orthogonal elastoplastic constitutive model of saturated soil, a non-orthogonal elastoplastic constitutive model for soil in the unsteady water environment is provided by determining the specific form of the unknown elements. Then, the constitutive relation of soil in the unsteady water environment is presented by the calculation of total strain increment. The capability of the proposed model to capture the mechanical behaviour for soil in the unsteady water environment is validated by simulating test results. Three typical paths of unsaturated soil are adopted to demonstrate the performance of the proposed model.

D. Lu et al., *Multi-physical and Non-orthogonal Geotechnical Plasticity Modeling and Algorithms*, https://doi.org/10.1007/978-981-96-2369-3_4

4.1 Modeling Approach for Soil in Unsteady Water Environment

The proposed modeling approach is mainly used to describe the mechanical behaviour of soil in the unsteady water environment. The saturated state can be regarded as a special case of the unsteady water environment. The proposed modeling approach takes a non-orthogonal elastoplastic constitutive model of saturated soil as a foundation. The influence of the water environment on the mechanical behaviour is considered from the perspective of the variation in the stress state and material parameters with the water environment state.

The mechanical behaviour for soil in the unsteady water environment can be presented by the calculation of the total strain increment $\mathrm{d}\boldsymbol{\varepsilon}$. In accordance with elastoplastic theory, $\mathrm{d}\boldsymbol{\varepsilon}$ can be expressed as the sum of a recoverable elastic part and an unrecoverable plastic part.

$$\mathrm{d}\boldsymbol{\varepsilon} = \mathrm{d}\boldsymbol{\varepsilon}^{\mathrm{e}} + \mathrm{d}\boldsymbol{\varepsilon}^{\mathrm{p}} \tag{4.1}$$

where $\mathrm{d}\varepsilon^{\mathrm{e}}$ is the elastic strain increment and $\mathrm{d}\boldsymbol{\varepsilon}^{\mathrm{p}}$ is the plastic strain increment.

At the saturated state, $\mathrm{d}\boldsymbol{\varepsilon}^{\mathrm{e}}$ can be reproduced through the generalized Hooke's law in Terzaghi's effective stress $\boldsymbol{\sigma}'$ space.

$$\mathrm{d}\boldsymbol{\varepsilon}^{\mathrm{e}} = \frac{1+\nu}{E}\mathrm{d}\boldsymbol{\sigma}' - \frac{\nu}{E}\left(\mathrm{d}\boldsymbol{\sigma}' : \mathbf{1}\right) \otimes \mathbf{1} \tag{4.2}$$

where ν is Poisson's ratio, E is the elastic Young's modulus, and $\mathbf{1}$ is the 2nd order unit tensor. E is stress-dependent and can be determined by the unloading–reloading line. As shown in Fig. 4.1, in the saturated state, the unloading–reloading line is a straight line with a slope of $-\kappa$ in the e–ln p' plane. Here, e is the void ratio, and p' ($=\boldsymbol{\sigma}' : \mathbf{1}/3$) is the mean stress of Terzaghi's effective stress. Thus, E can be expressed as:

$$E = \frac{3(1+e_0)(1-2\nu)}{\kappa}p' \tag{4.3}$$

where e_0 is the initial void ratio.

Under the unsteady water environment, changes in both the external load and the water environment lead to the accumulation of elastic deformation. The proposed modeling approach captures the elastic deformation in the unsteady water environment by adopting a specific form of Bishop's effective stress to describe the stress state in the unsteady water environment. The adopted Bishop's effective stress takes into account the influence of water environment on stress state and is written as:

$$\boldsymbol{\sigma} = \overline{\boldsymbol{\sigma}} + S_{\mathrm{e}} s \mathbf{1} \tag{4.4}$$

where s is the suction and is expressed as follows:

Fig. 4.1 Unloading-reloading line at the saturated state in the e–ln p' plane

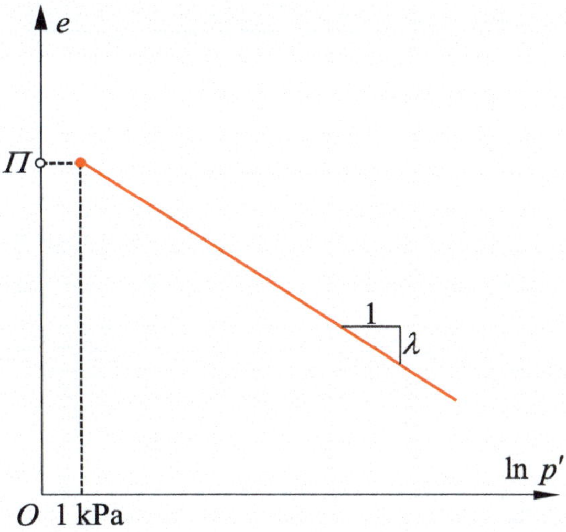

$$s = u_\mathrm{a} - u_\mathrm{w} \tag{4.5}$$

where u_a is the pore air pressure and u_w is the pore water pressure.

$\overline{\boldsymbol{\sigma}}$ is the net stress and can be expressed as:

$$\overline{\boldsymbol{\sigma}} = \boldsymbol{\sigma}^{\mathrm{total}} - u_\mathrm{a}\mathbf{1} \tag{4.6}$$

S_e is the effective degree of saturation, which considers the difference in water in macropores and micropores (Alonso et al., 2010; Zhou et al., 2012a), and is written as follows:

$$S_\mathrm{e} = \frac{S_\mathrm{r} - S_\mathrm{r}^{\mathrm{res}}}{S_\mathrm{r}^{0} - S_\mathrm{r}^{\mathrm{res}}} \tag{4.7}$$

where S_r is the degree of saturation, S_r^0 is the degree of saturation at zero suction, and $S_\mathrm{r}^{\mathrm{res}}$ is the residual degree of saturation.

Bishop's effective stress automatically degrades to Terzaghi's effective stress under saturated conditions and to the net stress under completely dry conditions. A smooth transition between the saturated state and the unsaturated state can be easily achieved by adopting Bishop's effective stress as a variable to describe the stress state in the unsteady water environment. The selection of Bishop's effective stress facilitates the establishment of the constitutive model in the unsteady water environment.

Substituting Eq. (4.3) into Eq. (4.2) and replacing $\boldsymbol{\sigma}'$ with $\boldsymbol{\sigma}$ provides the expression for calculating $\mathrm{d}\boldsymbol{\varepsilon}^\mathrm{e}$ in the unsteady water environment.

$$d\boldsymbol{\varepsilon}^{e} = \frac{(1+\nu)\kappa}{3(1+e_0)(1-2\nu)p}d\boldsymbol{\sigma} - \frac{\nu\kappa}{3(1+e_0)(1-2\nu)p}(d\boldsymbol{\sigma} : \mathbf{1}) \otimes \mathbf{1} \tag{4.8}$$

where p (=$\boldsymbol{\sigma} : \mathbf{1}/3$) is the mean stress of Bishop's effective stress.

Equation (4.8) indicates that the elastic deformation in the unsteady water environment is captured through the generalized Hooke's law in Bishop's effective stress space and that the unloading–reloading line in the unsteady water environment is a straight line with a slope of $-\kappa$ in the e–ln p plane. In addition, Eq. (4.8) automatically incorporates the elastic deformation due to changes in s through Bishop's effective stress.

Based on plasticity theory, $d\boldsymbol{\varepsilon}^{p}$ can be written as the product of the magnitude and direction.

$$d\boldsymbol{\varepsilon}^{p} = \Lambda \mathbf{m} \tag{4.9}$$

where Λ is a scalar, indicates the magnitude of $d\boldsymbol{\varepsilon}^{p}$, and is determined through the consistency condition of yield function. $\mathbf{m}$ is a tensor, denotes the direction of $d\boldsymbol{\varepsilon}^{p}$, and is obtained by applying the non-orthogonal plastic flow rule to the yield function. Thus, the specific form of the yield function in the unsteady water environment is the core for determining Λ and $\mathbf{m}$.

At the saturated state, the yield function establishes the relationship between Terzaghi's effective stress and the hardening parameter, and the general form of the yield function can be written as:

$$f = f\left(\boldsymbol{\sigma}', H\right) = 0 \tag{4.10}$$

where H is the hardening parameter and is a function of plastic strain $\boldsymbol{\varepsilon}^{p}$.

Since the proposed modeling approach adopts Bishop's effective stress as a variable describing the stress state in the unsteady water environment, the yield function in a steady water environment can be obtained by replacing Terzaghi's effective stress in Eq. (4.10) with Bishop's effective stress. The influence of water environment state on yield surface, which reflects the changes in the shape and magnitude of the yield surface in Bishop's effective stress space with respect to the water environment, is considered from the perspective of changes in the material parameters with respect to the water environment state. For the proposed modeling approach, S_{e} is taken as a variable to describe the water environment state, and the effect of the water environment on the yield function is captured by establishing the relationship between the material parameters and S_{e}. Thus, the general form of the yield function in the unsteady water environment can be expressed as:

$$f = f(\boldsymbol{\sigma}, N_x, H) = 0 \tag{4.11}$$

where N_x ($x = 1, 2, 3\ldots$) represents the parameters of the yield function in a steady water environment, which determines the shape and magnitude of yield surface in Bishop's effective stress space when H is a certain value, and is a function of S_{e}.

The proposed modeling approach determines **m** by applying the non-orthogonal plastic flow rule (Liang et al., 2019; Lu et al., 2019) to the yield function in the unsteady water environment (Eq. 4.11). The concept of fractional derivatives is adopted for obtaining the non-orthogonal gradient of the yield function in the unsteady water environment. Thus, **m** can be expressed as:

$$\mathbf{m}=\frac{\partial^{\mu}\mathbf{f}}{\partial\mathbf{p}^{\mu}}\frac{\partial\mathbf{p}}{\partial\boldsymbol{\sigma}}+\frac{\partial^{\mu}\mathbf{f}}{\partial\mathbf{q}^{\mu}}\frac{\partial\mathbf{q}}{\partial\boldsymbol{\sigma}} \tag{4.12}$$

where $q\,[=\sqrt{3/2(\boldsymbol{\sigma}-p\mathbf{1}):(\boldsymbol{\sigma}-p\mathbf{1})}]$ is the deviatoric stress of Bishop's effective stress. μ is the fractional order and is a function of S_e. $\partial^{\mu}f/\partial p^{\mu}$ and $\partial^{\mu}f/\partial q^{\mu}$ are the fractional partial derivative of f with respect to p and the fractional partial derivative f with respect to q, respectively. $\partial p/\partial\boldsymbol{\sigma}$ and $\partial q/\partial\boldsymbol{\sigma}$ are the partial derivatives of p with respect to $\boldsymbol{\sigma}$ and q with respect to $\boldsymbol{\sigma}$, respectively.

Taking the full differentiation of Eq. (4.11) and combining Eqs. (4.12) and (4.9) yields the expression for calculating Λ.

$$\Lambda=\frac{\frac{\partial f}{\partial\boldsymbol{\sigma}}:\mathrm{d}\boldsymbol{\sigma}+\frac{\partial f}{\partial N_x}\frac{\partial N_x}{\partial S_e}\mathrm{d}S_e}{-\frac{\partial f}{\partial H}\frac{\partial H}{\partial\boldsymbol{\varepsilon}^{\mathrm{p}}}:\left(\frac{\partial^{\mu}f}{\partial p^{\mu}}\frac{\partial p}{\partial\boldsymbol{\sigma}}+\frac{\partial^{\mu}f}{\partial q^{\mu}}\frac{\partial q}{\partial\boldsymbol{\sigma}}\right)} \tag{4.13}$$

By substituting Eqs. (4.8), (4.9), (4.12) and (4.13) into Eq. (4.1), the expression for calculating $\mathrm{d}\boldsymbol{\varepsilon}^{\mathrm{p}}$ can be obtained.

$$\begin{aligned}\mathrm{d}\boldsymbol{\varepsilon}=&\frac{(1+\nu)\kappa}{3(1+e_0)(1-2\nu)p}\mathrm{d}\boldsymbol{\sigma}-\frac{\nu\kappa}{3(1+e_0)(1-2\nu)p}(\mathrm{d}\boldsymbol{\sigma}:\mathbf{1})\otimes\mathbf{1}\\&+\frac{\frac{\partial f}{\partial\boldsymbol{\sigma}}:\mathrm{d}\boldsymbol{\sigma}+\frac{\partial f}{\partial N_x}\frac{\partial N_x}{\partial S_e}\mathrm{d}S_e}{-\frac{\partial f}{\partial H}\frac{\partial H}{\partial\boldsymbol{\varepsilon}^{\mathrm{p}}}:\left(\frac{\partial^{\mu}f}{\partial p^{\mu}}\frac{\partial p}{\partial\boldsymbol{\sigma}}+\frac{\partial^{\mu}f}{\partial q^{\mu}}\frac{\partial q}{\partial\boldsymbol{\sigma}}\right)}\left(\frac{\partial^{\mu}f}{\partial p^{\mu}}\frac{\partial p}{\partial\boldsymbol{\sigma}}+\frac{\partial^{\mu}f}{\partial q^{\mu}}\frac{\partial q}{\partial\boldsymbol{\sigma}}\right)\end{aligned} \tag{4.14}$$

It is worth noting that the expression for calculating $\mathrm{d}\boldsymbol{\varepsilon}$ involves the description of the unsteady water environment. The core of describing the unsteady water environment is to establish the relationship between S_e and s. Extensive experimental data show that a decrease in e leads to an increase in S_e and that the relationship between S_e and s is path dependent (D'Onza et al., 2011; Tarantino, 2009; Vanapalli et al., 1999; Wheeler et al., 2003). Therefore, the expression for describing the unsteady water environment can be written as:

$$\mathrm{d}S_e=\frac{\partial S_e}{\partial s}\mathrm{d}s+\frac{\partial S_e}{\partial e}\mathrm{d}e \tag{4.15}$$

The combination of Eqs. (4.14) and (4.15) provides the expression for the calculation of the total strain increment in the unsteady water environment. The proposed approach provides a novel strategy for establishing the constitutive models of soil in the unsteady water environment. The most prominent advantage of the proposed approach is that it can establish a new constitutive model of soil in the unsteady water environment by utilizing an arbitrary yield function of saturated soils.

4.2 Establishment of Constitutive Model

Based on the proposed modeling approach, the core of establishing a constitutive model for soil in the unsteady water environment is to determine the expressions for the following three unknown elements: (1) the yield function in a steady water environment; (2) the influence of water environment state on the yield function; and (3) the description of the unsteady water environment. In what follows, the above aspects are presented in detail. Finally, the constitutive equation of soil in the unsteady water environment is presented through the determined ingredients.

4.2.1 *Yield Function in Steady Water Environment*

A simple yield function for saturated clay (Liang et al., 2019), which has the same shape as the yield function of the modified Cam Clay model (Roscoe et al., 1963) and takes the plastic volumetric strain $\varepsilon_{\mathrm{v}}^{\mathrm{p}}$ as the hardening parameter, is employed to obtain the yield function in a steady water environment. Based on the proposed modeling approach, the yield function in a steady water environment is determined by replacing Terzaghi's effective stress with Bishop's effective stress.

$$
\begin{aligned}
f &= p^2 + \frac{q^2}{N^2} - pp_{\mathrm{x}} \\
&= p^2 + \frac{q^2}{N^2} - p\exp\left(\frac{1+e_0}{\lambda-\kappa}\varepsilon_{\mathrm{v}}^{\mathrm{p}} + \frac{\Pi - e_0 - \kappa \ln p_0}{\lambda - \kappa}\right) \\
&= 0
\end{aligned}
\tag{4.16}
$$

where N is a material parameter used to control the shape of f in a steady water environment in the p–q plane; p_x is an internal variable characterizing the magnitude of the yield surface in a steady water environment in $\boldsymbol{\sigma}$ space; Π and λ are the material parameters used to describe the evolution law of p_x with $\varepsilon_{\mathrm{v}}^{\mathrm{p}}$, and determine the position of the NCL in a steady water environment in the e–ln p plane; and p_0 is the initial value of p.

Figure 4.2 plots the yield function in a steady water environment in the p–q plane according to Eq. (4.16). As shown in Fig. 4.2, in the p–q plane, the yield function in a steady water environment is an ellipse, N is the ratio of the ordinate to the abscissa of the elliptic vertex (i.e., point A in Fig. 4.2), and p_x is the right intersection point of f with the p-axis (i.e., point B in Fig. 4.2). In Fig. 4.2, M is the critical stress ratio.

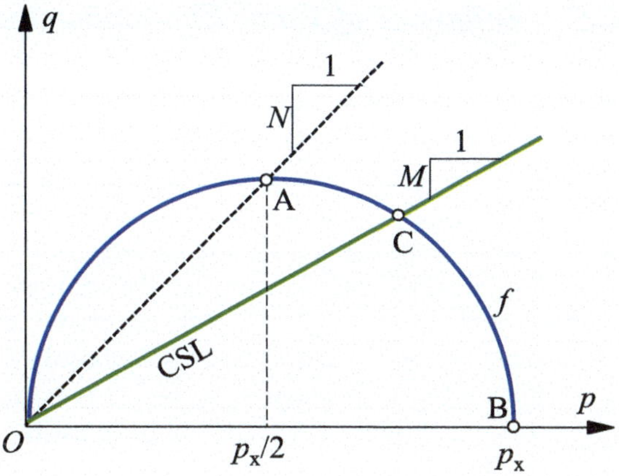

Fig. 4.2 Yield function with constant S_e in the p–q plane

It can be seen from Eq. (4.16) that the NCL in a steady water environment is a straight line with a slope of $-\lambda$ in the e–ln p plane. Figure 4.3 displays NCL in a steady water environment in the e–ln p plane. Π is the void ratio corresponding to $p = 1$ kPa on the NCL in a steady water environment. Thus, the NCL in a steady water environment can be expressed as:

$$e = \Pi - \lambda \ln p \tag{4.17}$$

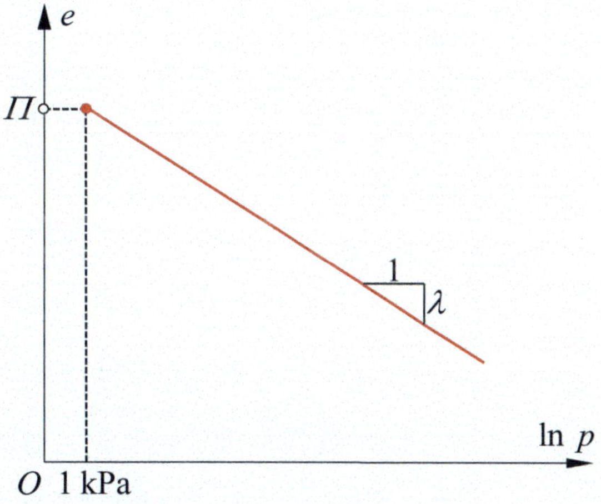

Fig. 4.3 NCL with constant S_e in the e–ln p plane

4.2.2 Influence of Water Environment on Yield Function

The proposed modeling approach takes into account the influence of water environment on yield function by establishing the relationship between the material parameter and S_e. The yield function in a steady water environment involves five material parameters (λ, κ, Π, M and N). The material parameters (λ and Π) are used to describe the isotropic compression behaviour and determine the position of NCL in the e–ln p plane. Experimental data indicate that the void ratio in the normal compression state increases with decreasing S_e. This characteristic can be captured by establishing the relationship between λ and S_e or the relationship between Π and S_e. In this study, λ is assumed to be independent of S_e, and Π is assumed to monotonically increase as S_e decreases and follows the following expression:

$$\Pi = \Pi_{sat} + (\lambda - \kappa)\ln\{1 + a_1 - a_1\exp[-a_2(1 - S_r)^{a_3}]\} \tag{4.18}$$

where a_1, a_2 and a_3 are the fitting parameters that describe the variation model of Π with S_e, and Π_{sat} is the value of Π at the saturated state.

Substituting Eq. (4.18) into Eq. (4.17) yields the following expression.

$$e = \Pi_{sat} + (\lambda - \kappa)\ln\{1 + a_1 - a_1\exp[-a_2(1 - S_r)^{a_3}]\} - \lambda\ln p \tag{4.19}$$

Equation (4.19) gives the normal compression surface in the e–p–S_e space and provides the relationship between e, p and S_e at the normal consolidated state. Based on Eq. (4.19), Fig. 4.4 plots the normal compression surface in the e–ln p–S_e space and the NCL with constant S_e in the e–ln p plane. The black solid line, the red solid line and the blue solid line represent NCL at the saturated state, NCL at $S_e = S_{e1}$ and NCL at $S_e = 0.0$, respectively. Here, $0.0 < S_{e1} < 1.0$. As shown in Fig. 4.4, in the e–ln p plane, the NCL with constant S_e is a straight line with a slope of $-\lambda$. With the decreasing S_e, the position of the NCL gradually moves up along the positive direction of the e-axis.

Based on Eq. (4.19) and taking ε_v^p as the hardening parameter, the yield curve in the unsteady water environment under isotropic compression conditions, which is also referred to as the loading-collapse (L-C) yield curve, can be written as:

$$\begin{aligned} f &= p - \{1 + a_1 - a_1\exp[-a_2(1 - S_e)^{a_3}]\} \\ &\quad \exp\left(\frac{\Pi_{sat} - e_0 - \kappa\ln p_0}{\lambda - \kappa} + \frac{1 + e_0}{\lambda - \kappa}\varepsilon_v^p\right) \\ &= p - \{1 + a_1 - a_1\exp[-a_2(1 - S_e)^{a_3}]\}p_{sat} = 0 \end{aligned} \tag{4.20}$$

where p_{sat} is the yield stress under isotropic compression conditions at the saturated state.

The effects of a_1, a_2 and a_3 on the L-C yield surface are shown in Fig. 4.5a–c, respectively. As shown in Fig. 4.5, with a decrease in S_e, p increases monotonically.

Fig. 4.4 Normal compression surface and NCL with constant S_e

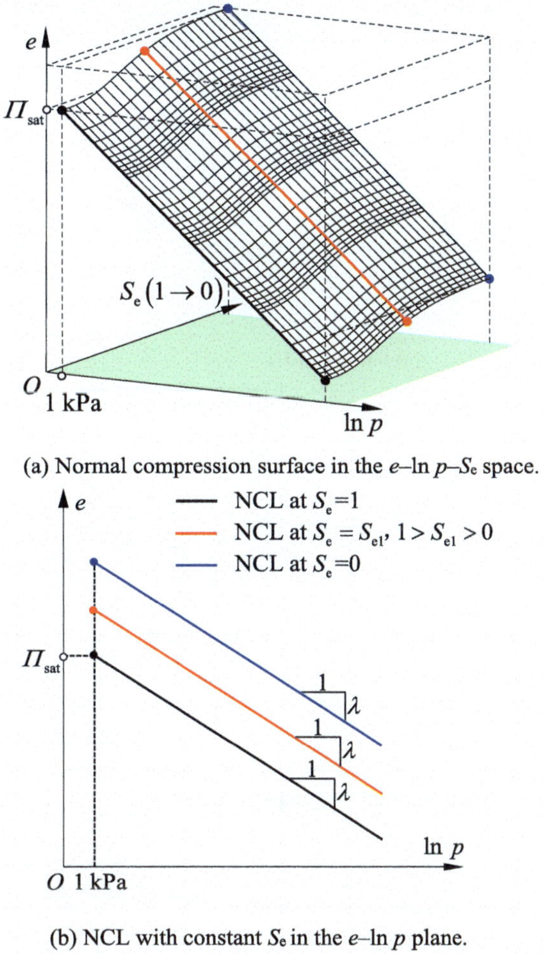

The ratio of p at $S_e = 0.0$ to p at $S_e = 1.0$ gradually increases with increasing a_1, and the shape of the L-C yield curve is basically independent of the value of a_1 as shown in Fig. 4.5a. The shape of the L-C yield curve is controlled by a_2 and a_3. An increase in a_2 changes the shape of the L-C yield curve from a concave curve to an S-shaped curve (Fig. 4.5b). An increase in a_3 causes the shape of the L-C yield curve to change from a convex curve to an S-shaped curve (Fig. 4.5c).

The material parameter N is used to control the shape of the yield surface in Bishop's effective stress space. Experimental data indicate that the volumetric strain produced by shearing is closely related to S_e and that the volumetric strain produced by shearing in unsaturated state may be small than that in saturated state (Cui & Delage, 1996; Estabragh & Javadi, 2008; Oka et al., 2010; Wang et al., 2002). The feature is closely related to the shape of yiled surface in Bishop's effective stress space and can be captured through the following simple expression:

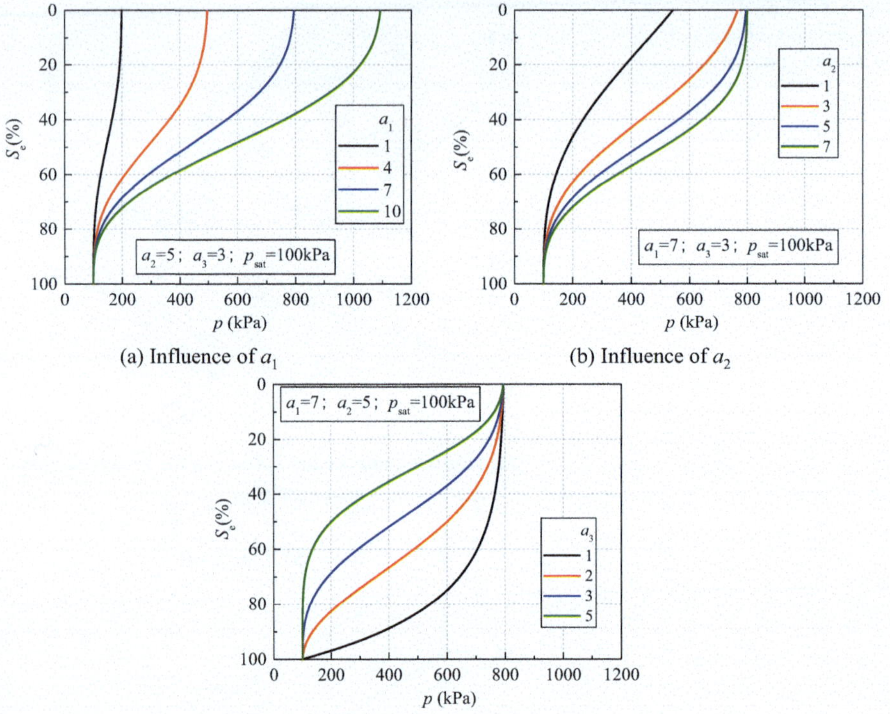

(a) Influence of a_1

(b) Influence of a_2

(c) Influence of a_3

Fig. 4.5 L-C yield curve in the p–S_e plane

$$N = \sqrt{2M^2\left(1 - S_e^d\right) + N_0^2 S_e^d} \tag{4.21}$$

where d is a material parameter used to describe the variation model of N with S_e, and N_0 is the value of N at $S_e = 1.0$.

Substituting Eqs. (4.21) and (4.20) into Eq. (4.16) and rearranging Eq. (4.16) yields the specific expression of f in the unsteady water environment.

$$\begin{aligned} f &= p^2 + \frac{q^2}{2M^2\left(1 - S_e^d\right) + N_0^2 S_e^d} - pp_x \\ &= p^2 + \frac{q^2}{2M^2\left(1 - S_e^d\right) + N_0^2 S_e^d} - p\left\{1 + a_1 - a_1 \exp\left[-a_2(1 - S_e)^{a_3}\right]\right\}p_{sat} \\ &= p^2 + \frac{q^2}{2M^2\left(1 - S_e^d\right) + N_0^2 S_e^d} - p\left\{1 + a_1 - a_1 \exp\left[-a_2(1 - S_e)^{a_3}\right]\right\} \\ &\quad \exp\left(\frac{\Pi_{sat} - e_0 - \kappa \ln p_0}{\lambda - \kappa} + \frac{1 + e_0}{\lambda - \kappa}\varepsilon_v^p\right) \\ &= 0 \end{aligned} \tag{4.22}$$

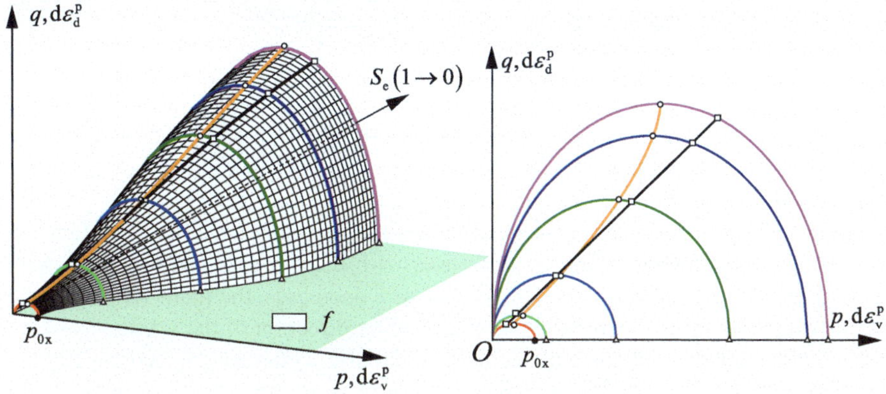

Fig. 4.6 Yield surface in the p–q–S space and in the p–q plane

According to Eq. (4.22), Fig. 4.6 plots the yield surface in the p–q–S_e space and the yield curve in the p–q plane. The orange curve indicates f at $q/p = N$. The black curve represents f at $q/p = M$. The red, green, blue, olive, royal and magenta curves represent f at $S_e = 1.0$, f at $S_e = 0.8$, f at $S_e = 0.6$, f at $S_e = 0.4$, f at $S_e = 0.2$ and f at $S_e = 0.0$, respectively. In the p–q–S_e space, the yield surface in the unsteady water environment is a continuous and smooth surface, and both the yield curve at $q/p = N$ (orange curve) and the yield curve at $q/p = M$ (black curve) are spatial curves. In the p–q plane, f with constant S_e is an ellipse, both the shape and magnitude of yield curve vary with S_e, and the yield curve at $q/p = M$ is a straight line with a slope of M. As S_e decreases, the ratio of the ordinate to the abscissa of the vertex of f and the magnitude of f gradually increases in the p–q plane.

4.2.3 Description of Unsteady Water Environment

The proposed modeling approach describes the unsteady water environment by determining the specific expressions of $\partial S_e/\partial s$ and $\partial S_e/\partial e$. For the proposed model, $\partial S_e/\partial s$ and $\partial S_e/\partial e$ are determined through the main drying curve (S_{ed}) and the main wetting curve (S_{ew}). Based on the van Genuchten soil–water characteristic curve (van Genuchten, 1980), the expressions of S_{ed} and S_{ew} are constructed to consider the influence of e on S_e.

$$S_{ed} = \left[1 + \left(\frac{s}{a_d}\right)^m \left(\frac{e}{\Pi_{sat}}\right)^b\right]^{1/m-1} \tag{4.23}$$

$$S_{\text{ew}} = \left[1 + \left(\frac{s}{a_{\text{w}}}\right)^{m} \left(\frac{e}{\Pi_{\text{sat}}}\right)^{b}\right]^{1/m-1} \tag{4.24}$$

where a_{w}, a_{d} and m are the fitting parameters for S_{ed} and S_{ew}; and b is a hydro-mechanical parameter to consider the mechanical behaviour on the water environment.

Figure 4.7 plots S_{ed} and S_{ew} in the S_{e}–ln s–e space and in the S_{e}–ln s plane. In the S_{e}–ln s–e space (Fig. 4.7a), S_{e} decreases with increasing s and increases with decreasing e. In the S_{e}–ln s plane (Fig. 4.7b), as e decreases, the main wetting curve and the main drying curve move along the positive direction of the s-axis.

$\partial S_{\text{e}}/\partial s$ is path dependent and includes hydraulic hysteresis. Based on Zhou's model (Zhou & Sheng, 2015), a simple boundary scanning rule is taken to obtain $\partial S_{\text{e}}/\partial s$.

$$\frac{\partial S_{\text{e}}}{\partial s} = \begin{cases} \left(\frac{S_{\text{e}}}{S_{\text{e,d}}}\right)^{c} \frac{\partial S_{\text{e,d}}}{\partial s} & \mathrm{d}s > 0(\text{drying}) \\ \left(\frac{S_{\text{e,w}}}{S_{\text{e}}}\right)^{c} \frac{\partial S_{\text{e,w}}}{\partial s} & \mathrm{d}s < 0(\text{wetting}) \end{cases} \tag{4.25}$$

where $S_{\text{e,d}}$ and $S_{\text{e,w}}$ are the degrees of saturation on S_{ed} and S_{ew}, respectively. The material parameter c reflects the nonlinearity of the scanning line. Based on the current state (s, S_{e}, e), as shown in Fig. 4.8, $S_{\text{e,d}}$ and $S_{\text{e,w}}$ can be written as:

$$S_{\text{e,d}} = \left[1 + \left(\frac{s}{a_{\text{d}}}\right)^{m} \left(\frac{e}{\Pi_{\text{sat}}}\right)^{b}\right]^{1/m-1} \tag{4.26}$$

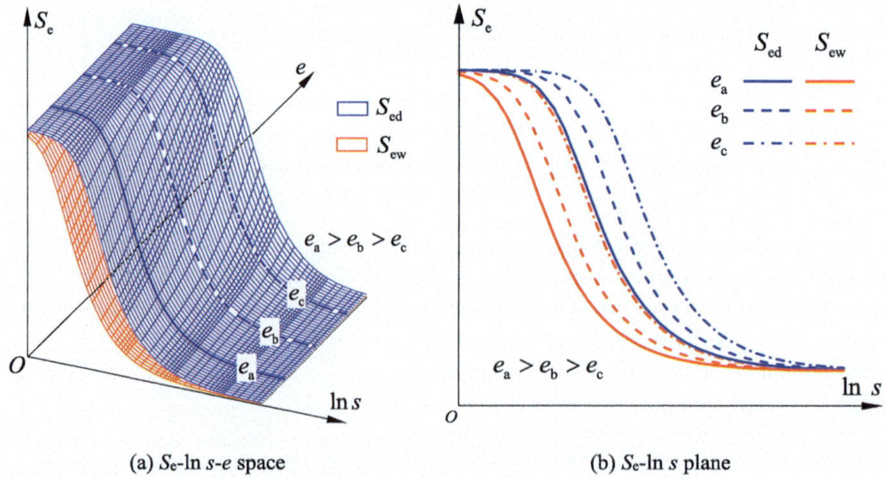

Fig. 4.7 Main drying surface and main wetting surface

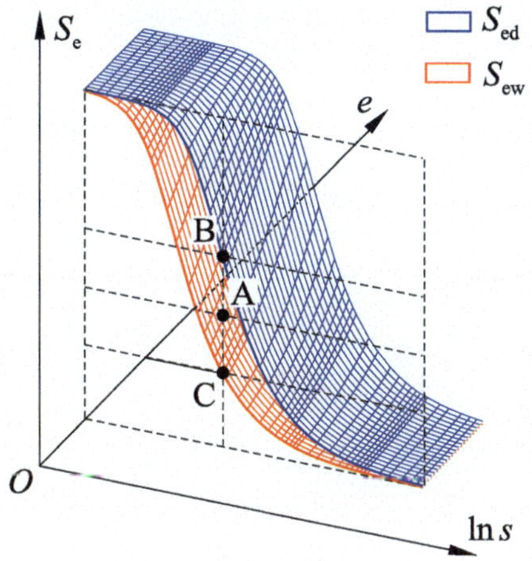

(a) S_e-ln s-e space

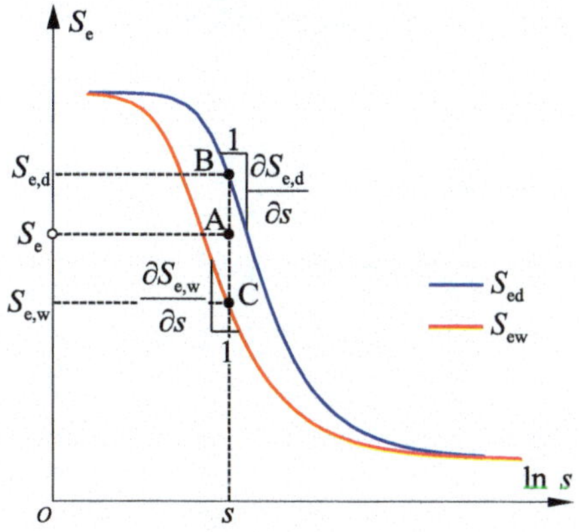

(b) S_e-ln s plane

Fig. 4.8 Determination of $\partial S_{e,d}/\partial s$ and $\partial S_{e,w}/\partial s$

$$S_{e,w} = \left[1 + \left(\frac{s}{a_w}\right)^m \left(\frac{e}{\Pi_{sat}}\right)^b\right]^{1/m-1} \tag{4.27}$$

$\partial S_{e,d}/\partial s$ and $\partial S_{e,w}/\partial s$ can be expressed as:

$$\frac{\partial S_{e,d}}{\partial s} = \frac{1-m}{a_d}\left[1 + \left(\frac{s}{a_d}\right)^m \left(\frac{e}{\Pi_{sat}}\right)^b\right]^{1/m-2} \left(\frac{e}{\Pi_{sat}}\right)^b \left(\frac{s}{a_d}\right)^{m-1} \tag{4.28}$$

$$\frac{\partial S_{e,w}}{\partial s} = \frac{1-m}{a_w}\left[1 + \left(\frac{s}{a_w}\right)^m \left(\frac{e}{\Pi_{sat}}\right)^b\right]^{1/m-2} \left(\frac{e}{\Pi_{sat}}\right)^b \left(\frac{s}{a_w}\right)^{m-1} \tag{4.29}$$

$\partial S_e/\partial e$ takes the following form to ensure that the hydraulic state is within the region surrounded by S_{ed} and S_{ew}.

$$\frac{\partial S_e}{\partial e} = \frac{S_{e,d} - S_e}{S_{e,d} - S_{e,w}} \frac{\partial S_{e,w}}{\partial e} + \frac{S_e - S_{e,w}}{S_{e,d} - S_{e,w}} \frac{\partial S_{e,d}}{\partial e} \tag{4.30}$$

Based on the current state (s, S_e, e), as shown in Fig. 4.9, $\partial S_{e,d}/\partial e$ and $\partial S_{e,w}/\partial e$ can be expressed as:

$$\frac{\partial S_{e,d}}{\partial e} = \frac{(1/m-1)b}{\Pi_{sat}}\left[1 + \left(\frac{s}{a_d}\right)^m \left(\frac{e}{\Pi_{sat}}\right)^b\right]^{1/m-2} \left(\frac{e}{\Pi_{sat}}\right)^{b-1} \left(\frac{s}{a_d}\right)^m \tag{4.31}$$

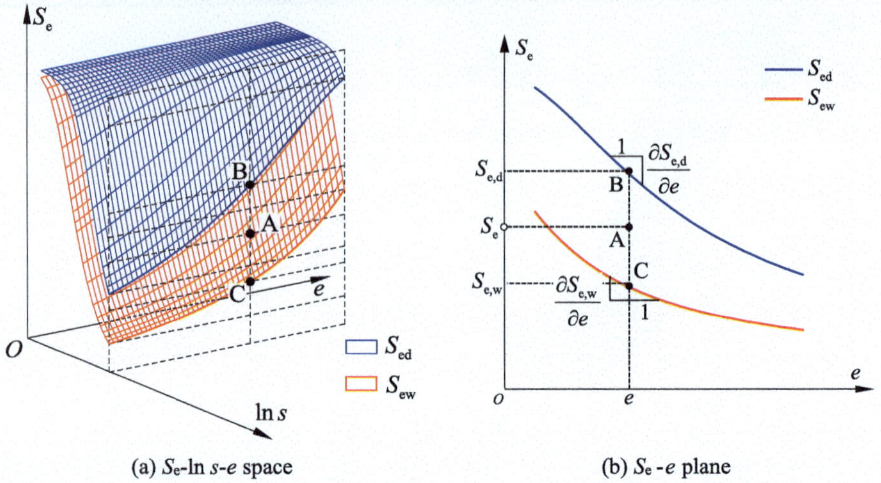

Fig. 4.9 Determination of $\partial S_{e,d}/\partial e$ and $\partial S_{e,w}/\partial e$

$$\frac{\partial S_{e,w}}{\partial s} = \frac{(1/m-1)b}{\Pi_{sat}}\left[1+\left(\frac{s}{a_w}\right)^m\left(\frac{e}{\Pi_{sat}}\right)^b\right]^{1/m-2}\left(\frac{e}{\Pi_{sat}}\right)^{b-1}\left(\frac{s}{a_w}\right)^m \tag{4.32}$$

4.3 Constitutive Equations

The constitutive equation of the proposed model is presented by the expressions for calculating the elastic strain increment and the plastic strain increment. The elastic strain increment is given through Eq. (4.8). Based on Eq. (4.9), the plastic strain increment can be determined by giving the specific form of Λ and $\mathbf{m}$.

The proposed modeling approach determines the specific form of $\mathbf{m}$ by applying the non-orthogonal plastic flow to f (Eq. 4.12). Thus, $\partial^\mu f/\partial p^\mu$, $\partial^\mu f/\partial q^\mu$, $\partial p/\partial \boldsymbol{\sigma}$, and $\partial q/\partial \boldsymbol{\sigma}$ are kernel elements to determine $\mathbf{m}$. For the proposed model, the Riemann–Liouville fractional derivative, which can be used to obtain the fractional derivative in the range of $\mu = 0$ to $\mu = 2$, is employed to determine $\partial^\mu f/\partial p^\mu$ and $\partial^\mu f/\partial q^\mu$. Thus, based on the specific expression of f in the unsteady water environment, $\partial^\mu f/\partial p^\mu$ and $\partial^\mu f/\partial q^\mu$ can be written as:

$$\frac{\partial^\mu f}{\partial p^\mu} = \frac{\mu\left[2M^2\left(1-S_e^d\right)+N_0^2S_e^d\right]p^2-\mu(2-\mu)q^2}{\left[2M^2\left(1-S_e^d\right)+N_0^2S_e^d\right]\Gamma(3-\mu)p^\mu} \tag{4.33}$$

$$\frac{\partial^\mu f}{\partial q^\mu} = \frac{q^{2-\mu}\left(3\mu-\mu^2\right)}{\left[2M^2\left(1-S_e^d\right)+N_0^2S_e^d\right]\Gamma(3-\mu)} \tag{4.34}$$

where $\Gamma(\mu)$ is the Gamma function, equals $\int_0^\infty e^{-\tau}\tau^{\mu-1}\mathrm{d}\tau$, and follows $\Gamma(\mu+1)=\mu\cdot\Gamma(\mu)$.

$\partial p/\partial \boldsymbol{\sigma}$ and $\partial q/\partial \boldsymbol{\sigma}$ are derived from the definitions of p and q:

$$\frac{\partial p}{\partial \boldsymbol{\sigma}}=\frac{\mathbf{1}}{3} \tag{4.35}$$

$$\frac{\partial q}{\partial \boldsymbol{\sigma}}=\frac{3(\boldsymbol{\sigma}-p\mathbf{1})}{2q} \tag{4.36}$$

Thus, $\mathbf{m}$ in the unsteady water environment can be expressed as:

$$\begin{aligned}\mathbf{m}=&\frac{\mu\left[2\mathbf{M}^2\left(1-\mathbf{S}_\mathrm{e}^\mathbf{d}\right)+\mathbf{N}_0^2\mathbf{S}_\mathrm{e}^\mathbf{d}\right]\mathbf{p}^2+\left(\mu^2-2\mu\right)\mathbf{q}^2}{\left[2\mathbf{M}^2\left(1-\mathbf{S}_\mathrm{e}^\mathbf{d}\right)+\mathbf{N}_0^2\mathbf{S}_\mathrm{e}^\mathbf{d}\right]\Gamma(3-\mu)\mathbf{p}^\mu}\frac{\mathbf{1}}{\mathbf{3}}\\&+\frac{3\mu(3-\mu)q^{1-\mu}p^\mu(\boldsymbol{\sigma}-p\mathbf{1})}{2\left[2M^2\left(1-S_\mathrm{e}^d\right)+N_0^2S_\mathrm{e}^d\right]\Gamma(3-\mu)p^\mu}\end{aligned} \tag{4.37}$$

Critical state theory defines that the plastic volumetric strain increment $\mathrm{d}\varepsilon_\mathrm{v}^\mathrm{p}$ is zero when the Bishop's effective stress ratio η ($= q/p$) is equal to M. Thus, the following expression needs to be satisfied.

$$\mu=\left(2-\frac{N_0^2}{M^2}\right)S_\mathrm{e}^d \tag{4.38}$$

Figure 4.10 depicts f and $\mathbf{m}$ in the p–q–S_e space and in the p–q plane. The blue arrow indicates the direction of $\mathrm{d}\boldsymbol{\varepsilon}^\mathrm{p}$. The black solid line indicates f at $\eta = M$. The direction of $\mathrm{d}\boldsymbol{\varepsilon}^\mathrm{p}$ on f at $\eta = M$ is parallel to the q-axis. The red solid line indicates f at $\eta = N$. Notably, the direction of $\mathrm{d}\boldsymbol{\varepsilon}^\mathrm{p}$ is always parallel to the p–q plane. With decreasing S_e, the position of the critical stress state moves along the positive direction of the p-axis. Except for the isotropic compression state and the critical state, the ratio of plastic shear strain increment to plastic volumetric strain increment $\mathrm{d}\varepsilon_\mathrm{d}^\mathrm{p}/\mathrm{d}\varepsilon_\mathrm{v}^\mathrm{p}$ gradually decreases with decreasing S_e.

For the determination of Λ, the first step is to obtain the total differential of f.

$$\frac{\partial f}{\partial \boldsymbol{\sigma}}:\mathrm{d}\boldsymbol{\sigma}+\frac{\partial f}{\partial \varepsilon_\mathrm{v}^\mathrm{p}}\mathrm{d}\varepsilon_\mathrm{v}^\mathrm{p}+\frac{\partial f}{\partial S_\mathrm{e}}\mathrm{d}S_\mathrm{e}=0 \tag{4.39}$$

where $\partial f/\partial \boldsymbol{\sigma}$, $\partial f/\partial \varepsilon_\mathrm{v}^\mathrm{p}$, and $\partial f/\partial S_\mathrm{e}$ are the partial derivatives of f and can be written as:

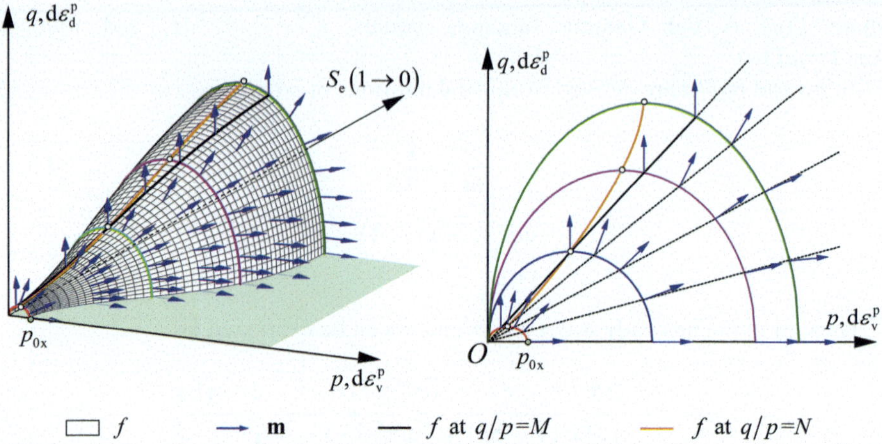

Fig. 4.10 f and **m** in the p–q–S_e space and in the p–q plane

$$\begin{cases} \dfrac{\partial f}{\partial \boldsymbol{\sigma}} = \dfrac{\mathbf{1}}{3}(2p - p_x) + \dfrac{3(\boldsymbol{\sigma} - p\mathbf{1})}{2M^2\left(1 - S_e^d\right) + N_0^2 S_e^d} \\ \dfrac{\partial f}{\partial \varepsilon_v^p} = -\dfrac{1 + e_0}{\lambda - \kappa} p p_x \\ \dfrac{\partial f}{\partial S_e} = \dfrac{p p_{sat} a_1 a_2 a_3 (1 - S_e)^{a_3 - 1}}{\exp\left[a_2 (1 - S_e)^{a_3}\right]} - \dfrac{q^2 d S_e^{d-1}\left(N_0^2 - 2M^2\right)}{\left[2M^2\left(1 - S_e^d\right) + N_0^2 S_e^d\right]^2} \end{cases} \tag{4.40}$$

Λ in the unsteady water environment is derived by substituting Eqs. (4.40) and (4.37) into Eq. (4.39), and rearranging Eq. (4.39).

$$\Lambda = \frac{\left[2M^2\left(1 - S_e^d\right) + N_0^2 S_e^d\right]\Gamma(3 - \mu)p^{\mu}}{\mu\left[2M^2\left(1 - S_e^d\right) + N_0^2 S_e^d\right]p^2 + \left(\mu^2 - 2\mu\right)q^2} \cdot$$
$$\frac{\lambda - \kappa}{(1 + e_0)p p_x}\left\{\begin{array}{l} \left[\dfrac{\mathbf{1}}{3}(2p - p_x) + \dfrac{3(\boldsymbol{\sigma} - p\mathbf{1})}{2M^2\left(1 - S_e^d\right) + N_0^2 S_e^d}\right] : d\boldsymbol{\sigma} \\ + \dfrac{p p_{sat} a_1 a_2 a_3 (1 - S_e)^{a_3 - 1}}{\exp\left[a_2 (1 - S_e)^{a_3}\right]} dS_e - \dfrac{q^2 d S_e^{d-1}\left(N_0^2 - 2M^2\right) dS_e}{\left[2M^2\left(1 - S_e^d\right) + N_0^2 S_e^d\right]^2} \end{array}\right\} \tag{4.41}$$

Substituting Eqs. (4.8), (4.9), (4.37) and (4.41) into Eq. (4.1) yields the specific constitutive relation, which is expressed by $d\boldsymbol{\sigma}$ and dS_e.

$$d\boldsymbol{\varepsilon} = \frac{(1 + \nu)\kappa}{3(1 + e_0)(1 - 2\nu)p} d\boldsymbol{\sigma} - \frac{\nu\kappa}{3(1 + e_0)(1 - 2\nu)p}(d\boldsymbol{\sigma} : \mathbf{1}) \otimes \mathbf{1}$$

$$
+\frac{\frac{\lambda-\kappa}{(1+e_0)p}\left[\frac{1}{3}+\frac{3(3-\mu)q^{1-\mu}p^{\mu}(\boldsymbol{\sigma}-p\mathbf{1})}{2\left[2M^2\left(1-S_e^d\right)+N_0^2S_e^d\right]p^2+(\mu-2)q^2}\right]}{\left\{1+a_1-a_1\exp\left[-a_2(1-S_e)^{a_3}\right]\right\}\exp\left(\frac{\Pi_{sat}-e_0-\kappa\ln p_0}{\lambda-\kappa}+\frac{1+e_0}{\lambda-\kappa}\varepsilon_v^p\right)}
\cdot\left\{\begin{aligned}&\left[\frac{\mathbf{1}}{3}(2p-p_x)+\frac{3(\boldsymbol{\sigma}-p\mathbf{1})}{\left(2-\mu_0S_e^d\right)M^2}\right]:\mathrm{d}\boldsymbol{\sigma}\\&+\frac{pp_{sat}a_1a_2a_3(1-S_e)^{a_3-1}}{\exp\left[a_2(1-S_e)^{a_3}\right]}\mathrm{d}S_e-\frac{q^2\mathrm{d}S_e^{d-1}\left(N_0^2-2M^2\right)}{\left[2M^2\left(1-S_e^d\right)+N_0^2S_e^d\right]^2}\mathrm{d}S_e\end{aligned}\right\} \tag{4.42}
$$

Equation (4.42) is used to calculate $\mathrm{d}\boldsymbol{\varepsilon}$ through $\mathrm{d}\boldsymbol{\sigma}$ and $\mathrm{d}S_e$. The combination of Eq. (4.42) and Eqs. (4.25)–(4.32) gives the expression for describing the mechanical behaviour in the unsteady water environment. From the expressions of Eq. (4.42) and Eqs. (4.25)–(4.32), it is clear that the proposed model requires a total of 15 material parameters-6 mechanical parameters ($\lambda, \kappa, \nu, M, \Pi_{sat}, N_0$), 4 hydraulic parameters (a_d, a_w, m, b) and 5 hydro-mechanical coupling parameters (a_1, a_2, a_3, c, d)-to describe the mechanical behaviour in the unsteady water environment. All the material parameters have clear physical meanings and can be easily determined through the element tests of soil. The mechanical parameters can be calibrated through conventional triaxial drained shearing tests and isotropic loading–unloading tests at saturated state. The hydraulic parameters can be determined through the drying test with constant e and the wetting test with constant e. For the hydro-mechanical parameters, a_1, a_2 and a_3 can be obtained by fitting the isotropic compression tests with constant s or the isotropic compression test with constant S_e in the e–ln p plane, and d can be determined by fitting the experimental data of the conventional triaxial drained shearing test with constant s. Notably, in the saturated state, the proposed model automatically degenerates to Liang's model.

4.4 Model Validation

The model ability in simulating the hydraulic and mechanical behaviour of unsaturated soil is validated against test data from silt sand (Cai et al., 2024), Pearl clay (Sun et al., 2007) and mixed soil (Almahbobi, 2018). The test data are indicated by hollow circles, and simulation results are represented by solid lines.

4.4.1 Silt Sand

Cai et al. (2024) reported test data from conventional triaxial drained shearing tests with constant s on a silt sand, which is taken from the excavation site of a station of Beijing subway line 16. The silty sand was composed of 75% sand, 20.5% silt and 4.5% clay. The liquid limit is 15.4%, the plastic limit is 9.9%, and the plastic index is 5.5%. The shape of the specimens is a cylinder with 80 mm in height and 39.1 mm in diameter. All the specimens are prepared by the moisture tamping method. The selected material parameters are summarized in Table 4.1. The initial state of the conventional triaxial drained shearing test with constant s is obtained through two steps. The first step is suction equalization, and the second step is isotropic consolidation with constant suction.

Figure 4.11 displays a comparison between test data from conventional triaxial drained shearing tests on saturated silty sand and corresponding simulation results. During the shearing process, as shown in Fig. 4.11, with the increase of the shear strain ε_d, q monotonically increases, tends to become stable, and reaches the critical state. The maximum value of q increases with the minior principal stress of Bishop's effective stress σ_3. It can be seen from Fig. 4.11 that the predictions (solid lines) of the proposed model agree well with the observed test results (hollow circle).

Figures 4.12, 4.13 and 4.14 show the test results and corresponding simulations of the conventional triaxial drained shearing test with constant s for unsaturated silt sand at $\overline{\sigma}_3 = 100$ kPa, $\overline{\sigma}_3 = 200$ kPa and $\overline{\sigma}_3 = 300$ kPa, respectively. Here, $\overline{\sigma}_3$ is the minor principal stress of net stress. During the drained shearing process with constant s, the variation of q with ε_d at unsaturated state is similar to that at the saturated state, and the evolution of S_r is consistent with the evolution of the volumetric strain ε_v. A comparison of the experimental data under different s-values

Table 4.1 Material parameters of silt sand

λ	κ	Π_{sat}	M	N_0	d	ν	a_1	a_2	a_3	a_w	a_d	m	b
0.064	λ/5	1.01	1.4	1.47	0.1	0.3	400	10	10	1 kPa	20 kPa	1.39	6.0

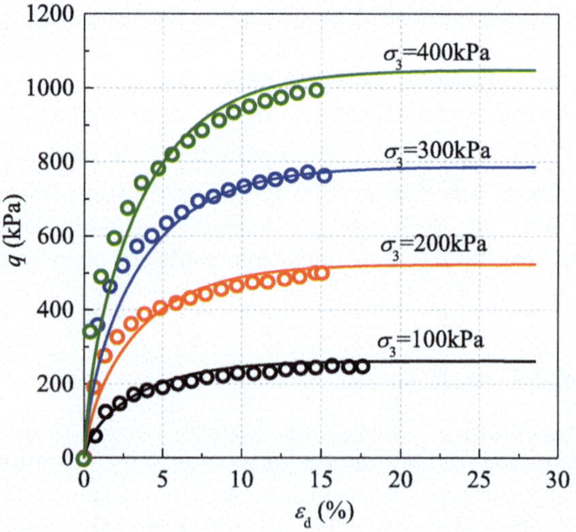

Fig. 4.11 Conventional triaxial drained shearing tests on saturated silt sand

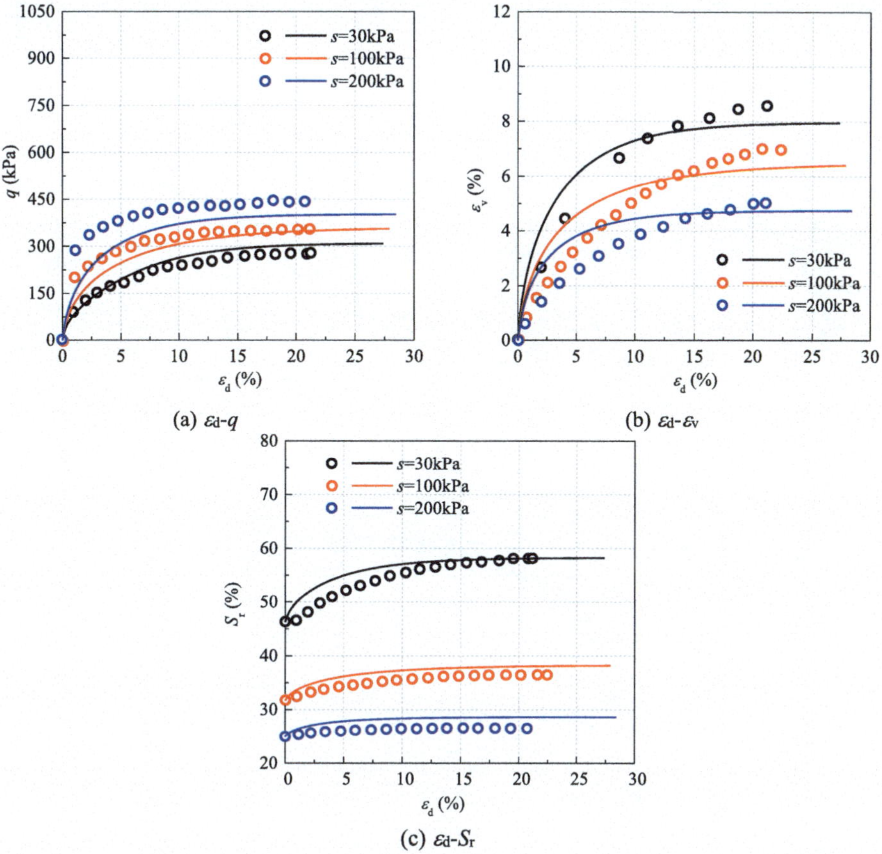

Fig. 4.12 Conventional triaxial drained shearing tests with $\overline{\sigma}_3 = 100$ kPa on unsaturated silt sand

reveals that q increases with increasing s and ε_v decreases with increasing s. This phenomenon can be effectively reproduced by the established model.

4.4.2 Pearl Clay

Sun et al. (2007) conducted extensive element tests on Pearl clay to investigate its hydro-mechanical behaviour. Pearl clay consists of 50% silt and 50% clay, and the content of expansive minerals is zero. The specific gravity is 2.71, the liquid limit is 49%, and the plasticity index is 22%. The shape of the specimens is a cylinder with a diameter of 35 mm and a height of 80 mm. $S_r^{res} = 0.0$ and $S_r^0 = 0.88$ are taken from the literature (Zhou et al., 2012b). The selected material parameters of Pearl clay are summarized in Table 4.2.

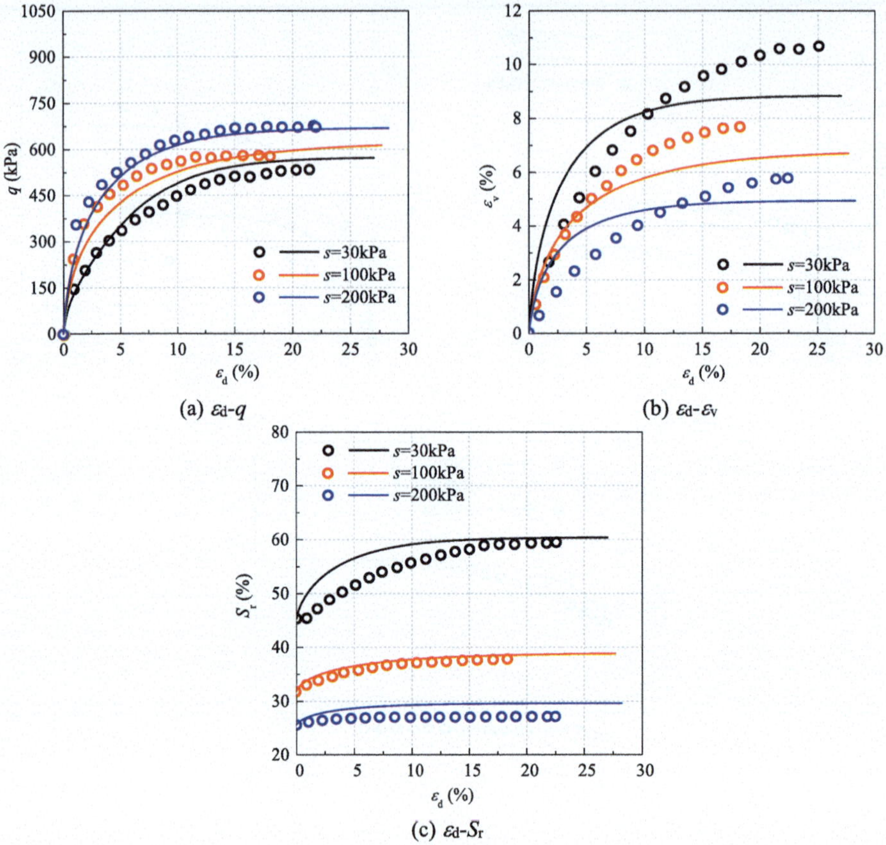

Fig. 4.13 Conventional triaxial drained shearing tests with $\overline{\sigma}_3 = 200$ kPa on unsaturated silt sand

Figure 4.15 compares the simulation results and observed experimental data of a wetting and drying test with $\overline{p} = 49$ kPa on Pearl clay. Here, $\overline{p}$ is the mean stress of net stress. During the wetting and drying test at $\overline{p} = 49$ kPa, s is first reduced from 245 kPa to 3.5 kPa, and subsequently is increased to 490 kPa (Sun et al., 2007). The experimental data presented in Fig. 4.15 indicate that both wetting process and drying process lead to a significant decrease in e. It can be seen from Fig. 4.15 that the proposed model can describe not only the variation of S_r but also the variation of e during wetting and drying tests under constant $\overline{p}$.

Figures 4.16, 4.17, 4.18, 4.19 illustrate the experimental data of the loading-wetting path on Pearl clay. For the loading-wetting path, e_0 is 1.27, the initial s is 147 kPa, and the initial $\overline{p}$ is 20 kPa. The loading-wetting path consists of two stages. The first stage is the isotropic compression path with $s = 147$ kPa, and the value of $\overline{p}$ increases from 20 to 98 kPa (see Fig. 4.16), 196 kPa (see Fig. 4.17), 392 kPa (see Fig. 4.18) and 588 kPa (see Fig. 4.18), respectively. The second stage is the wetting path with constant $\overline{p}$, and the value of s decreases from 147 to 0 kPa. The

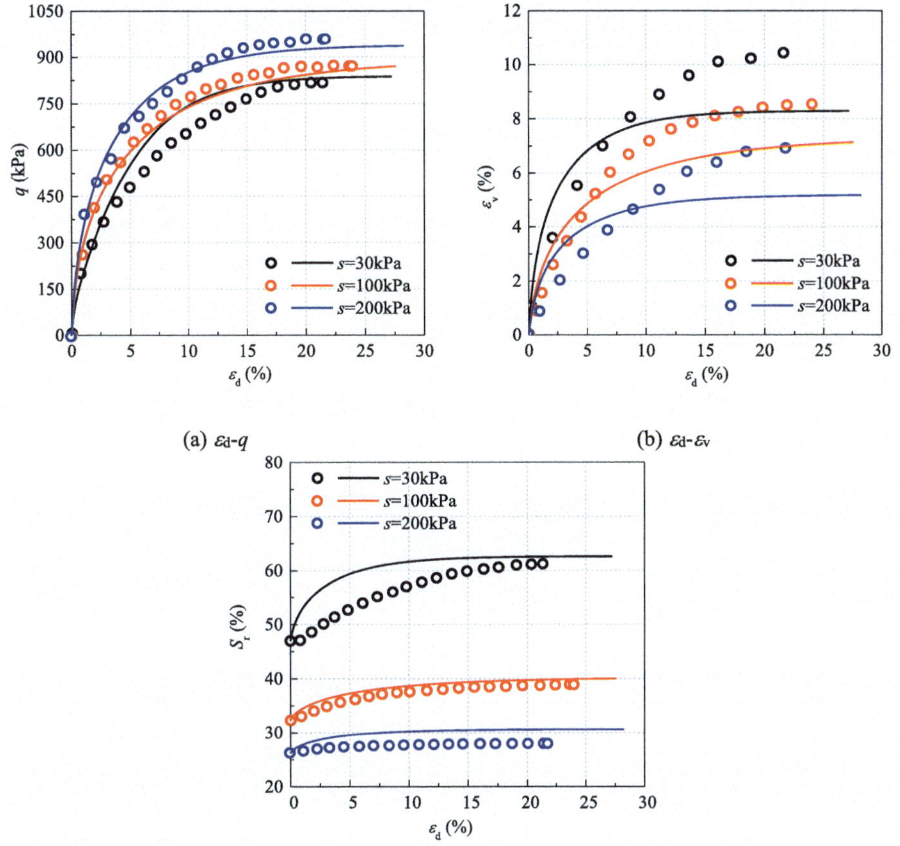

(c) ε_d-S_r

Fig. 4.14 Conventional triaxial drained shearing tests with $\overline{\sigma}_3 = 300$ kPa on unsaturated silt sand

Table 4.2 Material parameters of Pearl clay

λ	κ	Π_{sat}	M	N_0	ν	a_1	a_2	a_3	d	a_w	a_d	m	b	c
0.11	0.03	1.68	1.12	0.94	0.3	20	1.1	1.8	0.5	15 kPa	100 kPa	2.0	4.5	1.0

test data of the first stage indicate that the isotropic compression line with constant s is nonlinear in the e–ln $\overline{p}$ plane, and a decrease in e leads to an increase in S. The test data of the second stage show that a decrease in s results in a decrease in e. The proposed model is able to consider the interaction between mechanical behaviour and hydraulic behaviour.

Figures 4.20 and 4.21 display the experimental data on Pearl clay during the shearing-wetting-shearing path compared with the corresponding simulation results. For the shearing-wetting-shearing path (Figs. 4.20a and 4.21a), point A represents the

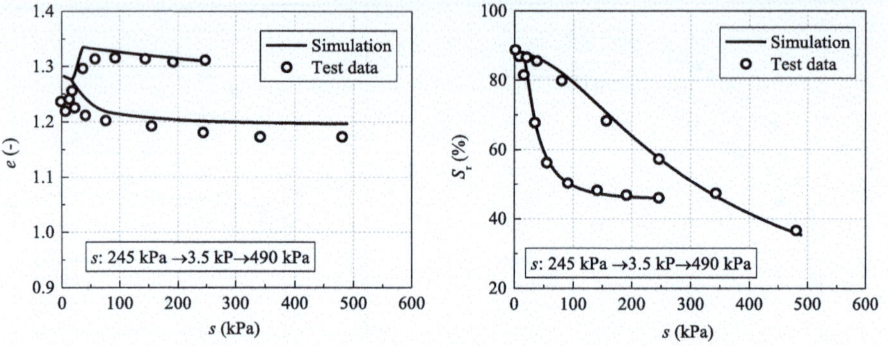

(a) S_r–s (b) e–s

Fig. 4.15 Wetting and drying test on Pearl clay at $\bar{p} = 49$ kPa

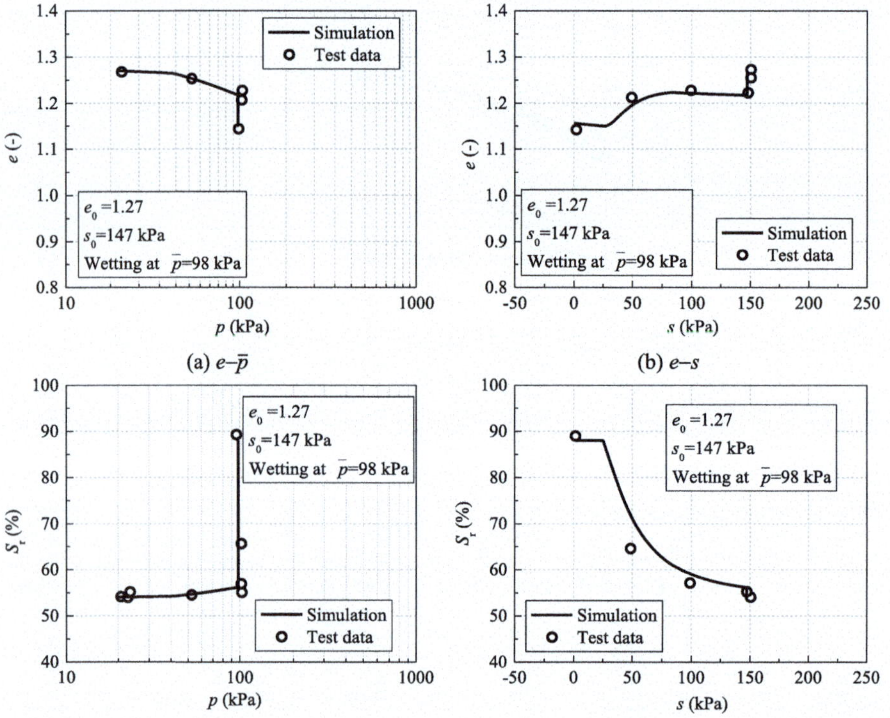

(c) S_r–$\bar{p}$ (d) S_r–s

Fig. 4.16 Hydro-mechanical behaviour of Pearl clay during the isotropic compression path with $s = 147$ kPa and the wetting path at $\bar{p} = 98$ kPa

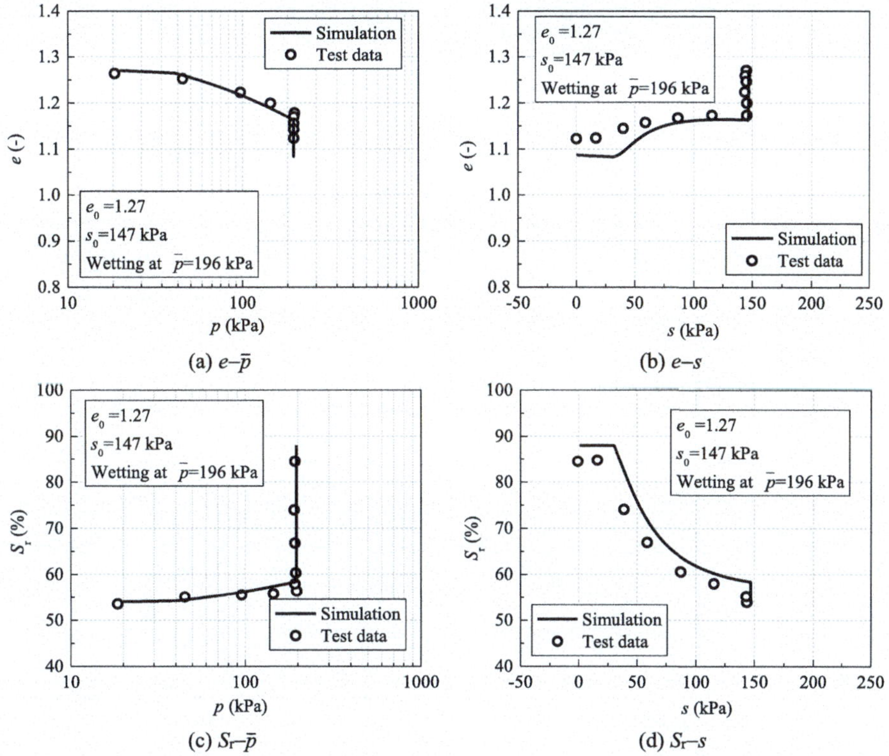

Fig. 4.17 Hydro-mechanical behaviour of Pearl clay during the isotropic compression path with s = 147 kPa and wetting path at $\bar{p}$ = 196 kPa

initial state, point D indicates the critical stress state, path AB represents the triaxial drained shearing path with $\bar{p}$ = 196 kPa and s = 147 kPa, path BC indicates that the wetting path with constant net stress state, and path CD refers to the triaxial drained shearing path with p = 196 kPa and s = 0 kPa. For the shearing-wetting-shearing path in Fig. 4.20a, $e_0 = 1.41$, $S_{r0} = 53.0\%$, and s_0 = 147 kPa; the initial net stress state is $\overline{\sigma}_1=\overline{\sigma}_2=\overline{\sigma}_3$=196 kPa; and the wetting path is conducted at q = 84 kPa. Here, $\overline{\sigma}_1$ is the major principal stress of net stress, and $\overline{\sigma}_2$ is the intermediate principal stress of net stress. For the shearing-wetting-shearing path in Fig. 4.21a, $e_0 = 1.35$, S_{r0} = 59.2%, and s_0 = 147 kPa; the initial net stress state is $\overline{\sigma}_1=\overline{\sigma}_2=\overline{\sigma}_3$=196 kPa; and the wetting path is conducted at q = 147 kPa.

It can be seen from Figs. 4.20 and 4.21 that the wetting path at q = 147 kPa causes a larger ε_1 than that at q = 84 kPa. During the triaxial drained shearing path with constant $\bar{p}$ and s (path AB), S_r increases with the increase of ε_v. By comparing the test data with the simulation results in Figs. 4.20 and 4.21, we may conclude that the proposed model provides a reasonable description of the hydraulic and mechanical behaviour of unsaturated soil during the shearing path with constant suction, the wetting path with constant net stress, and the shearing path with zero suction.

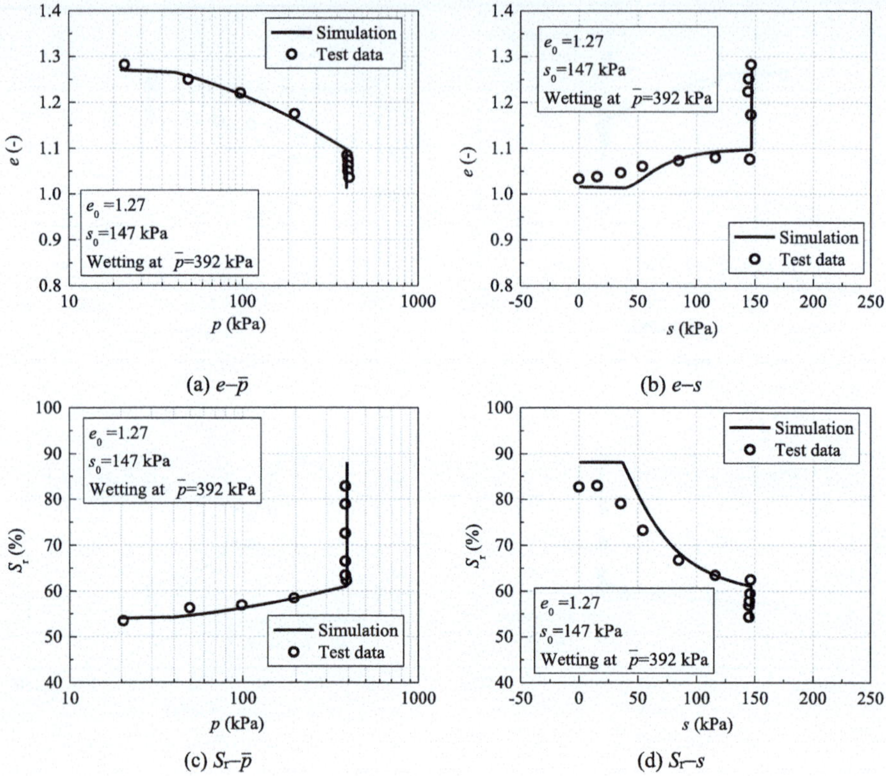

Fig. 4.18 Hydro-mechanical behaviour of Pearl clay during the isotropic compression path with $s = 147$ kPa and wetting path at $\bar{p} = 392$ kPa

4.4.3 Mixed Soil

Almahbobi (2018) conducted a series of conventional triaxial drained shearing tests with constant s on mixed soil. The mixed soil consists of 20% Speswhite kaolin, 40% Leighton Buzzard sand and 40% M400 silt. The specific gravity value is 2.65. The plasticity limit is 16%, and the liquid limit is 24%. The shape of the specimens for the triaxial drained shearing tests is a cylinder with 100 mm in height and 50 mm in diameter. $s = 0$ kPa represents the saturated state, and $s > 0$ kPa indicates the unsaturated state. $S_r^{res} = 0.06$ and $S_r^0 = 1.0$ are taken from the original literature. Thus, S_e can be calculated through Eq. (4.7). Because s remains unchanged in conventional triaxial shearing tests, the change of S_e can be determined by the change of e, and the material parameter c is not needed. The adopted material parameters are given in Table 4.3.

Figure 4.22 compares the simulation results and the experimental data for conventional triaxial drained shearing tests on saturated mixed soil. For the conventional triaxial drained shearing tests on saturated mixed soil, e_0 is 0.529 ($\sigma_3 = 100$ kPa),

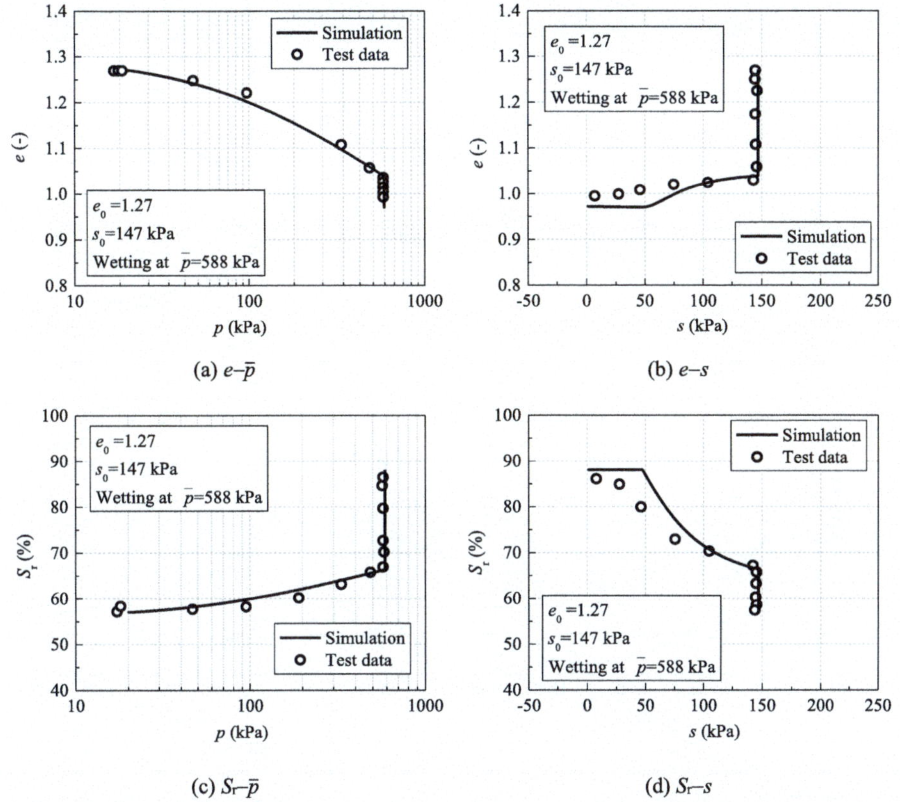

Fig. 4.19 Hydro-mechanical behaviour of Pearl clay during the isotropic compression path with $s = 147$ kPa and wetting path at $\bar{p} = 588$ kPa

0.476 ($\sigma_3 = 250$ kPa) and 0.438 ($\sigma_3 = 400$ kPa). At the saturated state, q and ε_v increase with ε_1 at different constant σ_3 values of 100, 250 and 400 kPa. The comparison between model predictions and experimental data shows that the proposed model can reasonably reproduce the above phenomenon.

Figure 4.23 displays model simulations and test data of the conventional triaxial drained shearing test with constant s. Table 4.4 summarizes the initial states of the conventional triaxial drained shearing test with constant s, including s, e_0, the initial degree of saturation S_{r0} and the initial mean net stress $\bar{p}_0$. During the conventional triaxial drained shearing path with constant s, q and ε_v increase with ε_1. The comparisons between simulation results and experimental data indicate that the proposed model provides a suitable description of the mechanical behaviour of saturated and unsaturated mixed soils.

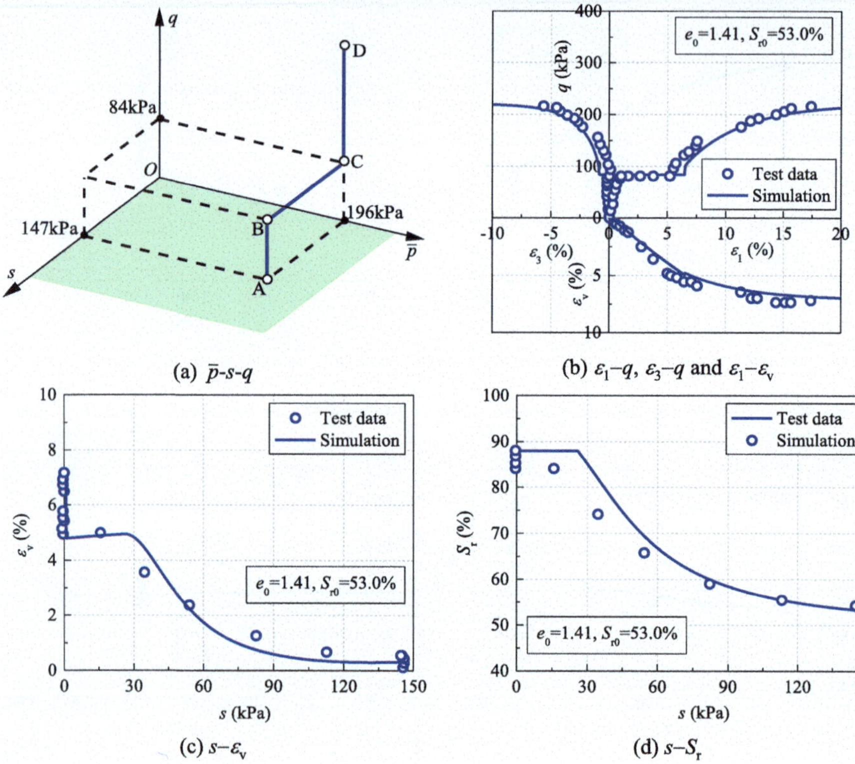

Fig. 4.20 Hydro-mechanical behaviour of Pearl clay during the shearing-wetting-shearing path with wetting path at $q = 84$ kPa and $\bar{p} = 196$ kPa

4.5 Model Performance Analysis

The performance of the proposed model is illustrated through three typical stress paths of unsaturated soil, namely, the isotropic compression path with constant suction, the wetting path with constant net stress, and the conventional triaxial drained shearing path with constant suction. The material parameters of Pearl clay are taken to describe the mechanical behaviour and the hydraulic behaviour of the selected typical stress paths. In addition, for convenience, $S_r^{res} = 0.0$ and $S_r^0 = 1.0$. Thus, based on Eq. (4.7), $S_e = S_r$. In this section, the initial state of the hypothetical soil is in the normally consolidated state.

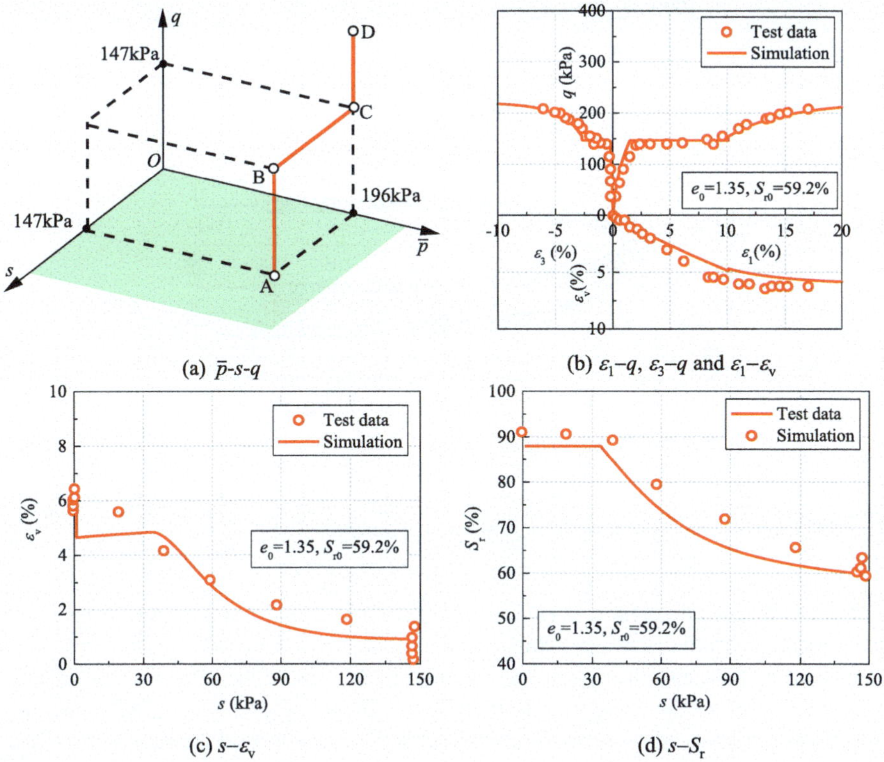

Fig. 4.21 Hydro-mechanical behaviour of Pearl clay during the shearing-wetting-shearing path with wetting path at q = 147 kPa and $\bar{p}$= 196 kPa

Table 4.3 Material parameters of mixed soil

λ	κ	Π_{sat}	M	N_0	d	v	a_1	a_2	a_3	a_w	a_d	m	b
0.1	0.03	1.04	1.08	1.25	0.15	0.3	150	0.2	1.5	10 kPa	44 kPa	1.5	1.0

4.5.1 Isotropic Compression Path with Constant Suction

For the isotropic compression path with constant s, the initial net stress state is taken as $\overline{\sigma}_1 = \overline{\sigma}_2 = \overline{\sigma}_3 = 10$ kPa; the initial hydraulic state is taken as $S_r = 0.3$ and $s =$ 200 kPa; and e_0 is calculated based on Eq. (4.19) and is equal to 1.395. The isotropic compression with constant s is terminated when the value of $\overline{p}$ reaches 800,000 kPa. Figure 4.24 displays the simulation results of the isotropic compression path with constant s.

As shown in Fig. 4.23, the initial state is located on the NCL with $S_e = 0.3$, and the isotropic compression line with constant s is a spatial curve lying on the

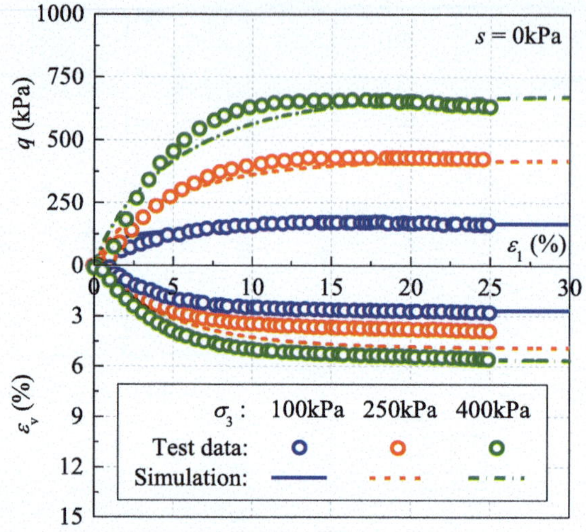

Fig. 4.22 Conventional triaxial shearing tests on saturated mixed soil

normal compression surface. With the increase of $\overline{p}$, e decreases, S_e increases, and the stress state gradually moves from the NCL at $S_e = 0.3$ to the NCL at $S_e = 1.0$. The increase of S_e indicates that the proposed model can capture the effect of mechanical behaviour on hydraulic behaviour. In both the $e - \ln p$ plane and the $e - \ln \overline{p}$ plane, the NCL with constant s shows obvious nonlinearity. The proposed model can describe the nonlinearity of the NCL with constant s in the e–$\ln p$ plane and in the $e - \ln \overline{p}$ plane.

4.5.2 *Wetting Path with Constant Net Stress*

For the wetting path with constant net stress, the initial net stress state is taken as $\overline{\sigma}_1 = \overline{\sigma}_2 = \overline{\sigma}_3 = 100$ kPa and the initial hydraulic state is taken as $S_e = 0.3$ and $s =$ 200 kPa. e_0 is calculated through Eq. (4.19) and equals to 1.304. Figure 4.25 shows the simulation results corresponding to the wetting path when s decreases from 200 to 0 kPa at $\overline{p} = 100$ kPa. In Fig. 4.25, point A represents the initial state.

Based on the changes in f (Fig. 4.25e) and S_r (Fig. 4.25a), the wetting process can be classified into four stages. At the stage AB, a decrease in s leads to an increase in S_r (Fig. 4.25a) and a decrease in p (Fig. 4.25b); the wetting path points to the inside of f in the p-S_r plane (Fig. 4.25e); and elastic expansive deformation occurs due to the decrease in p (Fig. 4.25d). Thus, e gradually increases (Fig. 4.25c) at the stage AB. At the stage BC, as s decreases, S_r increases and p decreases; the wetting path points to the outside of f in the p-S_r plane (Fig. 4.25e); plastic contraction deformation occurs due to the extension of f, and elastic expansive deformation occurs due to the decrease in p. Note that the absolute value of plastic contraction deformation is larger

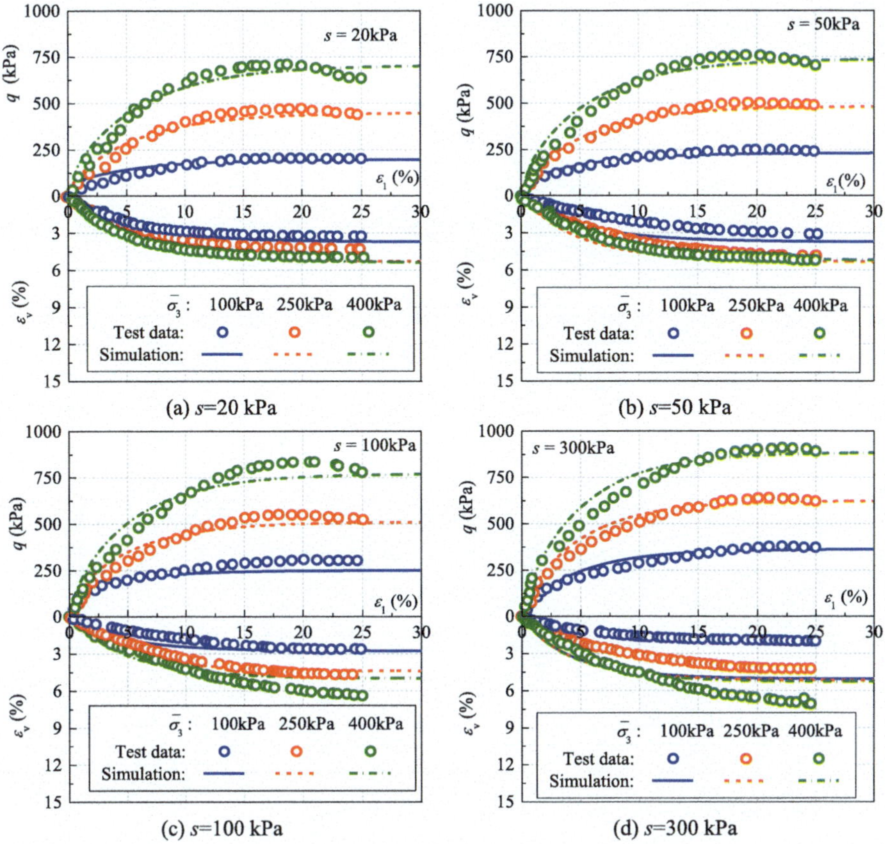

(a) s=20 kPa (b) s=50 kPa (c) s=100 kPa (d) s=300 kPa

Fig. 4.23 Conventional triaxial shearing tests with constant s on unsaturated mixed soil

Table 4.4 Initial states of the conventional triaxial drained shearing test with constant s on unsaturated mixed soil

s/kPa	$\bar{p}_0$/kPa	e_0/–	S_{r0}/–
20	100	0.589	0.881
	250	0.545	0.851
	400	0.470	0.906
50	100	0.628	0.755
	250	0.584	0.742
	400	0.530	0.759
100	100	0.668	0.511
	250	0.617	0.545
	400	0.578	0.574
300	100	0.702	0.407
	250	0.661	0.422
	400	0.626	0.436

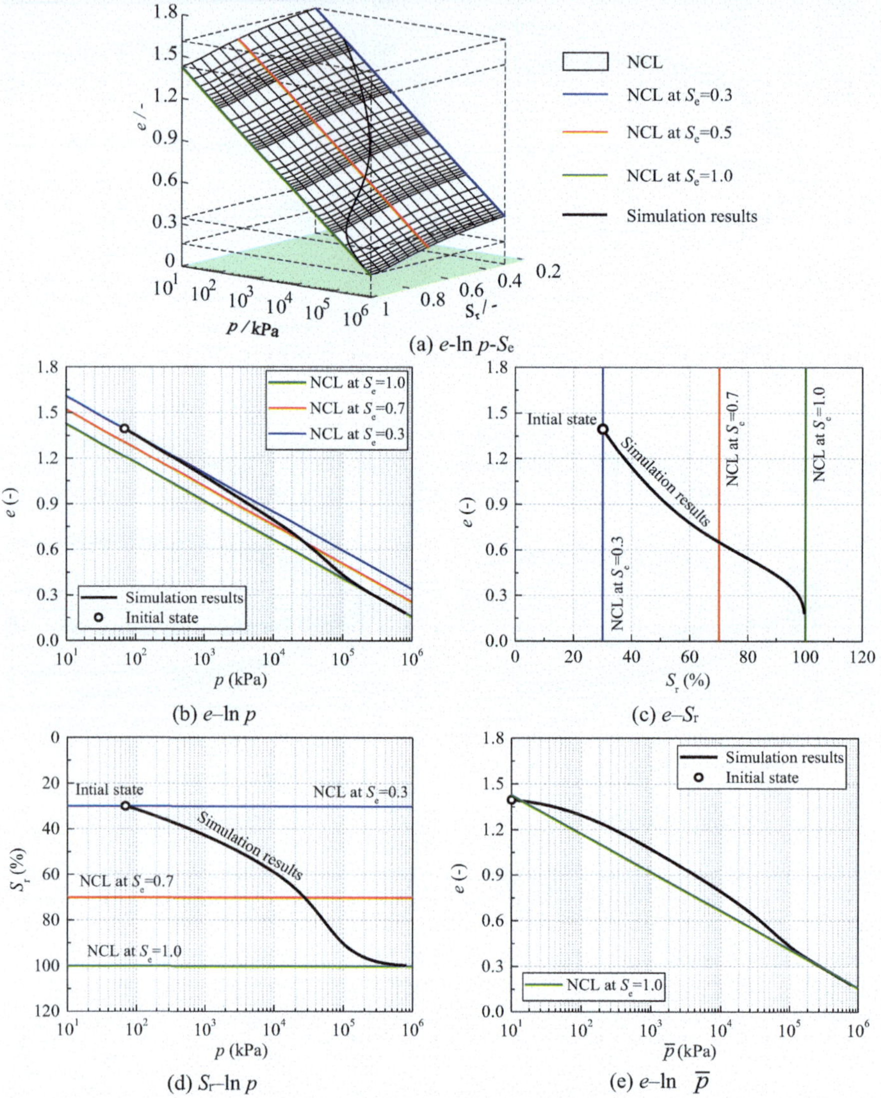

Fig. 4.24 Simulation of isotropic compression with constant s

than that of elastic expansive deformation. *e* decreases significantly during the stage BC (Fig. 4.25c). The characteristics of the stage CD are consistent with those of the stage AB. At the stage DE, *p* decreases linearly as *s* decreases, and elastic expansive deformation occurs. Thus, *e* at the stage DE increases. The most prominent feature of the DE stage is $S_r = 1.0$. The simulation results presented in Fig. 4.25 indicate that the wetting path with constant net stress causes the soil being in an over consolidated

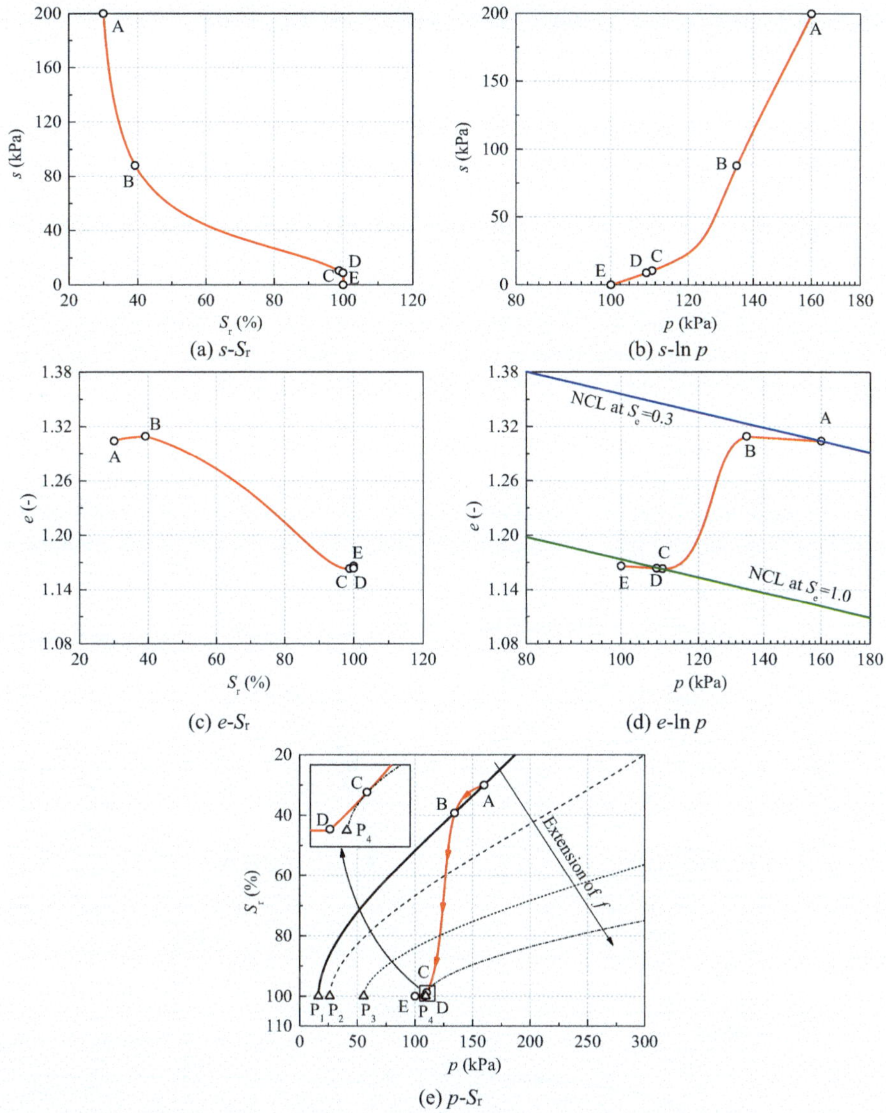

Fig. 4.25 Simulation of the wetting path at $\overline{p} = 100$ kPa

state. The proposed model is capable of describing the phenomenon in which e decreases significantly during wetting process.

4.5.3 Conventional Triaxial Drained Shearing Path with Constant Suction

The conventional triaxial drained shearing path with constant suction is employed to demonstrate the performance of the proposed model in terms of shearing behaviour. The initial net stress state is taken as $\overline{\sigma}_1 = \overline{\sigma}_2 = \overline{\sigma}_3 = 100$ kPa. The initial hydraulic state is taken as $s = 0$ kPa ($S_r = 100\%$), $s = 100$ kPa ($S_r = 87.95\%$), $s = 200$ kPa ($S_r = 67.86\%$), $s = 400$ kPa ($S_r = 41.93\%$), $s = 800$ kPa ($S_r = 22.50\%$) and $s = 1600$ kPa ($S_r = 11.47\%$). e_0 is calculated based on Eq. (4.19) and is 1.173, 1.136, 1.183, 1.229, 1.252, and 1.262. Figure 4.26 displays the simulation results of the conventional triaxial drained shearing path with constant s. In Fig. 4.26, $\overline{\eta}$ $(= \overline{q}/\overline{p})$ is the net stress ratio, and $s = 0$ kPa represents the saturated state.

At the saturated state ($s = 0$ kPa), as shown by the dashed line in Fig. 4.26, $\overline{\eta}$, η and ε_v increase monotonically with increasing ε_1, $\overline{\eta}$ is equal to η, and $S_r = 1.0$. During conventional triaxial shearing path with constant s, the variations of $\overline{\eta}$, η and ε_v at the unsaturated state ($s > 0$ kPa) are similar to those at the saturated state ($s = 0$ kPa). Comparing Fig. 4.26c with Fig. 4.26d, it can be found that the evolution

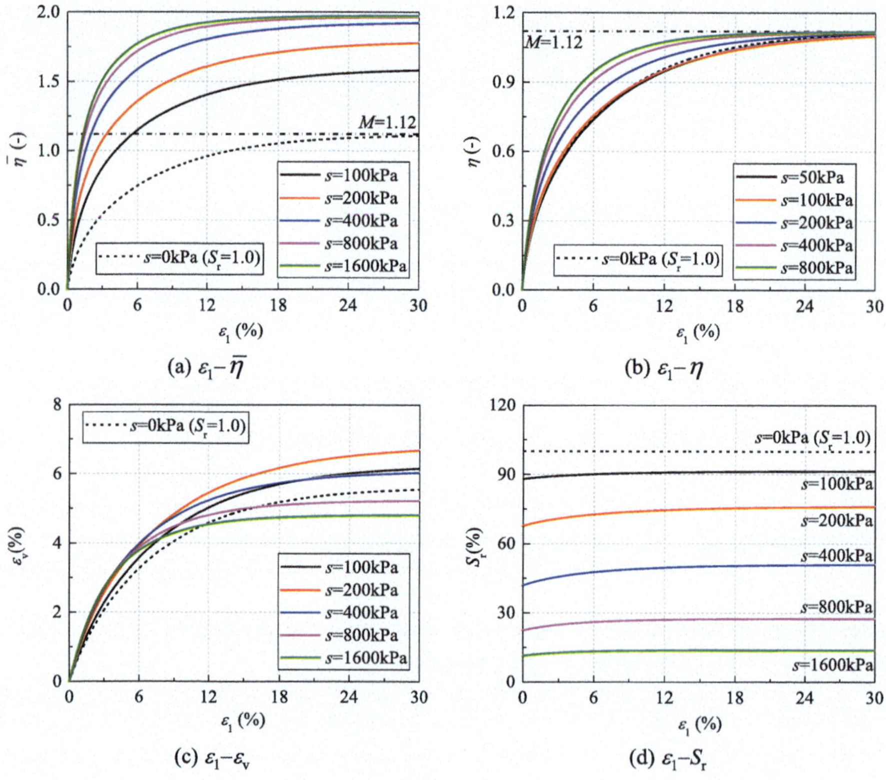

Fig. 4.26 Simulations of the conventional triaxial drained shearing path with constant s

of S_r is consistent with the evolution of ε_v at the unsaturated state. The increase of S_r is caused by the accumulation of ε_v. This phenomenon reflects the influence of the mechanical behaviour on the hydraulic behaviour. Based on the simulation results presented in Fig. 4.26c, it can be seen that ε_v generated during conventional triaxial drained shearing path with constant s may be greater than that generated under saturated conditions or less than that generated under saturated conditions. ε_v generated during conventional triaxial drained shearing path is closely related to the hydraulic state before shearing. This phenomenon is consistent with existing experimental results (Cui & Delage, 1996; Estabragh & Javadi, 2008; Oka et al., 2010).

4.6 Summary

A new approach to modeling the mechanical behaviour for soil in the unsteady water environment is presented from the perspectively of stress state and material parameters changing with water environment. The proposed modeling approach takes a non-orthogonal constitutive model of saturated soil as a foundation. Bishop's effective stress is adopted to describe the stress state in the unsteady water environment, and automatically considers the influence of water environment state. The elastic deformation is described through the generalized Hooke's law, in which Terzaghi's effective stress is replaced with Bishop's effective stress. For the plastic deformation, the direction of the plastic strain increment is obtained by applying the non-orthogonal plastic flow rule to yield function in the unsteady water environment, and the magnitude of the plastic strain increment is determined through the consistency conditions of yield function in the unsteady water environment. The yield function in the unsteady water environment is obtained by taking the same form of yield function for saturated soil, replacing Terzaghi's effective stress with Bishop's effective, and establishing the relationship between the material parameters and water environment state. A hydraulic constitutive equation is provided to describe the unsteady water environment. The proposed approach provides a novel strategy for establishing a constitutive model for soil in the unsteady water environment state. The most prominent advantage of the proposed approach is that it can establish a new constitutive model for soil in the unsteady water environment state by utilizing an arbitrary yield function of saturated soils.

Based on the proposed modeling approach, a non-orthogonal elastoplastic constitutive model for soil in the unsteady water environment is presented by determining the specific expressions for the unknown elements. For the proposed model, the specific form of the yield function in a steady water environment state is determined by taking the same form of the yield function of Liang's model and replacing Terzaghi's effective stress with Bishop's effective stress; and the influence of water environment state on yield surface is captured by establishing the relationship between Π and S_e and the relationship between N and S_e. The Riemann–Liouville fractional derivative is employed to obtain the non-orthogonal gradients of yield function in the unsteady water environment. A specific hydraulic constitutive relationship is provided to describe the unsteady water environment. The proposed model involves 15 material parameters, which includes 6 mechanical parameters (λ, κ, ν, M, Π_{sat}, N_0), 4 hydraulic parameters (a_d, a_w, m, b) and 5 hydro-mechanical coupling parameters (a_1, a_2, a_3, c, d). All material parameters have clear physical meanings and can be easily calibrated via conventional laboratory tests. At the saturated state, the proposed model degenerates to Liang's model (Liang et al., 2019).

The proposed model simultaneously considers the influence of the water environment state on the magnitude and direction of the plastic strain increment. The nonlinearity of the NCL with constant s in the e–ln p plane and e–ln $\overline{p}$ plane, the volume contraction during wetting process, and the increases of S_e due to the reduction of e can be reasonably reproduced by the proposed model. Notably, the proposed model is also capable of describing the situation where the volumetric strain of the triaxial drained shearing path with constant s under unsaturated conditions is less than that under saturated conditions. The comparisons between the model simulations and test data from the unsaturated soils indicate that the proposed model can provide a reasonable description of the mechanical behaviour of unsaturated soil, and validate the effectiveness of the proposed model.

References

Almahbobi, S. (2018). *Experimental study of volume change and shear strength behaviour of statically compacted collapsible soil*. Cardiff University.

Alonso, E. E., Pereira, J. M., Vaunat, J., & Olivella, S. (2010). A microstructurally based effective stress for unsaturated soils. *Géotechnique., 60*(12), 913–925.

Cai, G., Han, B., Asreazad, S., Liu, C., Zhou, A., Li, J., & Zhao, C. (2024). Experimental study on critical state behaviour of unsaturated silty sand under constant matric suctions. *Géotechnique,* 74(5), 409–430.

Cui, Y. J., & Delage, P. (1996). Yielding and plastic behaviour of an unsaturated compacted silt. *Géotechnique, 46*(2), 291–311.

D'Onza, F., Gallipoli, D., Wheeler, S., Casini, F., Vaunat, J., Khalili, N., Laloui, L., Mancuso, C., Mašín, D., Nuth, M., Pereira, J. M., & Vassallo, R. (2011). Benchmark of constitutive models for unsaturated soils. *Géotechnique, 61*(4), 283–302.

Estabragh, A. R., & Javadi, A. A. (2008). Critical state for overconsolidated unsaturated silty soil. *Canadian Geotechnical Journal, 45*(3), 408–420.

Liang, J. Y., Lu, D. C., Zhou, X., Du, X. L., & Wu, W. (2019). Non-orthogonal elastoplastic constitutive model with the critical state for clay. *Computers and Geotechnics, 116*, 103200.

Lu, D. C., Liang, J. Y., Du, X. L., Ma, C., & Gao, Z. W. (2019). Fractional elastoplastic constitutive model for soils based on a novel 3D fractional plastic flow rule. *Computers and Geotechnics, 105*, 277–290.

Oka, F., Kodaka, T., Suzuki, H., Kim, Y. S., Nishimatsu, N., & Kimoto, S. (2010). Experimental study on the behavior of unsaturated compacted silt under triaxial compression. *Soils and Foundations, 50*(1), 27–44.

Roscoe, K. H., Schofield, A. N., & Thurairajah, A. (1963). Yielding of clays in states wetter than critical. *Géotechnique, 13*(3), 211–240.

Sun, D. A., Sheng, D. C., Cui, H. B., & Sloan, S. W. (2007). A density-dependent elastoplastic hydro-mechanical model for unsaturated compacted soils. *International Journal for Numerical and Analytical Methods in Geomechanics, 31*(11), 1257–1279.

Tarantino, A. (2009). A water retention model for deformable soils. *Géotechnique, 59*(9), 751–762.

van Genuchten, M. T. (1980). A closed-form equation for predicting the hydraulic conductivity of unsaturated soils. *Soil Science Society of America Journal, 44*(5), 892–898.

Vanapalli, S. K., Fredlund, D. G., & Pufahl, D. E. (1999). The influence of soil structure and stress history on the soil-water characteristics of a compacted till. *Géotechnique, 49*(2), 143–159.

Wang, Q., Pufahl, D. E., & Fredlund, D. G. (2002). A study of critical state on an unsaturated silty soil. *Canadian Geotechnical Journal, 39*(1), 213–218.

Wheeler, S. J., Sharma, R. S., & Buisson, M. S. R. (2003). Coupling of hydraulic hysteresis and stress-strain behaviour in unsaturated soils. *Géotechnique, 53*(1), 41–54.

Zhou, A. N., & Sheng, D. C. (2015). An advanced hydro-mechanical constitutive model for unsaturated soils with different initial densities. *Computers and Geotechnics, 63*, 46–66.

Zhou, A. N., Sheng, D. C., Sloan, S. W., & Gens, A. (2012a). Interpretation of unsaturated soil behaviour in the stress-saturation space, I: Volume change and water retention behaviour. *Computers and Geotechnics, 43*, 178–187.

Zhou, A. N., Sheng, D. C., Sloan, S. W., & Gens, A. (2012b). Interpretation of unsaturated soil behaviour in the stress-saturation space II: Constitutive relationships and validations. *Computers and Geotechnics, 43*, 111–123.

Chapter 5
Non-orthogonal Plasticity Damage Model for Concrete

The irreversible plastic deformation and degradation of unloading stiffness are the most typical nonlinear mechanical behaviors of concrete materials, which can be described by the plasticity damage model that couples plastic mechanics and damage mechanics, as shown in Fig. 5.1. The methods to develop the plasticity damage model can be classified roughly into two categories: the nominal stress method and the effective stress method. The main difference between these two methods lies in whether the effective stress corresponding to the undamaged configuration or the nominal stress corresponding to the damaged configuration drives the development of plastic deformation. Researchers generally prefer to use effective stress method (Jason et al., 2006; Ju, 1989; Lee & Fenves, 1998; Simo & Ju, 1987; Zhang et al., 2016) for establishing the plasticity damage models because the decoupling of plastic and damage calculations makes models more stable in numerical implementation. However, the difficulty of this method lies in the difficulty of reasonably describing the plastic deformation behavior in the effective stress space, because the vast majority of concrete tests are based on nominal stress.

In this chapter, a 3D non-orthogonal plasticity damage model for concrete in steady environments is established based on the combination of the effective stress method and multi-physical and non-orthogonal geotechnical plasticity theory. The proposed model consists of the damage part driven by plastic strain and the plastic part based on effective stress. In the damage characterization, the degeneration law of elastic stiffness is described by an S-type damage variable. In the plastic characterization, the direction of plastic strain increment is obtained directly by the non-orthogonal gradient of yield surface and the need for the construction of the plastic potential function is bypassed. The magnitude of plastic strain increment is calculated based on the effective hardening function and the closed-form yield function, where the effective hardening function corresponding to the undamaged material is derived by both the nominal hardening/softening function corresponding to the damaged material and the damage variable. The ability of the model to capture the

D. Lu et al., *Multi-physical and Non-orthogonal Geotechnical Plasticity Modeling and Algorithms*, https://doi.org/10.1007/978-981-96-2369-3_5

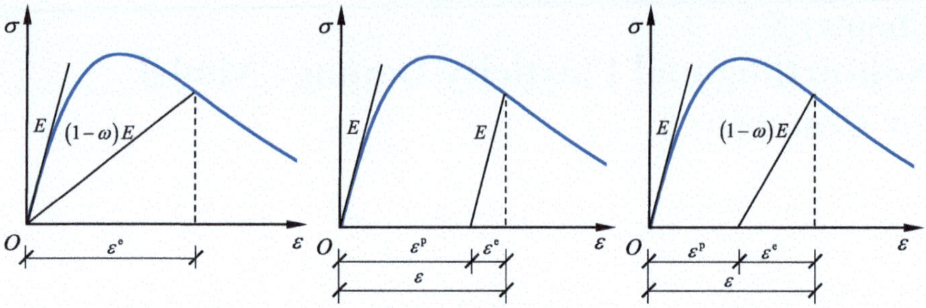

(a) Damage model (b) Plasticity model (c) Plasticity damage model

Fig. 5.1 Calculation results of different constitutive models

complex mechanical behavior of concrete is assessed by the monotonous and cyclic test data from the literature.

5.1 Theoretical Framework of Plasticity Damage Model

For the plasticity damage model utilizing the effective stress method, the mechanical evolution of material under the given load can be divided into the plastic part and the damage part. The characterization of plastic part is implemented in the effective stress space and can be completely decoupled from that of the damage part.

5.1.1 Damage Formulation

The stiffness degeneration of concrete material at the macroscopic scale can be attributed to the initiation, growth, and coalescence of both microcracks and microvoids. The concept of effective stress was first proposed by Kachanov (1958) to describe the material damage and further developed in subsequent studies (Dougill, 1976; Mazars, 1984; Rabotnov, 1969). Here, the effective stress represents the average stress which acts on the undamaged material where micro-defects are removed (Al-Rub & Kim, 2010). Correspondingly, the nominal stress, which is also known as Cauchy stress, represents the average stress which acts on the damaged material. Utilizing Hooke's law, the effective stress $\overline{\boldsymbol{\sigma}}$ and the nominal stress $\boldsymbol{\sigma}$ can be expressed as follows:

$$\overline{\boldsymbol{\sigma}} = \mathbb{D}_0 : \boldsymbol{\varepsilon}^{\mathrm{e}} = \mathbb{D}_0 : (\boldsymbol{\varepsilon} - \boldsymbol{\varepsilon}^{\mathrm{p}}) \tag{5.1}$$

$$\boldsymbol{\sigma} = \mathbb{D} : \boldsymbol{\varepsilon}^{\mathrm{e}} = \mathbb{D} : (\boldsymbol{\varepsilon} - \boldsymbol{\varepsilon}^{\mathrm{p}}) \tag{5.2}$$

where $\boldsymbol{\varepsilon}^{e}$, $\boldsymbol{\varepsilon}^{p}$, and $\boldsymbol{\varepsilon}$ denote the elastic strain, the plastic strain, and the total strain, respectively. The evolution for $\boldsymbol{\varepsilon}^{p}$ can be obtained by the plasticity theory implemented in the effective stress space. $\mathbb{D}_0$ and $\mathbb{D}$ denote the undamaged and the damaged elastic stiffness matrix, respectively. Herein, the degradation law of elastic stiffness is described by the scalar damage variable as follows:

$$\mathbb{D} = (1 - \omega)\mathbb{D}_0 \tag{5.3}$$

where ω denotes the damage variable driven by plastic strain. Strictly connected to the effective stress and damage concept, the strain equivalence hypothesis (Lemaitre, 1985; Simo & Ju, 1987), i.e., the strain caused by the effective stress is equivalent to the strain caused by the nominal stress, is employed to relate the effective stress to the nominal stress. By combining Eqs. (5.1), (5.2), and (5.3), the nominal stress can be derived by mapping the effective stress from the undamaged material into the damaged material as follows:

$$\boldsymbol{\sigma} = (1 - \omega)\overline{\boldsymbol{\sigma}} \tag{5.4}$$

In the process of characterizing the damage response, the material damage is defined as the degeneration of elastic stiffness. The damage variable is defined as the function of plastic strain and can be determined explicitly once the plastic evolution process is completed. The effective stress can be transformed into the nominal stress by the damage variable utilizing the strain equivalence hypothesis.

5.1.2 Plasticity Formulation

The standard increment plasticity theory is employed in the effective stress space. Taking the full differential for Eq. (5.1) leads to:

$$\mathrm{d}\overline{\boldsymbol{\sigma}} = \mathbb{D}_0 : (\mathrm{d}\boldsymbol{\varepsilon} - \mathrm{d}\boldsymbol{\varepsilon}^{p}) \tag{5.5}$$

where $\mathrm{d}\boldsymbol{\varepsilon}^{p}$ is the plastic strain increment and can be expressed as the product of magnitude and direction

$$\mathrm{d}\boldsymbol{\varepsilon}^{p} = \mathrm{d}\phi\mathbf{r} \tag{5.6}$$

where $\mathrm{d}\phi$ is the plastic multiplier and denotes the magnitude of $\mathrm{d}\boldsymbol{\varepsilon}^{p}$. $\mathbf{r}$ is the direction of $\mathrm{d}\boldsymbol{\varepsilon}^{p}$ and characterize the proportion relationship among components of $\mathrm{d}\boldsymbol{\varepsilon}^{p}$.

In the process of determining magnitude of $\mathrm{d}\boldsymbol{\varepsilon}^{p}$, the two core elements are required, i.e., the yield function and the effective hardening function. The former limits the current admissible stresses and the later describes the hardening law of undamaged material. The general expression of yield function f can be written as:

$$f\left[\overline{\boldsymbol{\sigma}}, \overline{H}(\boldsymbol{\varepsilon}^{\mathrm{p}})\right] = 0 \tag{5.7}$$

where $\overline{H}$ is the effective hardening function and controls evolution process of yield surface. During plastic loading process, the material produces $\mathrm{d}\boldsymbol{\varepsilon}^{\mathrm{p}}$ under the action of the given load increment. The yield surface expands outwards with the increase of $\overline{H}$. The new stress point is still located on the subsequent yield surface. This compatible relationship between the stress state and the accumulated plastic strain is known as the consistency condition, which can be expressed as follows:

$$\mathrm{d}f = f_{\overline{\boldsymbol{\sigma}}} : \mathrm{d}\overline{\boldsymbol{\sigma}} + f_{\overline{H}} \overline{H}_{\boldsymbol{\varepsilon}^{\mathrm{p}}} : \mathrm{d}\boldsymbol{\varepsilon}^{\mathrm{p}} = 0 \tag{5.8}$$

where $f_{\overline{\boldsymbol{\sigma}}} = \partial f / \partial \overline{\boldsymbol{\sigma}}$, $f_{\overline{H}} = \partial f / \partial \overline{H}$ and $\overline{H}_{\boldsymbol{\varepsilon}^{\mathrm{p}}} = \partial \overline{H} / \partial \boldsymbol{\varepsilon}^{\mathrm{p}}$ denote the partial derivatives of yield function with respect to the effective stress tensor and the effective hardening function, as well as the partial derivative of the effective hardening function to the plastic strain tensor, respectively. Substituting Eqs. (5.5) and (5.6) into Eq. (5.8), the expression for $\mathrm{d}\phi$ is derived as follows:

$$\mathrm{d}\phi = \frac{f_{\overline{\boldsymbol{\sigma}}} : \mathbb{D}_0 : \mathrm{d}\boldsymbol{\varepsilon}}{f_{\overline{\boldsymbol{\sigma}}} : \mathbb{D}_0 : \mathbf{r} - f_{\overline{H}} \overline{H}_{\boldsymbol{\varepsilon}^{\mathrm{p}}} : \mathbf{r}} \tag{5.9}$$

Herein, the non-orthogonal flow rule implemented by the fractional derivative (Lu et al., 2019a, 2019b) is employed. The direction of $\mathrm{d}\boldsymbol{\varepsilon}^{\mathrm{p}}$ is determined directly by the fractional gradient of the yield surface. Thus, the need for the construction of the plastic function is bypassed. The direction of $\mathrm{d}\boldsymbol{\varepsilon}^{\mathrm{p}}$ based on non-orthogonal flow rule is expressed as follows:

$$\mathbf{r} = \left(\frac{\partial^{\mu} f}{\partial S_1^{\mu}}, \frac{\partial^{\mu} f}{\partial S_2^{\mu}}, \frac{\partial^{\mu} f}{\partial S_3^{\mu}} \right) : \frac{\partial \mathbf{S}}{\partial \overline{\boldsymbol{\sigma}}} \tag{5.10}$$

where the non-orthogonal flow rule is used in the effective stress space. $\mathbf{S} = (S_1, S_2, S_3)$ is a set of basic stress invariants for the effective stress; μ is the fractional order. $\partial^{\mu} f / \partial \mathbf{S}^{\mu}$ represents the fractional gradient of yield surface in $\mathbf{S}$ coordinate system. Based on the coordinate transformation $\partial \mathbf{S} / \partial \overline{\boldsymbol{\sigma}}$, the general form for direction of $\mathrm{d}\boldsymbol{\varepsilon}^{\mathrm{p}}$ is determined by transforming $\partial^{\mu} f / \partial \mathbf{S}^{\mu}$ from $\mathbf{S}$ coordinate system into $\overline{\boldsymbol{\sigma}}$ coordinate system. A more detailed description concerning the non-orthogonal flow rule is presented in Chap. 2.

When the magnitude and direction of $\mathrm{d}\boldsymbol{\varepsilon}^{\mathrm{p}}$ are determined, submitting Eqs. (5.6) and (5.9) into Eq. (5.5), the plastic constitutive equation established in effective stress space can be determined as follows:

$$\mathrm{d}\overline{\boldsymbol{\sigma}} = \left[\mathbb{D}_0 - \frac{(\mathbb{D}_0 : \mathbf{r}) \otimes (f_{\overline{\boldsymbol{\sigma}}} : \mathbb{D}_0)}{f_{\overline{\boldsymbol{\sigma}}} : \mathbb{D}_0 : \mathbf{r} - f_{\overline{H}} \overline{H}_{\boldsymbol{\varepsilon}^{\mathrm{p}}} : \mathbf{r}} \right] : \mathrm{d}\boldsymbol{\varepsilon} = \mathbb{D}_0^{\mathrm{ep}} : \mathrm{d}\boldsymbol{\varepsilon} \tag{5.11}$$

Taking the full differential for Eq. (5.4), we obtains

$$\mathrm{d}\boldsymbol{\sigma} = (1-\omega)\mathrm{d}\overline{\boldsymbol{\sigma}} - \overline{\boldsymbol{\sigma}}\mathrm{d}\omega = (1-\omega)\mathbb{D}_0^{\mathrm{ep}} : \mathrm{d}\boldsymbol{\varepsilon} - \mathrm{d}\phi\overline{\boldsymbol{\sigma}}(\omega_{\boldsymbol{\varepsilon}^{\mathrm{p}}} : \mathbf{r}) \tag{5.12}$$

where $\omega_{\boldsymbol{\varepsilon}^{\mathrm{p}}} = \partial\omega/\partial\boldsymbol{\varepsilon}^{\mathrm{p}}$. Substituting Eq. (5.9) into Eq. (5.12), the plastic constitutive relationship from the perspective of nominal stress can be derived as follows:

$$\mathrm{d}\boldsymbol{\sigma} = \left[(1-\omega)\mathbb{D}_0^{\mathrm{ep}} - \frac{(\omega_{\boldsymbol{\varepsilon}^{\mathrm{p}}} : \mathbf{r})\overline{\boldsymbol{\sigma}} \otimes (f_{\overline{\sigma}} : \mathbb{D}_0)}{f_{\overline{\sigma}} : \mathbb{D}_0 : \mathbf{r} - f_{\overline{H}}\overline{H}_{\boldsymbol{\varepsilon}^{\mathrm{p}}} : \mathbf{r}}\right] : \mathrm{d}\boldsymbol{\varepsilon} = \mathbb{D}^{\mathrm{epd}} : \mathrm{d}\boldsymbol{\varepsilon} \tag{5.13}$$

5.2 Specifics Adopted for Concrete Model

The non-orthogonal plasticity damage model is developed based on four basic elements including the damage variable, the effective hardening function, the yield function, and the non-orthogonal flow rule. The degeneration law of unloading stiffness is characterized by the damage variable. The magnitude of $\mathrm{d}\boldsymbol{\varepsilon}^{\mathrm{p}}$ is calculated by the effective hardening function and the yield function. The non-orthogonal flow rule is used to describe the direction of $\mathrm{d}\boldsymbol{\varepsilon}^{\mathrm{p}}$.

5.2.1 Damage Variable

The degradation of unloading stiffness is the most intuitive damage behavior in the cyclic loading experiments of concrete. In this chapter, an S-type damage variable proposed by Wang et al. (2020) is adopted to describe the degeneration law of unloading stiffness as follows:

$$\omega = 1 - \exp\left[-\chi(x)^k\right] \tag{5.14}$$

where χ and k are material parameters controlling the damage growth rate. $x = \varepsilon_{\mathrm{d}}^{\mathrm{p}}/\varepsilon_{\mathrm{ds}}^{\mathrm{p}}$ and $\varepsilon_{\mathrm{ds}}^{\mathrm{p}}$ is the value of $\varepsilon_{\mathrm{d}}^{\mathrm{p}}$ at the normal peak stress, which can be determined by Eq. (5.38) in Sect. 5.3.1. The expression of $\varepsilon_{\mathrm{d}}^{\mathrm{p}}$ is

$$\varepsilon_{\mathrm{d}}^{\mathrm{p}} = \left\|\frac{2}{3}\left(\boldsymbol{\varepsilon}^{\mathrm{p}} - \frac{1}{3}\varepsilon_{\mathrm{v}}^{\mathrm{p}}\mathbf{1}\right)\right\|_2 \tag{5.15}$$

where $\varepsilon_{\mathrm{v}}^{\mathrm{p}}$ is the plastic volume strain. $\|\cdot\|_2$ is the 2-norm.

The test results have shown that the non-negligible stiffness degradation phenomenon can be observed in the pre-peak loading stage for concrete (Okamoto et al., 1976; Tanigawa & Uchida, 1979). However, the pre-peak damage behavior of concrete has been rarely captured in the existing plasticity damage models where damage is initiated after the normal peak stress (Grassl et al., 2013; Paliwal et al.,

2017). For the damage variable used in this chapter, the damage degree of the material at the normal peak stress is determined by the parameter χ. Figure 5.2 depicts the influence of χ on the damage rule with the fixed k value, where $x = 1$ represents that the loading stress reaches the normal peak stress. As the χ value increases, the damage ω_s at the normal peak stress is larger and the pre-peak damage behavior of concrete is more obvious. The influence of k on the damage rule with the fixed χ value is shown in Fig. 5.3, as the k value increases, the damage growth rate near the normal peak stress gradually increases. Note that ω_s depends only on the value of χ no matter what the parameter k changes. Substituting $x = 1$ and $\chi = 0.4$ into Eq. (5.14), $\omega_s \approx 0.33$ and the point (1, 0.33) in Fig. 5.3 can be obtained.

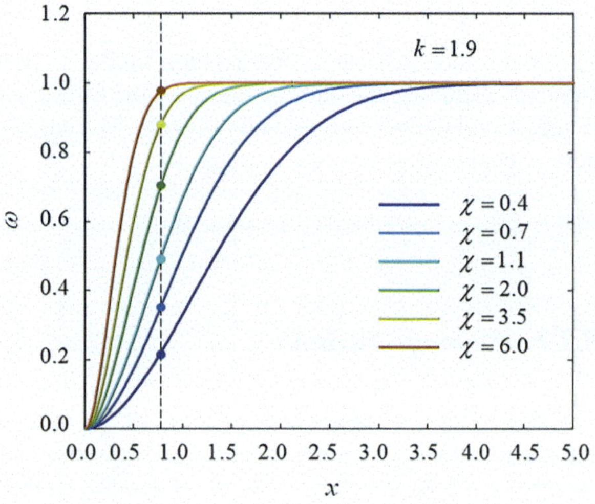

Fig. 5.2 Influence law of the parameter χ on the damage variable

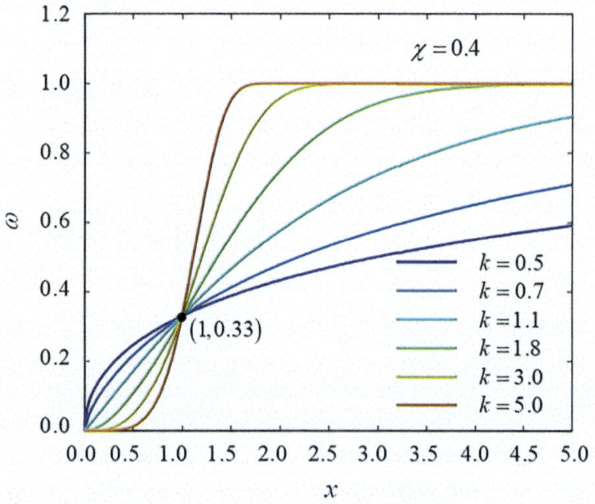

Fig. 5.3 Influence law of the parameter k on the damage variable

5.2.2 Effective Hardening Function

The evolution process of yield stress during the plastic loading process is characterized by the hardening function. For the plasticity damage model utilizing the effective stress method, the plastic process is driven by the effective stress. The hardening function defines the evolution law of effective yield stress during the loading process. The hardening function is generally obtained by the stress–strain curve of concrete. However, it must be pointed out that the effective stress–strain curve cannot be determined directly by conventional experiments of concrete. The reason is that the undamaged area of the specimen that is used to calculate the effective stress is difficult to measure accurately. Figure 5.4 demonstrates the two representative uniaxial compression stress–strain curves of concrete from the perspective of undamaged and damaged materials. Here, $\overline{\sigma}_1^*$ and σ_1^* are the effective axial yield stress and normal axial yield stress at any axial strain ε_1^*, respectively. f_c is the normal uniaxial compression strength. ε_{1s} denotes the axial strain which corresponds to f_c. $\overline{f}_c$ is the effective axial stress at ε_{1s} and $\overline{f}_c = f_c / (1 - \omega_s)$. From the perspective of undamaged material, concrete material mainly exhibits the strain-hardening behavior and the unloading stiffness of concrete remains constant. For the nominal stress–strain law that can be obtained directly in concrete experiments, the strain-hardening and strain-softening behaviors can be observed successively. The unloading stiffness of material gradually decreases with of devolvement of damage process.

The normal axial yield stress σ_1 can be expressed by the nominal hardening/softening function normalized by f_c as follows:

$$\sigma_1 = H(\varepsilon_1^p) f_c \tag{5.16}$$

Fig. 5.4 Uniaxial compression stress–strain curves

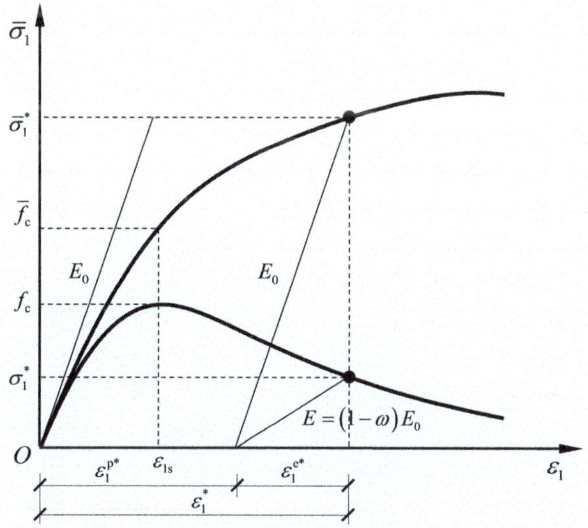

where $\varepsilon_1^{\mathrm{p}}$ denotes the axial plastic strain. H is the nominal hardening/softening in the uniaxial condition. Similarly, the effective axial yield stress $\overline{\sigma}_1$ can be expressed as the product of the dimensionless effective hardening function and $\overline{f}_{\mathrm{c}}$ as follows:

$$\overline{\sigma}_1 = \overline{H}\left(\varepsilon_1^{\mathrm{p}}\right)\overline{f}_{\mathrm{c}} \tag{5.17}$$

where $\overline{H}$ is the effective hardening in the uniaxial condition.

According to Eq. (5.4) and Fig. 5.4, the relationship between $\overline{\sigma}_1$ and σ_1 at the same strain can be established by the damage variable as follows:

$$\sigma_1 = (1 - \omega)\overline{\sigma}_1 \tag{5.18}$$

Substituting Eqs. (5.16) and (5.17) into the Eq. (5.18) and considering $\overline{f}_{\mathrm{c}} = f_{\mathrm{c}} / (1 - \omega_{\mathrm{s}})$, $\overline{H}$ can be derived based on H and ω as follows:

$$\overline{H}\left(\varepsilon_1^{\mathrm{p}}\right) = \frac{1 - \omega_{\mathrm{s}}}{1 - \omega} H\left(\varepsilon_1^{\mathrm{p}}\right) \tag{5.19}$$

The effective hardening curve and the nominal hardening/softening curve are demonstrated simultaneously in Fig. 5.5. Here, $\varepsilon_{1\mathrm{s}}^{\mathrm{p}}$ is the axial plastic strain corresponding to f_{c}. It can be found from the figure that, during the initial loading period, the values of both $\overline{H}$ and H increase with the accumulation of plastic strain. $\overline{H}$ and H increase to 1 simultaneously when $\varepsilon_1^{\mathrm{p}}$ reaches $\varepsilon_{1\mathrm{s}}^{\mathrm{p}}$. With the further development of plastic strain, $\overline{H}$ continues to increase but H starts to decrease, which results in two distinct mechanical behavior, i.e., the strain-hardening in the undamaged material and the strain-softening in the damaged material.

The method defined in Eq. (5.19) is further used to derive the effective hardening function suitable for the multiaxial stress condition. A nominal hardening/softening function proposed by Wang et al. (2018), which considers the influence of stress sate on the hardening/softening behavior of concrete, is adopted to derive the effective

Fig. 5.5 Effective hardening curve and nominal hardening/softening curve

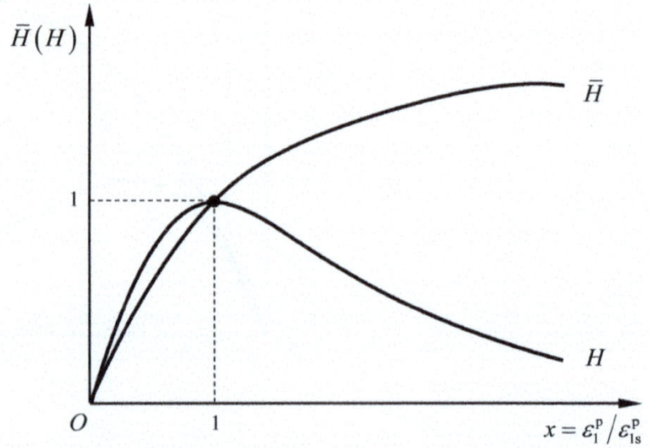

hardening function of the proposed model and can be expressed as follows:

$$H = \frac{Ax + (D - 1)x^2}{1 + (A - 2)x + Dx^2} \tag{5.20}$$

where the variable $x = \varepsilon_{\mathrm{d}}^{\mathrm{p}}/\varepsilon_{\mathrm{ds}}^{\mathrm{p}}$. $\varepsilon_{\mathrm{d}}^{\mathrm{p}}$ is also chosen as the plastic internal variable. The failure of concrete materials are mainly caused by the shear load. $\varepsilon_{\mathrm{d}}^{\mathrm{p}}$ defined by Eq. (5.15) can well reflect the mechanical evolution process of concrete. Therefore, $\varepsilon_{\mathrm{d}}^{\mathrm{p}}$ is used as both the damage internal variable and the plastic internal variable to record the loading history of material herein. A and D are the parameters that are used to adjust the nominal hardening/softening law of concrete. It is noted that the parameters $\varepsilon_{\mathrm{ds}}^{\mathrm{p}}$ and A are related to the stress state and their stress dependence will be described by two empirical equations in Sect. 5.3.1. The function form of H is developed from the semi-empirical stress–strain curve of concrete proposed by Sargin (1971). Then, this function form is used to develop the nominal hardening/softening function of concrete in Wang et al. (2018). When $\varepsilon_{\mathrm{d}}^{\mathrm{p}} < \varepsilon_{\mathrm{ds}}^{\mathrm{p}}$, H increases from 0 to 1 with increases in $\varepsilon_{\mathrm{d}}^{\mathrm{p}}$ and describes the hardening behavior. When $\varepsilon_{\mathrm{d}}^{\mathrm{p}} > \varepsilon_{\mathrm{ds}}^{\mathrm{p}}$, H decreases with increases in $\varepsilon_{\mathrm{d}}^{\mathrm{p}}$ and describes the softening behavior.

Then, substituting H defined in Eq. (5.20) and ω defined in Eq. (5.14) into the Eq. (5.19), $\overline{H}$ suitable for the multiaxial stress condition can be derived as follows:

$$\overline{H} = \frac{1 - \omega_{\mathrm{s}}}{1 - \omega} H = (1 - \omega_{\mathrm{s}}) \exp\left[\chi (x)^k\right] \frac{Ax + (D - 1)x^2}{1 + (A - 2)x + Dx^2} \tag{5.21}$$

When $\overline{H}$ is determined by Eq. (5.21), H and ω are mutually independent but coordinated. The hardening/softening law of damaged material is defined individually by H and is independent of ω. Therefore, the hardening/softening behavior of damaged material and the degradation law of unloading stiffness can be captured by the proposed model simultaneously.

5.2.3 *Yield Function*

The yield function is first proposed in triaxial compression condition and then is developed to the 3D stress condition by the transformed stress (TS) method (Du et al., 2010; Yao et al., 2004). The yield function expressed by effective stress can turn into the strength criterion expressed by nominal stress when $\overline{H}$ increases to 1.

The yield function of concrete can be generally constructed by combining the hardening function and the strength criterion. In this way, the hardening behavior defined by the hardening function can be extended to multiaxial stress conditions. Based on the expression of extended Leon criterion (Etse & Willam, 1994) under the condition of triaxial compression and $\overline{H}$ defined in Eq. (5.21), a yield function suitable for triaxial compression condition is presented as follows:

$$f = \left[\left(1-\overline{H}'\right)\left(\frac{\overline{q}}{3\overline{f}_{c}}-\frac{\overline{p}}{\overline{f}_{c}}\right)^{2}+\frac{\overline{q}}{\overline{f}_{c}}\right]^{2}+m_{0}\overline{H}'\overline{H}\left(\frac{\overline{q}}{3\overline{f}_{c}}-\frac{\overline{p}}{\overline{f}_{c}}\right)-\overline{H}^{2} \tag{5.22}$$

where, $\overline{H}'$ is defined as $\overline{H}$ in pre-peak stage (i.e., $x \leq 1$) and as 1 in post-peak stage (i.e., $x > 1$). $\overline{q}$ and $\overline{p}$ represents respectively the generalized shear stress and the mean stress of $\overline{\boldsymbol{\sigma}}$. m_0 is the friction parameter and $m_0 = \overline{f}_c/\overline{\sigma}_0 = f_c/\sigma_0$. $\overline{\sigma}_0$ is the effective yield stress which corresponds to the normal 3D tensile strength σ_0 and $\overline{\sigma}_0 = \sigma_0/(1-\omega_s)$. When $\overline{H}$ increases to 1 and ω reaches ω_s, $\overline{q} = q/(1-\omega_s)$ and $\overline{p} = p/(1-\omega_s)$. Substituting the above conditions into Eq. (5.22), the yield function defined in Eq. (5.22) can be reduced to the extended Leon criterion under the condition of triaxial compression as follows:

$$f = \left(\frac{q}{f_c}\right)^2 + m_0\left(\frac{q}{3f_c}-\frac{p}{f_c}\right) - 1 \tag{5.23}$$

where q and p represents the generalized shear stress and the mean stress of $\boldsymbol{\sigma}$, respectively.

The yield function defined in Eq. (5.22) is only suitable for triaxial compression condition, where the strength of concrete except for triaxial compression is overestimated when $\overline{H}$ reaches 1. The TS method proposed by Yao et al. (2004) provides an indirect manner to develop the yield function suitable for the 3D stress condition. Utilizing the TS tensor defined by the TS method, the yield function is also established in triaxial compression conditions, but it is capable of reflecting the mechanical characteristic of concrete in the 3D stress condition (Lu et al., 2019b; Wang et al., 2018). The yield function in Eq. (5.22) can be reconstructed by the TS as follows:

$$f = \left[\left(1-\overline{H}'\right)\left(\frac{\tilde{q}}{3\overline{f}_{c}}-\frac{\tilde{p}-\overline{\sigma}_0}{\overline{f}_{c}}\right)^{2}+\frac{\tilde{q}}{\overline{f}_{c}}\right]^{2}+m_{0}\overline{H}'\overline{H}\left(\frac{\tilde{q}}{3\overline{f}_{c}}-\frac{\tilde{p}-\overline{\sigma}_0}{\overline{f}_{c}}\right)-\overline{H}^{2} \tag{5.24}$$

where $\tilde{p}$ and $\tilde{q}$ are the mean stress and the generalized shear stress of the TS tensor $\tilde{\boldsymbol{\sigma}}$. $\tilde{\boldsymbol{\sigma}}$ is generally constructed in conjunction with a specific strength criterion.

$$\tilde{\boldsymbol{\sigma}} = (\overline{p}+\overline{\sigma}_0)\mathbf{1} + \frac{\hat{q}_{\alpha}^{*}}{\overline{q}}(\overline{\boldsymbol{\sigma}}-\overline{p}\mathbf{1}) \tag{5.25}$$

where $\hat{q}_{\alpha}^{*}$ can be expressed as follows:

$$q_{\alpha}^{*} = \alpha\sqrt{\hat{I}_1^2-3\hat{I}_2} + \frac{2(1-\alpha)\hat{I}_1}{3\sqrt{(\hat{I}_1\hat{I}_2-\hat{I}_3)/(\hat{I}_1\hat{I}_2-9\hat{I}_3)}-1} \tag{5.26}$$

where Eq. (5.26) is the expression of nonlinear unified strength criterion (Du et al., 2010; Gao et al., 2010; Wang et al., 2018; Yao et al., 2004) on the deviatoric plane, which is used to derive the TS tensor herein. α is the triaxial extension compression strength ratio. $\hat{I}_1$, $\hat{I}_2$, and $\hat{I}_3$ are respectively the first, second and third stress invariants of the stress tensor $\hat{\boldsymbol{\sigma}}$.

$$\hat{I}_1 = \mathrm{trace}(\hat{\boldsymbol{\sigma}}) \tag{5.27}$$

$$\hat{I}_2 = \frac{1}{2}\left[\mathrm{trace}(\hat{\boldsymbol{\sigma}})^2 - \mathrm{trace}(\hat{\boldsymbol{\sigma}} \cdot \hat{\boldsymbol{\sigma}})\right] \tag{5.28}$$

$$\hat{I}_3 = \det(\hat{\boldsymbol{\sigma}}) \tag{5.29}$$

where the "·" denotes the single dot product of tensor operation. The "trace(**A**)" represents the operation of obtaining the trace of matrix **A**. The expression of $\hat{\boldsymbol{\sigma}}$ is

$$\hat{\boldsymbol{\sigma}} = \overline{\boldsymbol{\sigma}} + \overline{\sigma}_0 \mathbf{1} \tag{5.30}$$

On the deviatoric plane, the nonlinear unified strength criterion defined by Eq. (5.30) is a series of continuous and smooth strength curves where the Matsuoka-Nakai curved triangle is the lower bound and the Drucker-Prager circle is the upper bound, as shown in Fig. 5.6. The nonlinear unified strength criterion can describe the lode angle effect of different geomaterials. The 3D mechanical characteristic of concrete can be described reasonably when the TS tensor in conjunction with the nonlinear unified strength criterion is used for the construction of yield function defined by Eq. (5.24). The expression defined in Eq. (5.26) is constructed by the effective stress. It can also turn into the expression constructed by the nominal stress by substituting $\overline{\sigma}_0 = \sigma_0/(1-\omega_s)$, $\overline{\boldsymbol{\sigma}} = \boldsymbol{\sigma}/(1-\omega)$ and Eq. (5.30) into Eq. (5.26).

The isotropic hardening law is employed in the proposed model. Figure 5.7 presents the evolution law of yield surface from the perspectives of the effective stress and the nominal stress, respectively. When $\overline{H} < 1$, the yield surface in the effective stress space and in the nominal stress space expand toward with the increase of $\overline{H}$, in which the material plasticization caused by hydrostatic loading is captured by the closed form elastic domain. When $\overline{H}$ reaches 1, the yield surface evolves from the closed-form to the open form. The yield surface exactly turns into the strength surface in the nominal stress space. As $\overline{H}$ increases, the yield surface continues to expand outward in the effective stress space. However, the yield surface gradually shrinks inward in the nominal stress space, which corresponds to the softening behavior of damaged material.

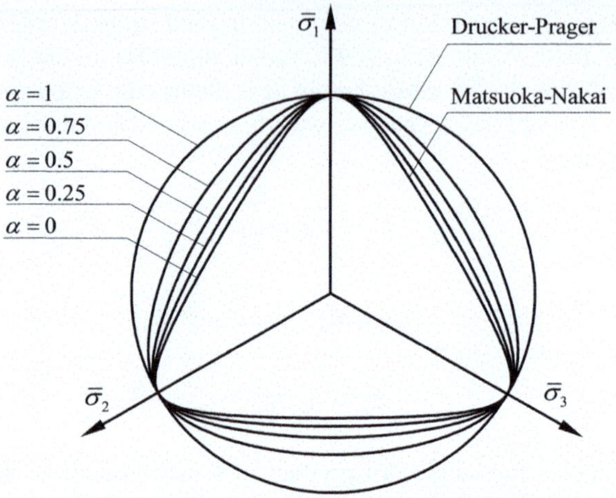

Fig. 5.6 Strength curves of the nonlinear unified strength criterion on the deviatoric plane

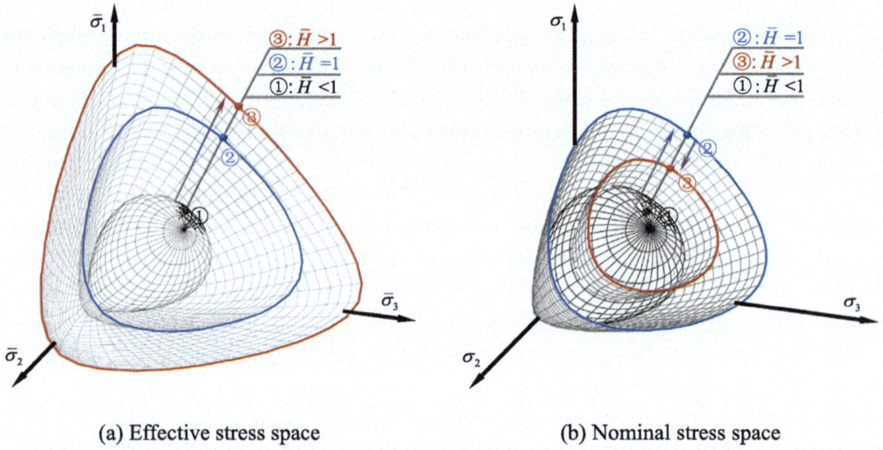

Fig. 5.7 Evolution law of the yield surface

5.2.4 Non-orthogonal Flow Rule

The non-orthogonal flow rule is employed herein due to its flexibility to obtain the non-orthogonal direction of yield surface. For the proposed model, the yield function in the TS space is composed of stress invariants $\tilde{p}$ and $\tilde{q}$, independent of $\tilde{\theta}$. Thus, the direction of $d\boldsymbol{\varepsilon}^p$ based on the non-orthogonal flow rule is expressed as follows:

$$\mathbf{r} = \frac{\partial^\mu f}{\partial \tilde{p}^\mu}\frac{\partial \tilde{p}}{\partial \tilde{\boldsymbol{\sigma}}} + \frac{\partial^\mu f}{\partial \tilde{q}^\mu}\frac{\partial \tilde{q}}{\partial \tilde{\boldsymbol{\sigma}}} \tag{5.31}$$

The Riemann–Liouville (RL) fractional derivative is used to derive the non-orthogonal gradient in Eq. (5.31), which is defined as:

$$ {}_{0}^{RL}D_t^{\mu} f(t) = \frac{\partial^{\mu} f(t)}{\partial t^{\mu}} = \frac{1}{\Gamma(n-\mu)} \frac{\mathrm{d}^n}{\mathrm{d}t^n} \left(\int_0^t (t-\tau)^{n-\mu-1} f(\tau) d\tau \right) \tag{5.32} $$

where ${}_{0}^{RL}D_t^{\mu}$ is the fractional partial differential operator; n is the smallest integer larger than the fractional order μ; t denotes the current stress located on the yield surface herein; τ is the integration variable defined in the interval $[0, t]$; $\Gamma(t) = \int_0^{\infty} e^{-\tau} \tau^{t-1} \mathrm{d}\tau$ is gamma function and $\Gamma(t+1) = t\Gamma(t)$.

In order to facilitate fractional derivative operation, the yield function is further expanded in the form of sums of power functions for $\tilde{p}$ and $\tilde{q}$, respectively.

$$ \begin{cases} f(\tilde{p}) = a_0 + a_1 \tilde{p} + a_2 \tilde{p}^2 + a_3 \tilde{p}^3 + a_4 \tilde{p}^4 \\ f(\tilde{q}) = b_0 + b_1 \tilde{q} + b_2 \tilde{q}^2 + b_3 \tilde{q}^3 + b_4 \tilde{q}^4 \end{cases} \tag{5.33} $$

where the expressions of coefficients a_k and b_k ($k = 0, 1, 2, 3, 4$) are given as follows:

$$ \begin{cases} a_0(\tilde{q}) = \left[\dfrac{\tilde{q}}{\overline{f}_c} + \dfrac{\left(1-\overline{H}'\right)(\tilde{q}+3\overline{\sigma}_0)^2}{9\overline{f}_c^2} \right]^2 + \dfrac{m_0 \overline{H}' \overline{H} (\tilde{q}+3\overline{\sigma}_0)}{3\overline{f}_c} - \overline{H}^2 \\ a_1(\tilde{q}) = \dfrac{-4(\tilde{q}+3\overline{\sigma}_0)^3 \left(1-\overline{H}'\right)^2}{27\overline{f}_c^4} - \dfrac{4\left(1-\overline{H}'\right)(\tilde{q}+3\overline{\sigma}_0)\tilde{q}}{3\overline{f}_c^3} - \dfrac{m_0 \overline{H}' \overline{H}}{\overline{f}_c} \\ a_2(\tilde{q}) = \dfrac{2\left(1-\overline{H}'\right)^2 (\tilde{q}+3\overline{\sigma}_0)^2}{3\overline{f}_c^4} + 2\left(1-\overline{H}'\right)\dfrac{\tilde{q}}{\overline{f}_c^3} \\ a_3(\tilde{q}) = -\dfrac{4\left(1-\overline{H}'\right)^2 (\tilde{q}+3\overline{\sigma}_0)}{3\overline{f}_c^4} \\ a_4(\tilde{q}) = \dfrac{\left(1-\overline{H}'\right)^2}{\overline{f}_c^4} \end{cases} \tag{5.34} $$

$$\begin{cases} b_0(\tilde{p}) = \dfrac{(\tilde{p}-\overline{\sigma}_0)^4\left(1-\overline{H}'\right)^2}{\overline{f}_c^4} - \dfrac{m_0\overline{H}'\overline{H}(\tilde{p}-\overline{\sigma}_0)}{\overline{f}_c} - \overline{H}^2 \\ b_1(\tilde{p}) \dfrac{-4\left(1-\overline{H}'\right)^2(\tilde{p}-\overline{\sigma}_0)^3 + 6\overline{f}_c\left(1-\overline{H}'\right)(\tilde{p}-\overline{\sigma}_0)^2}{3\overline{f}_c^4} + \dfrac{m_0\overline{H}'\overline{H}}{3\overline{f}_c} \\ b_2(\tilde{p}) = \dfrac{2(\tilde{p}-\overline{\sigma}_0)^2\left(1-\overline{H}'\right)^2 - 4\overline{f}_c(\tilde{p}-\overline{\sigma}_0)\left(1-\overline{H}'\right)}{3\overline{f}_c^4} + \dfrac{1}{\overline{f}_c^2} \\ b_3(\tilde{p}) = \dfrac{2\left(1-\overline{H}'\right)}{9\overline{f}_c^3} - \dfrac{4(\tilde{p}-\overline{\sigma}_0)\left(1-\overline{H}'\right)^2}{27\overline{f}_c^4} \\ b_4(\tilde{p}) = \dfrac{\left(1-\overline{H}'\right)^2}{81\overline{f}_c^4} \end{cases} \tag{5.35}$$

Substituting Eq. (5.33) into Eq. (5.32) and utilizing the basic properties of RL fractional derivatives listed in Appendix A, the fractional gradient components of the yield function are obtained as follows:

$$\begin{cases} \dfrac{\partial^\mu f}{\partial \tilde{p}^\mu} = \dfrac{a_0\tilde{p}^{-\mu}}{\Gamma(1-\mu)} + \dfrac{a_1\Gamma(2)\tilde{p}^{1-\mu}}{\Gamma(2-\mu)} + \dfrac{a_2\Gamma(3)\tilde{p}^{2-\mu}}{\Gamma(3-\mu)} + \dfrac{a_3\Gamma(4)\tilde{p}^{3-\mu}}{\Gamma(4-\mu)} + \dfrac{a_4\Gamma(5)\tilde{p}^{4-\mu}}{\Gamma(5-\mu)} \\ \dfrac{\partial^\mu f}{\partial \tilde{q}^\mu} = \dfrac{b_0\tilde{q}^{-\mu}}{\Gamma(1-\mu)} + \dfrac{b_1\Gamma(2)\tilde{q}^{1-\mu}}{\Gamma(2-\mu)} + \dfrac{b_2\Gamma(3)\tilde{q}^{2-\mu}}{\Gamma(3-\mu)} + \dfrac{b_3\Gamma(4)\tilde{q}^{3-\mu}}{\Gamma(4-\mu)} + \dfrac{b_4\Gamma(5)\tilde{q}^{4-\mu}}{\Gamma(5-\mu)} \end{cases} \tag{5.36}$$

$d\varepsilon_v^p$ and $d\varepsilon_d^p$ under the given load increment can be determined by the Eq. (5.36) and the plastic multiplier defined by Eq. (5.9). The dilatancy law of concrete reflects essentially the change law of proportional relationship between $d\varepsilon_v^p$ and $d\varepsilon_d^p$, which can be described geometrically by the direction of $d\boldsymbol{\varepsilon}^p$ on the meridian plane. Based on the non-orthogonal flow rule, Table 5.1 shows the evolution law of direction of $d\boldsymbol{\varepsilon}^p$ in hardening and softening stages on the meridian plane, where loading paths are set to the uniaxial compression, the biaxial compression, and the triaxial compression. It can be found from the figure that, in the hardening stage, the direction of $d\boldsymbol{\varepsilon}^p$ firstly points to the compression direction of hydrostatic axis and then turns toward the tension direction of hydrostatic axis near the normal peak stress. In the softening stage, the direction of $d\boldsymbol{\varepsilon}^p$ always points to the tension direction of hydrostatic axis. It shows that, under different loading paths, the dilatancy law of concrete from contraction to dilation can be captured satisfactorily by the non-orthogonal flow rule.

For the classical plasticity theory, the non-associated flow rule is employed to obtain the non-orthogonal direction of the yield surface by constructing the additional plastic potential function. Compared with the non-associated flow rule, the advantage

Table 5.1 Evolution law for the direction of $d\varepsilon^{p}$

Loading paths	Hardening stage	Softening stage
Uniaxial compression	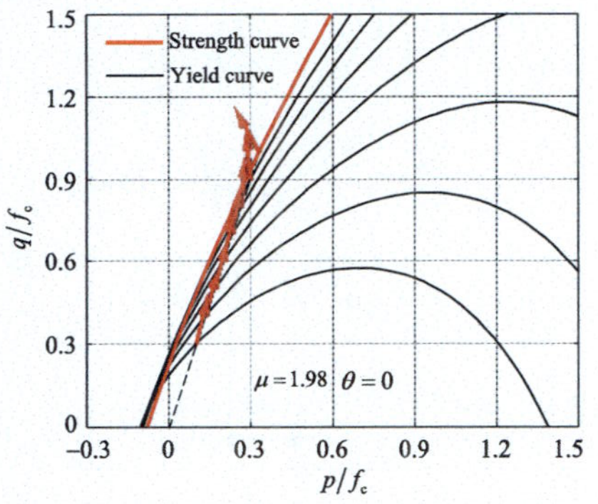	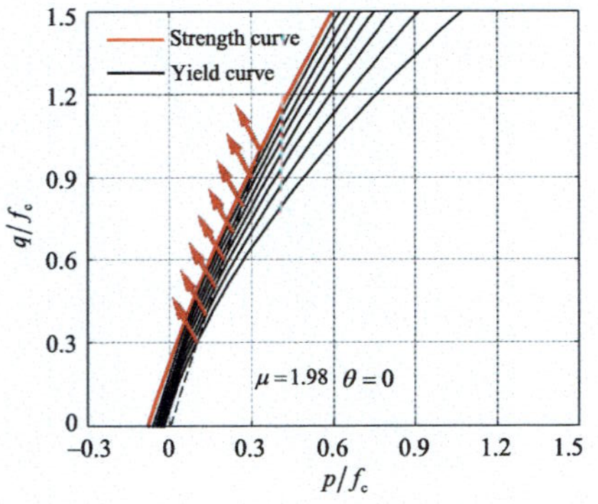

(continued)

Table 5.1 (continued)

Loading paths	Hardening stage	Softening stage
Biaxial compression	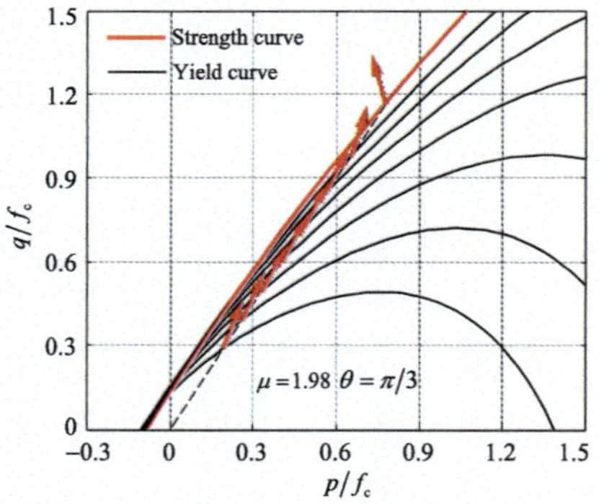	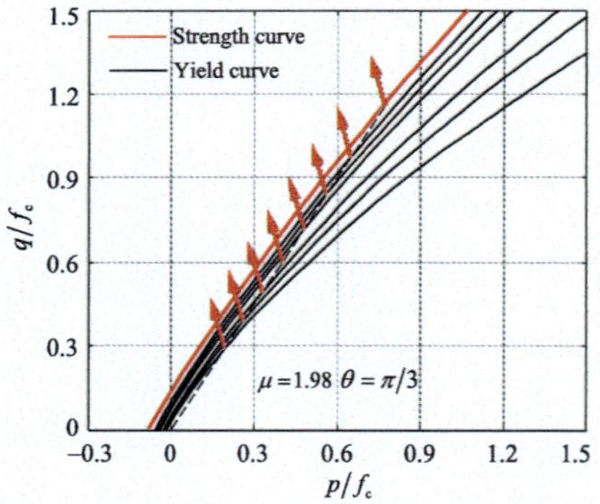
Triaxial compression	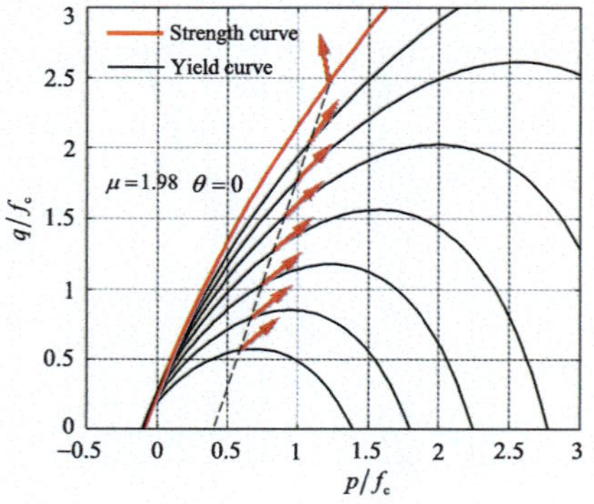	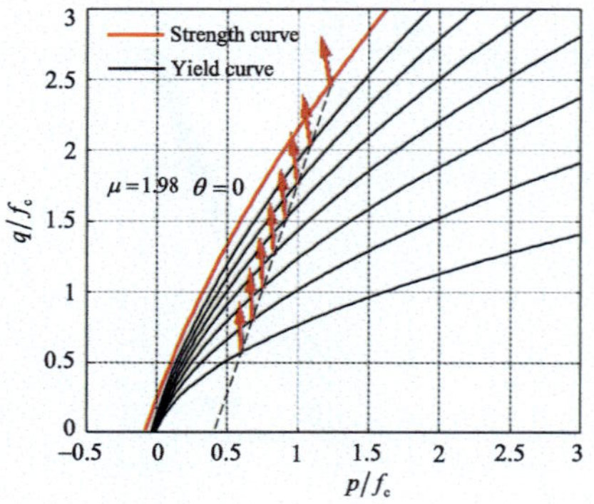

of non-orthogonal flow rule is that the direction of $d\boldsymbol{\varepsilon}^{p}$ is obtained directly by the yield function and the need for the construction of the plastic potential function is bypassed. However, it is undeniable that the formulas of the non-orthogonal flow rule are complex due to the use of the fractional derivative. This flaw may be solved by the numerical methods of fractional derivatives.

5.3 Parameter Calibration and Model Validation

5.3.1 Parameter Calibration

The proposed model has eleven material parameters that can be calibrated by the uniaxial and biaxial experiments of concrete. There are two elastic constants (E_0, ν), two damage parameters (χ, k), three material parameters $(A,\ D,\ \varepsilon_{\text{ds}}^{\text{p}})$ in the nominal hardening/softening function, three strength parameters (f_c, σ_0, α), and the fractional order μ. The calibration methods for these parameters are presented as follows.

The elastic constants (E_0, ν) can be determined easily by the uniaxial compression test of concrete. Noted that E_0 is generally considered to be independent of pressure herein.

The parameters (χ, k) define the damage evolution characteristics of unloading stiffness, among which χ can be calibrated by ω_s as follows:

$$\chi = \ln \frac{1}{1-\omega_s} \tag{5.37}$$

where $\omega_s = 1 - E_s/E_0$. E_0 denotes the undamaged Young's modulus and E_s denotes the damaged Young's modulus at f_c. Once the parameter χ is determined, the parameter k can be obtained by substituting ω and x at any test point into Eq. (5.14).

Three material parameters $(A,\ D,\ \varepsilon_{\text{ds}}^{\text{p}})$ controls the hardening/softening behavior of damaged material. The existing test data (Du et al., 2010; Gao et al., 2010; Wang et al., 2018; Yao et al., 2004) indicated that the parameters $\varepsilon_{\text{ds}}^{\text{p}}$ and A dependent on the stress state. The two linear expressions are used to describe the stress dependence of $\varepsilon_{\text{ds}}^{\text{p}}$ and A as follows:

$$\varepsilon_{\text{ds}}^{\text{p}} = \left(28.12\frac{\overline{\sigma}_3+\overline{\sigma}_0}{\overline{f}_c+\overline{\sigma}_0} - 1.56\right)\varepsilon_{1,\text{ds}}^{\text{p}} \tag{5.38}$$

$$A = 10.997\frac{\overline{\sigma}_3+\overline{\sigma}_0}{\overline{f}_c+\overline{\sigma}_0} + A_s - 1 \tag{5.39}$$

where $\overline{\sigma}_3$ denotes the effective minor principle stress. $\varepsilon_{1,\text{ds}}^{\text{p}}$ is the value of $\varepsilon_{\text{ds}}^{\text{p}}$ at f_c. A_s is the value of A in uniaxial compression. Equations (5.38) and (5.39) are the two empirical equations proposed by Wang et al. (2018) and are obtained by fitting the test data from the literature (Candappa et al., 2001; Kotsovos, 1980; Kupfer et al., 1969; Lu & Hsu, 2007). The parameters $\left(A_s,\ D,\ \varepsilon_{1,\mathrm{ds}}^{\mathrm{p}}\right)$ can be calibrated by the uniaxial compression test, as illustrated in Wang et al. (2018).

The strength parameters $(f_c,\ \sigma_0,\ \alpha)$ define the failure shape of the strength surface. According to Lu et al. (2019b), σ_0 and α can be determined using the basic strength indexes as follows:

$$\sigma_0 = \frac{f_c^2 f_{bt}}{f_c^2 - f_{bt}^2} \tag{5.40}$$

$$\begin{aligned}\alpha = {} & \frac{m_0^2 - m_0\sqrt{(m_0^2 + 24\rho m_0 + 36)} + 18}{6\rho m_0} \\ & + \frac{m_0 - \sqrt{(m_0^2 + 24\rho m_0 + 36)}}{2\rho^2 m_0} + 2\end{aligned} \tag{5.41}$$

where $\rho = f_{bc}/f_c$. f_{bc} and f_{bt} denote the normal biaxial compression strength and the normal biaxial tensile strength, respectively.

The parameter μ controls the volume change law of concrete. When the concrete is loaded to the phase transformation state (i.e., the plastic volume strain of concrete changes from shrinkage to expansion), the plastic volume strain increment $d\varepsilon_v^p$ is zero.

$$\begin{aligned}d\varepsilon_v^p &= d\phi\tilde{p}^{-\mu}\left[\frac{a_0}{\Gamma(1-\mu)} + \frac{a_1\Gamma(2)\tilde{p}}{\Gamma(2-\mu)} + \frac{a_2\Gamma(3)\tilde{p}^2}{\Gamma(3-\mu)} + \frac{a_3\Gamma(4)\tilde{p}^3}{\Gamma(4-\mu)} + \frac{a_4\Gamma(5)\tilde{p}^4}{\Gamma(5-\mu)}\right] \\ &= 0\end{aligned} \tag{5.42}$$

where μ in Eq. (5.42) is determined based on the stress and strain condition at the phase transformation state. Under uniaxial compression condition, H represents the ratio of normal axial yield stress to normal uniaxial compression strength σ_1/f_c. Utilizing σ_1 at the phase transformation state, the corresponding H can be obtained and the same time x can be solved by Eq. (5.20). Then, ω, $\overline{H}$, and $\overline{\sigma}_1$ at the phase transformation state can be determined successively by Eqs. (5.21), (5.14) and (5.18). The effective stress invariants $\overline{p}$ and $\overline{q}$ can be expressed as $\overline{p} = \overline{\sigma}_1/3$ and $\overline{q} = \overline{\sigma}_1$. Considering the relationship between the TS and the effective stress under the triaxial compression condition, the stress invariants of TS appeared in Eq. (5.42) can be determined by $\tilde{p} = \overline{p} + \overline{\sigma}_0$ and $\tilde{q} = \overline{q}$. Finally, the value of μ can be solved by substituting $\tilde{p}$, $\tilde{q}$, and $\overline{H}$ at the phase transformation state into Eq. (5.42).

5.3.2 Model Validation for Monotonic Loading

The predictive capability of the non-orthogonal plasticity damage model is assessed based on the monotonic and cyclic test data of concrete from the available literature. In the following figures, the model predictions are depicted by solid lines and the test data are indicated by hollow circles. σ_1 and σ_3 denote the normal axial stress and the lateral stress, respectively. ε_1, ε_3, and ε_v denote the axial strain, the lateral strain, and the volume strain, respectively. Table 5.2 lists all model parameters used for predictions.

First, the two monotonic triaxial compression test of concrete are used for the validation of the non-orthogonal plasticity damage model. Figure 5.8 presents the triaxial compression test results of concrete measured by Candappa et al. (2001), where the confinement pressure was respectively set to 0, 4, 8 and 12 MPa, and its excellent simulations by the proposed model. In Fig. 5.8a and b, the comparisons between test data and prediction results are shown in the form of both, σ_1 *vs.* ε_v, as well as σ_1 *vs.* ε_1 and ε_3, respectively. Figure 5.9 depicts the model predictions and the triaxial compression test data of concrete reported by Lu and Hsu (2007) where the confinement pressure varied from 0 to 56 MPa. In Fig. 5.9a and b, the calculated stress–strain curves on σ_1 versus ε_v, as well as σ_1 versus ε_1 and ε_3 agree well with the test results of concrete with different lateral confinement pressure. As can be seen from Figs. 5.8 and 5.9, the proposed model is capable of capturing the enhancement in strength and ductility for concrete subjected to low, normal, and high lateral confinement pressures. In particular, the dilatancy behavior for concrete under various confinement pressures can be reasonably captured by the proposed model owing to the use of the non-orthogonal flow rule.

Table 5.2 Evolution law for the direction of $d\varepsilon^p$

Researchers	E_0/MPa	ν	f_c/ MPa	σ_0/ MPa	α	A_s	D	ε^p_{ds}	μ	χ	k
Candappa et al. (2001)	32,500	0.15	41.9	3.37	0.345	9.3	0.50	0.0015	1.92	0.30	0.70
Lu and Hsu (2007)	33,000	0.14	68.6	6.58	0.378	10.7	0.50	0.001	1.90	0.30	0.70
Sinha et al. (1964)	19,560	0.15	26.5	2.63	0.383	6.9	0.55	0.001	1.93	0.30	0.88
Karsan and Jirsa (1969)	30,000	0.15	28.2	2.42	0.357	4.9	0.50	0.008	1.95	0.36	0.96
Okamoto et al. (1976)	25,500	0.15	30	3.03	0.388	7.1	0.7	0.0007	1.90	0.33	0.67
	39,000	0.15	40	4.04	0.388	6.3	0.75	0.0007	1.85	0.44	0.54
Tanigawa and Uchida (1979)	40,000	0.15	40	4.04	0.388	6.9	0.80	0.0009	1.80	0.40	0.54
Imran (1994)	29,570	0.22	45	4.86	0.400	10.7	0.50	0.0012	1.95	0.30	0.88

Fig. 5.8 Comparison with test results reported by Candappa et al. (2001)

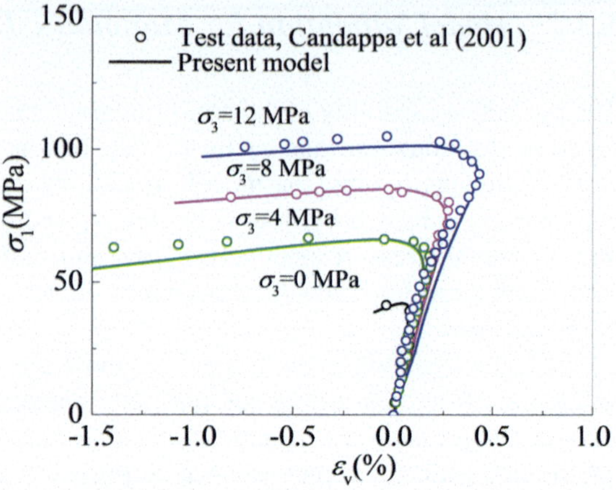

(a) σ_1 *vs.* ε_v

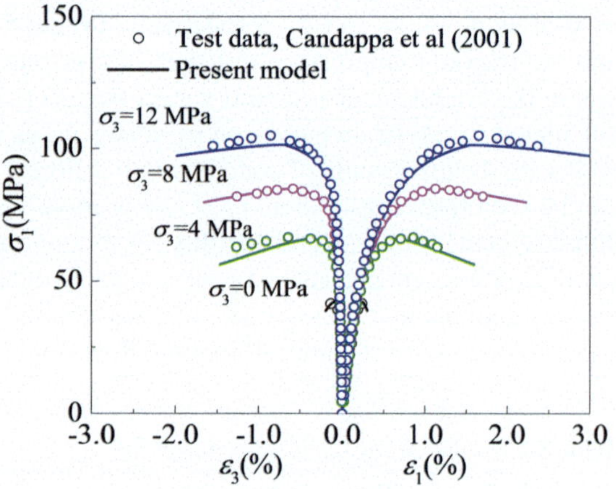

(b) σ_1 *vs.* ε_1 and ε_3

5.3.3 Model Validation for Cyclic Loading

The experimental study of plastic and damage properties of concrete in the uniaxial stress condition is quite sufficient. Then, five sets of cyclic uniaxial compression tests of concrete, reported in (Karsan & Jirsa, 1969; Okamoto et al., 1976; Sinha et al., 1964; Tanigawa & Uchida, 1979), are used to examine the predictive capacity of the proposed model. Figure 5.10 presents the predicted stress–strain curves and the cyclic uniaxial compression test data of concrete measured by Sinha et al. (1964). Figure 5.11 depicts the cyclic uniaxial compression test data from (Karsan & Jirsa, 1969) and its predictions by the proposed model.

In Fig. 5.12a and b, two sets of cyclic uniaxial compression tests of C30 and C40 concrete conducted by Okamoto et al. (1976) are simulated with the proposed model.

Fig. 5.9 Comparison with test results reported by Lu and Hsu (2007)

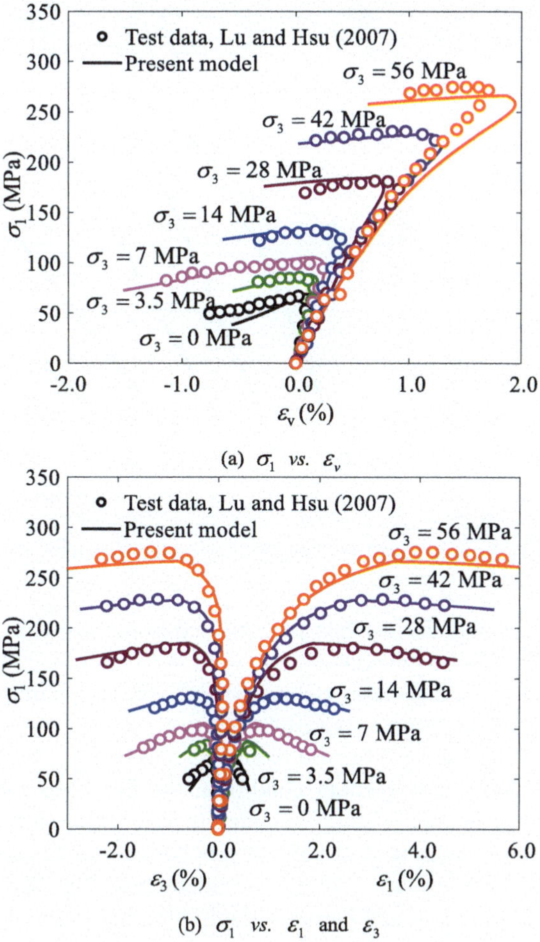

Fig. 5.10 Comparison with test results reported by Sinha et al. (1964)

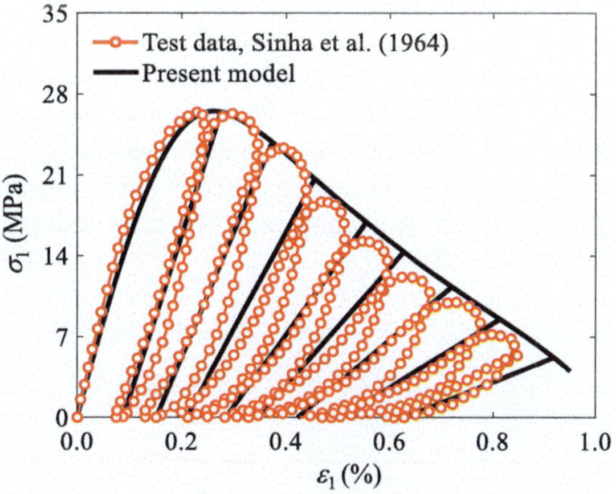

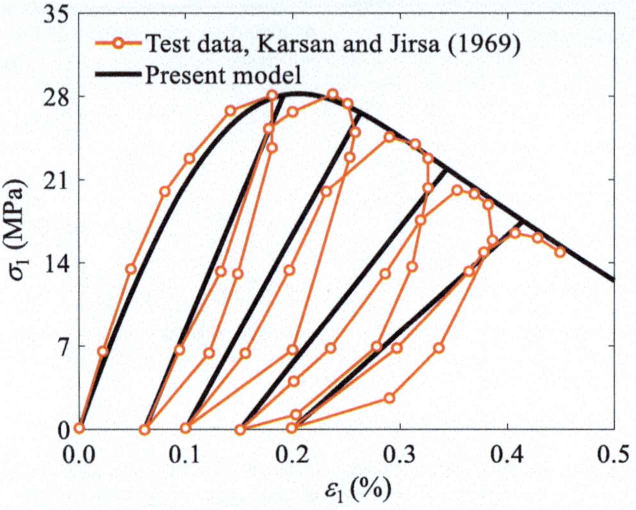

Fig. 5.11 Comparison with test results reported by Karsan and Jirsa (1969)

Note that the test data are collected from the literature reported by Sima et al. (2008). The cyclic uniaxial compression test data reported in Tanigawa and Uchida (1979) and its excellent model predictions can be found in Fig. 5.13. Note that the test data are from Sima et al. (2008).

As can be observed from Figs. 5.10, 5.11, 5.12 and 5.13, in the cyclic uniaxial compression loading condition, the proposed model realistically reproduces the experimental findings of concrete from different literature. Particularly, the pre-peak hardening and post-softening behavior and the stiffness degradation phenomenon of concrete are captured simultaneously by the proposed model. The reason is that the nominal hardening/softening behavior of concrete can be captured by the nominal hardening/softening function and is independent of the damage variable in establishing the non-orthogonal plasticity damage model.

Finally, a set of cyclic triaxial compression tests is adopted to further assess the performance of the model. The test was conducted by Imran (1994) with 17.2 MPa confinement pressure. The test data from the literature (Sarikaya & Erkmen, 2019) and the model prediction curves are plotted in Fig. 5.14. The results show that model predictions excellently predict the test results. The ductility behavior and the stiffness reduction phenomenon observed in the concrete experiments are very well reproduced.

Fig. 5.12 Comparison with test results reported by Okamoto et al. (1976)

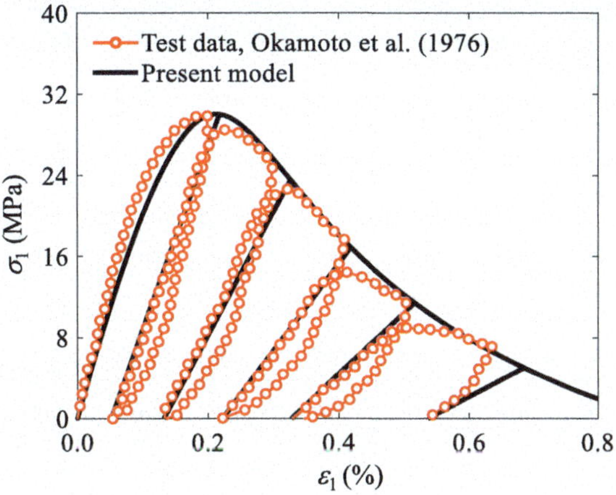

(a) C30 concrete

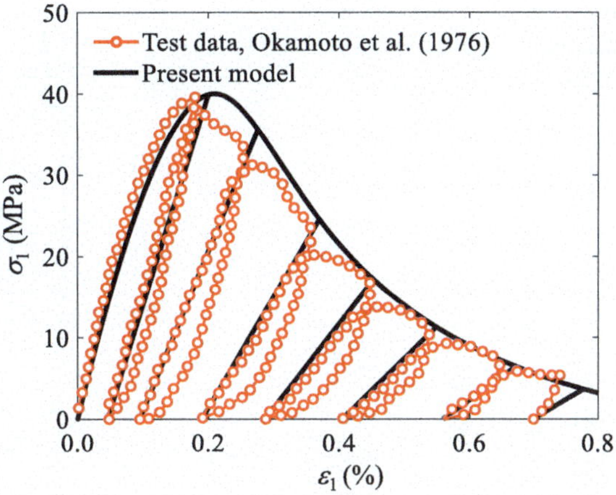

(b) C40 concrete

5.4 Summary

A 3D non-orthogonal plasticity damage model is developed to characterize the typical nonlinear mechanical behavior of concrete including the irreversible deformation and the stiffness degradation. In the proposed model, the plastic characterization is formulated in the effective stress space and is independent of damage characterization. The damage process is driven by plastic strain and is described by an S-type damage variable.

In the proposed model, the effective hardening function suitable for the multi-axial stress conditions is derived by the nominal hardening/softening function and the damage variable. The hardening/softening behavior of damaged material and the degeneration law of unloading stiffness can be thus captured separately. The

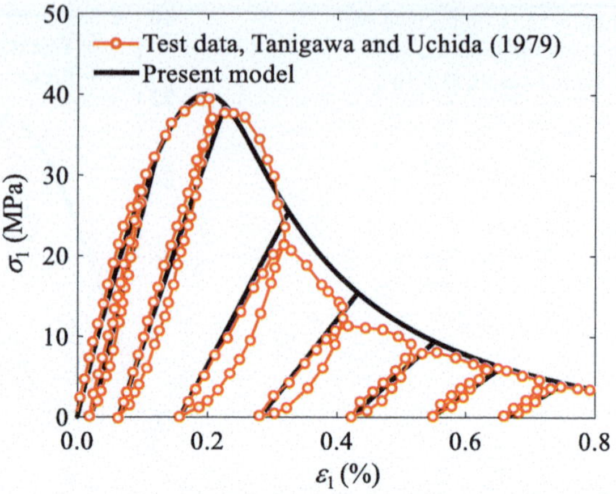

Fig. 5.13 Comparison with test results reported by Tanigawa and Uchida (1979)

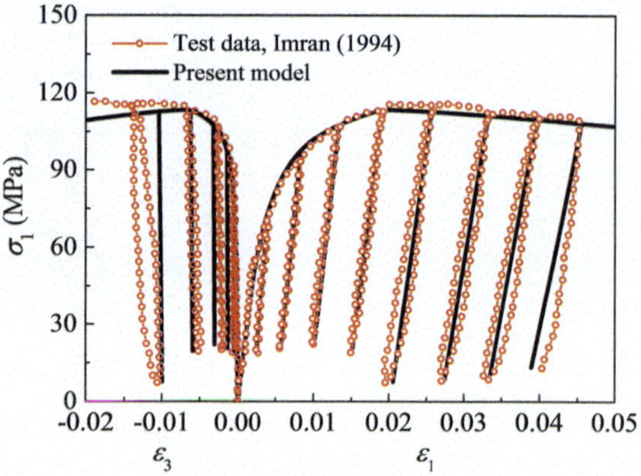

Fig. 5.14 Comparison with test results reported by Imran (1994)

direction of the plastic strain increment is determined by the non-orthogonal flow rule utilizing the RL fractional derivative. The need for the construction of plastic potential function is bypassed but the ability to reasonably capture the dilatancy law is achieved. The calibration method for model parameters is discussed in detail. The comparison of model predictions with test results from the available literature shows that the proposed model can reasonably describe the complex and diverse mechanical behaviors of concrete under different loading.

References

Al-Rub, R. K. A., & Kim, S. (2010). Computational applications of a coupled plasticity-damage constitutive model for simulating plain concrete fracture. *Engineering Fracture Mechanics, 77*(10), 1577–1603.

Candappa, D. C., Sanjayan, J. G., & Setunge, S. (2001). Complete triaxial stress-strain curves of high-strength concrete. *Journal of Materials in Civil Engineering, 13*(3), 209–215.

Dougill, J. W. (1976). On stable progressively fracturing solids. *Journal of Applied Mechanics and Technical Physics, 27*(4), 423–437.

Du, X. L., Lu, D. C., Gong, Q. M., & Zhao, M. (2010). Nonlinear unified strength criterion for concrete under three-dimensional stress states. *Journal of Engineering Mechanics, 136*(1), 51–59.

Etse, G., & Willam, K. (1994). Fracture energy formulation for inelastic behavior of plain concrete. *Journal of Engineering Mechanics, 120*(9), 1983–2011.

Gao, Z. W., Zhao, J. D., & Yao, Y. P. (2010). A generalized anisotropic failure criterion for geomaterials. *International Journal of Solids and Structures, 47*(22–23), 3166–3185.

Grassl, P., Xenos, D., Nyström, U., Rempling, R., & Gylltoft, K. (2013). CDPM2: A damage-plasticity approach to modelling the failure of concrete. *International Journal of Solids and Structures, 50*(24), 3805–3816.

Imran, A. (1994). Applications of non-associated plasticity in modelling the mechanical response of concrete. Ph.D. thesis. University of Toronto.

Jason, L., Huerta, A., Pijaudier-Cabot, G., & Ghavamian, S. (2006). An elastic plastic damage formulation for concrete: Application to elementary tests and comparison with an isotropic damage model. *Computer Methods in Applied Mechanics and Engineering, 195*(52), 7077–7092.

Ju, J. W. (1989). On energy-based coupled elastoplastic damage theories: constitutive modeling and computational aspects. *International Journal of Solids and Structures, 25*(7), 803–833.

Kachanov, L. M. (1958). Time of the rupture process under creep conditions. *Izvestija Akademii Nauk Sojuza Sovetskich Socialisticeskich Respubliki, 8*, 26–31.

Karsan, I. D., & Jirsa, J. O. (1969). Behavior of concrete under compressive loadings. *Journal of the Structural Division, 95*, 2543–2563.

Kotsovos, M. D. (1980). A mathematical model of the deformational behavior of concrete under generalised stress based on fundamental material properties. *Matériaux Et Construction, 13*(4), 289–298.

Kupfer, H., Hilsdorf, H. K., & Rusch, H. (1969). Behavior of concrete under biaxial stresses. *Journal of American Concrete Institute, 66*, 656–666.

Lee, J., & Fenves, G. L. (1998). Plastic-damage model for cyclic loading of concrete structures. *Journal of Engineering Mechanics, 124*(8), 892–900.

Lemaitre, J. (1985). Coupled elasto-plasticity and damage constitutive equations. *Computer Methods in Applied Mechanics and Engineering, 51*(1–3), 31–49.

Lu, X., & Hsu, C. T. (2007). Tangent poisson's ratio of high-strength concrete in triaxial compression. *Magazine of Concrete Research, 59*(1), 69–77.

Lu, D. C., Liang, J. Y., Du, X. L., Ma, C., & Gao, Z. W. (2019a). Fractional elastoplastic constitutive model for soils based on a novel 3D fractional plastic flow rule. *Computers and Geotechnics, 105*, 277–290.

Lu, D. C., Zhou, X., Du, X. L., & Wang, G. S. (2019b). A 3D fractional elastoplastic constitutive model for concrete material. *International Journal of Solids and Structures, 165*, 160–175.

Mazars, J. (1984). Application de la mécanique de l'endommagement au comportement non linéaire et à la rupture du béton de structure, Ph.D. thesis. Université Paris VI, France.

Okamoto, S., Shiomi, S., & Yamabe, K. (1976). Earthquake resistance of prestressed concrete structures. In: *Proceedings of annual architectural institute of Japan (AIJ) convention, Japan* (pp. 1251–1252).

Paliwal, B., Hammi, Y., Moser, R. D., & Horstemeyer, M. F. (2017). A three-invariant cap-plasticity damage model for cementitious materials. *International Journal of Solids and Structures, 108*, 186–202.

Rabotnov, Y. N. (1969). Creep rupture. In Applied mechanics: Proceedings of the Twelfth International Congress of Applied Mechanics, Stanford University, August 26–31, 1968 (pp. 342–349). Berlin, Heidelberg: Springer Berlin Heidelberg.

Sargin, M. (1971). Stress-strain relationship for concrete and the analysis of structural concrete sections. University of Waterloo, Solid Mechanics Division.

Sarikaya, A., & Erkmen, R. E. (2019). A plastic-damage model for concrete under compression. *International Journal of Mechanical Sciences, 150*, 584–593.

Sima, J. F., Roca, P., & Molins, C. (2008). Cyclic constitutive model for concrete. *Engineering Structures, 30*(3), 695–706.

Simo, J.C., & Ju, J. W. (1987). Strain-and stress-based continuum damage models—I. Formulation. *International Journal of Solids and Structures*, *23*(7), 821–840.

Sinha, B. P., Gerstle, K. H., & Tulin, L. G. (1964). Stress-strain relations for concrete under cyclic loading. *Journal of American Concrete Institution, 61*(2), 195–211.

Tanigawa, Y., & Uchida, Y. (1979). Hysteretic characteristics of concrete in the domain of high compressive strain. In: *Proceedings of annual architectural institute of Japan (AIJ) convention, Japan* (pp. 449–450).

Wang, G.S., Lu, D.C., Du, X.L., Zhou, X., & Cao, S.T. (2018). A true 3D frictional hardening elastoplastic constitutive model of concrete based on a unified hardening/softening function. *Journal of the Mechanics and Physics of Solids, 119*, 250–273.

Wang, G. S., Lu, D. C., Zhou, X., Wu, Y. F., Du, X. L., & Xiao, Y. (2020). A stress-path-independent damage variable for concrete under multiaxial stress conditions. *International Journal of Solids and Structures, 206*, 59–74.

Yao, Y. P., Lu, D. C., Zhou, A. N., & Bo, Z. (2004). Generalized non-linear strength theory and transformed stress space. *Science China Technological Sciences, 47*(6), 691–709.

Zhang, J., Li, J., & Ju, J. W. (2016). 3D elastoplastic damage model for concrete based on novel decomposition of stress. *International Journal of Solids and Structures, 94–95*, 125–137.

Chapter 6
Non-orthogonal Elastoplastic Model for Concrete in Variable Strain Rate Environment

A large number of test results show that strain rate significantly affects the mechanical properties of concrete materials (Kong et al., 2018; Shkolnik, 2008). There are generally two methods of establishing the dynamic constitutive model to reflect the mechanical characteristics affected by the strain rate. One simplified method is to introduce the strain rate u_{d} into the material parameters in the stiffness matrix, i.e., $\mathrm{d}\boldsymbol{\sigma} = \mathbb{D}(u_{\mathrm{d}}) : \mathrm{d}\boldsymbol{\varepsilon}$. When performing the consistency condition in the differential form to determine the plasticity multiplier, it is assumed that the strain rate is constant (Bai et al., 2020). Here, the strain rate is considered a factor influencing material parameters rather than a constitutive variable. Several popular concrete models, e.g., the Holmquist–Johnson–Cook model (Holmquist & Johnson, 2011), the Riedel–Hiermaier–Thoma model (Borrvall & Riedel, 2011), and the Karagozian and Case model (Kong et al., 2017), have been developed based on this modeling method. The advantage of this method is that the existing damage model or plasticity model can be directly developed into a dynamic constitutive model by developing model parameters into functions of strain rate, but the mechanical response caused by the strain rate change cannot be captured. Another method is to treat the strain rate as the constitutive variable, similar to stress and strain, i.e., $\mathrm{d}\boldsymbol{\sigma} = \mathbb{D}(u_{\mathrm{d}}) : \mathrm{d}\boldsymbol{\varepsilon} + \mathbf{D}_{u_{\mathrm{d}}}\mathrm{d}u_{\mathrm{d}}$. The constitutive model is established in the form of incremental stress–strain–strain rate to describe the mechanical response of concrete under the coupled action of load and environment (Lu et al., 2020). Especially, the mechanical response, i.e., $\mathbf{D}_{u_{\mathrm{d}}}\mathrm{d}u_{\mathrm{d}}$, caused by the changes in the strain rate can be captured (Hossain & Weiss, 2004; Rahnavard et al., 2022; Torelli et al., 2018). The viscoplastic model is a representative model established by this method. Here, the viscoplastic strain rate is considered as a constitutive variable in calculating the viscoplastic multiplier (Aráoz & Luccioni, 2015; Kang & Willam, 2000), so the relaxation or creep process caused by a change in viscoplastic strain rate can be described (Naghdi & Murch, 1963). However, neglecting the elastic strain rate prevents the relaxation process (creep process) of the stress state beginning within the yield surface from being captured

D. Lu et al., *Multi-physical and Non-orthogonal Geotechnical Plasticity Modeling and Algorithms*, https://doi.org/10.1007/978-981-96-2369-3_6

(Heeres et al., 2002; Qiao et al., 2016). There are few reports on establishing the incremental stress–strain–strain rate constitutive model by introducing the total strain rate into the yield function.

In this context, a dynamic elastoplastic constitutive model for concrete is established based on the multi-physical and non-orthogonal geotechnical plasticity modeling method. The total strain rate is regarded as a constitutive variable when determining the magnitude of plastic strain, which can describe the plastic deformation behavior under its independent action. The direction of the plastic strain increment is determined through the non-orthogonal gradient of the dynamic yield function. A loading and unloading criterion is established in the stress–strain rate coordinate space to determine the conditions for plastic deformation of concrete.

6.1 Specifics Adopted for Concrete Model

According to a general expression of incremental stress–strain–environment relation, to establish a model with strain rate as a constitutive variable, the expressions for yield function, hardening parameter, and flow rule need to be provided, in which strain rate is introduced into strength parameters. The yield function, hardening parameter, and consistency criterion are combined to determine $d\phi$. The flow rule is used to calculate $\mathbf{r}$.

6.1.1 *Yield Function*

The shape of the yield surface depends on the characteristics of the material. Bridgeman (1964) demonstrated through experiments that the mechanical properties of metal materials are independent of hydrostatic pressure. However, the stress–strain behavior of friction materials such as soil, concrete, and rock depends on hydrostatic pressure. The advantage of the closed yield function is that it can more completely describe the stress state of concrete including under shear loads and hydrostatic pressure loads. A dynamic closed yield function is developed by referring to the function form in the literature (Etse & Willam, 1994; Grassl & Jirásek, 2006; Lu et al., 2022; Zhou et al., 2020) as follows:

$$
\begin{aligned}
f = & \left[(1-H)\left(\frac{q}{3f_{\mathrm{c}}^{\mathrm{d}}} - \frac{p}{f_{\mathrm{c}}^{\mathrm{d}}}\right)^2 + \frac{q}{f_{\mathrm{c}}^{\mathrm{d}}}\right]^2 \\
& + m_0^{\mathrm{d}} H^2 \left[\frac{q}{3f_{\mathrm{c}}^{\mathrm{d}}} r(\cos\theta) - \frac{p}{f_{\mathrm{c}}^{\mathrm{d}}}\right] - H^2
\end{aligned}
\tag{6.1}
$$

where f_c^d and m_0^d indicate the dynamic uniaxial compressive strength and dynamic friction parameter respectively. p, q, and θ are the hydrostatic pressure, the generalized shear stress, and the stress Lode angle respectively. $p = \boldsymbol{\sigma} : \mathbf{1}/3$, $q = \sqrt{3J_2}$, and $\theta = \arccos\left(3\sqrt{3}J_3/2J_2^{3/2}\right)/3$, in which J_2 and J_3 are the second and third invariants, respectively, of the deviatoric stress tensor $\mathbf{s}$. $J_2 = \mathbf{s} : \mathbf{s}/2$, $J_3 = \det(\mathbf{s})$, and $\mathbf{s} = \boldsymbol{\sigma} - p\mathbf{1}$. $\mathbf{1}$ is 2nd order unit tensor. The function $r(\cos\theta)$ is expressed by:

$$r(\cos\theta) = \frac{4\left[1 - \left(e^d\right)^2\right]\cos^2\theta + \left[2e^d - 1\right]^2}{2\left[1 - \left(e^d\right)^2\right]\cos\theta + \left[2e^d - 1\right]\sqrt{4\left[1 - \left(e^d\right)^2\right]\cos^2\theta + 5\left(e^d\right)^2 - 4e^d}} \tag{6.2}$$

where e^d is the dynamic eccentricity parameter. The three material parameters in the yield function, namely f_c^d, m_0^d, and e^d, can be obtained by the basic dynamic strength parameters, namely f_c^d, f_t^d, and f_{bc}^d:

$$m_0^d = 3\frac{\left(f_c^d\right)^2 - \left(f_t^d\right)^2}{f_c^d f_t^d}\frac{e^d}{e^d + 1} \tag{6.3}$$

$$e^d = \frac{1 + \epsilon^d}{2 - \epsilon^d} \tag{6.4}$$

$$\epsilon^d = \frac{f_t^d}{f_{bc}^d}\frac{\left(f_{bc}^d\right)^2 - \left(f_c^d\right)^2}{\left(f_c^d\right)^2 - \left(f_t^d\right)^2} \tag{6.5}$$

where f_t^d denotes dynamic uniaxial tensile strength and f_{bc}^d indicates dynamic equibiaxial compressive strength. It is possible to write f_{bc}^d as a function of f_c^d, i.e., $f_{bc}^d = h \cdot f_c^d$, where h is a constant (Wang et al., 2018a). The strain rate is introduced into the yield function as a constitutive variable by strength parameters. Experimental data (Yu et al., 2013) show that at very high strain rates, the dynamic strength will inevitably approach a limit value. Therefore, the S-shaped dynamic increase factor (DIF) is employed to describe the dynamic strength property of concrete (Lu et al., 2017).

The yield function is a four-dimensional (4D) hypersurface in the (p, q, θ, u_d) coordinate space. u_d is the logarithm of the generalized shear strain rate $\dot{\varepsilon}_d = \sqrt{2/3}\,\|(\dot{\boldsymbol{\varepsilon}} - \dot{\varepsilon}_v\mathbf{1}/3)\|_2$. The 4D hypersurface can be decomposed into two 3D yield surfaces in the (u_d, p, q) and $(u_d, x_\sigma = q \cdot \sin\theta, y_\sigma = q \cdot \cos\theta)$ coordinate spaces. Figure 6.1a presents the 3D yield surface in the (u_d, p, q) coordinate space by setting θ equal to a constant. The yield surface's variation trend with the strain rate in the meridian plane is compatible with the evolution law of the adopted dynamic strength criterion, which has an S-shaped shape and will not expand indefinitely. As illustrated in Fig. 6.1b, the $(u_d, x_\sigma, y_\sigma)$ coordinate space can present the evolution of the yield curve with the strain rate in the deviatoric space when p is constant.

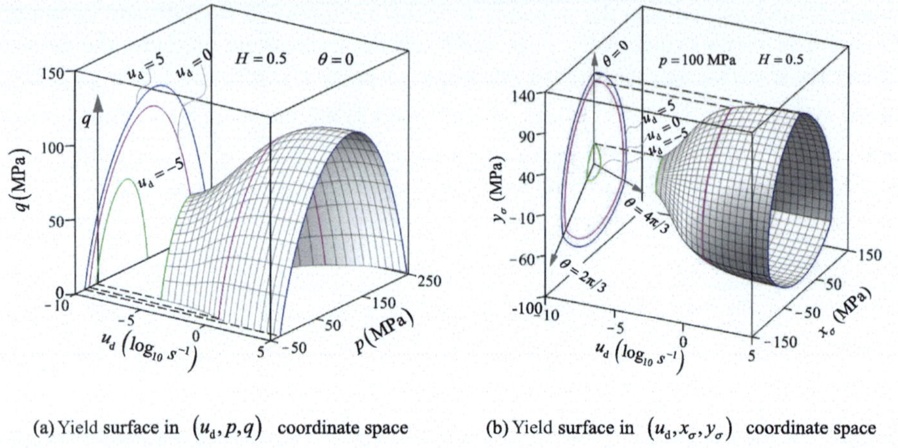

Fig. 6.1 Evolution of yield surface with strain rate

6.1.2 *Hardening Parameter*

The hardening–softening parameter (H) establishes the relationship between stress and strain and is used to control the shape of the yield surface. For concrete materials with obvious peak points, H needs to describe both nonlinear hardening and post-peak softening behaviors. The H represented by Eq. (6.6) is adopted (Wang et al., 2018b):

$$H = \frac{Ax + (D-1)x^2}{1 + (A-2)x + Dx^2} \tag{6.6}$$

where a continuous function simultaneously describes the deformation behavior of concrete in the hardening and softening stages. When $H = 1$, the yield surface in Eq. (6.1) is degenerated into an open strength surface, as shown in Fig. 6.2. A and D are the model parameters. x is a plastic internal variable, and $x = \varepsilon_d^p/\varepsilon_{ds}^p$. ε_d^p indicates the equivalent plastic shear strain. ε_{ds}^p is the value of ε_d^p at peak stress. The test data (Candappa et al., 2001; Imran & Pantazopoulou, 1996; Kupfer et al., 1969; Lu & Hsu, 2007) reveal that parameters ε_{ds}^p and A (cf. Fig. 6.3) correlate with the stress state as follows:

$$\frac{\varepsilon_{ds}^p}{\varepsilon_{1,ds}^p} = \frac{14.70}{1 + e^{\ln(13.70)+2.30-2.30\cdot\left(\frac{q_{max}}{f_c}\right)}} \tag{6.7}$$

$$\frac{A}{A_1} = \frac{3.88}{1 + e^{\ln(2.88)+2.22-2.22\cdot\left(\frac{q_{max}}{f_c}\right)}} \tag{6.8}$$

where $\varepsilon^{p}_{1,ds}$ and A_1 are the value of ε^{p}_{ds} and A under uniaxial compressive conditions respectively. The physical meaning of A can be found in ref (Wang et al., 2018b). q_{max} is the largest q in loading history.

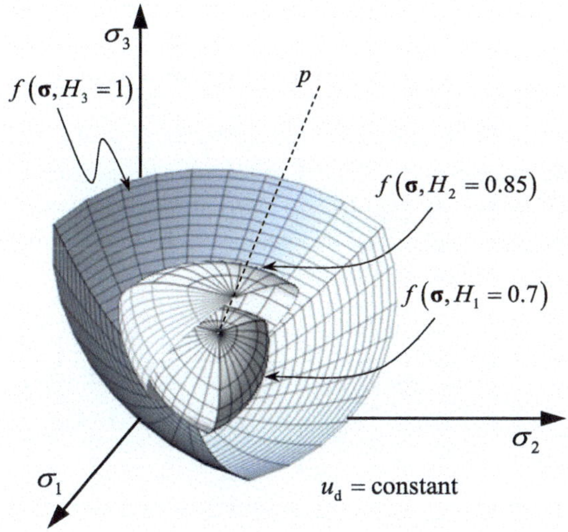

Fig. 6.2 Evolution of yield surface with H when u_d is constant

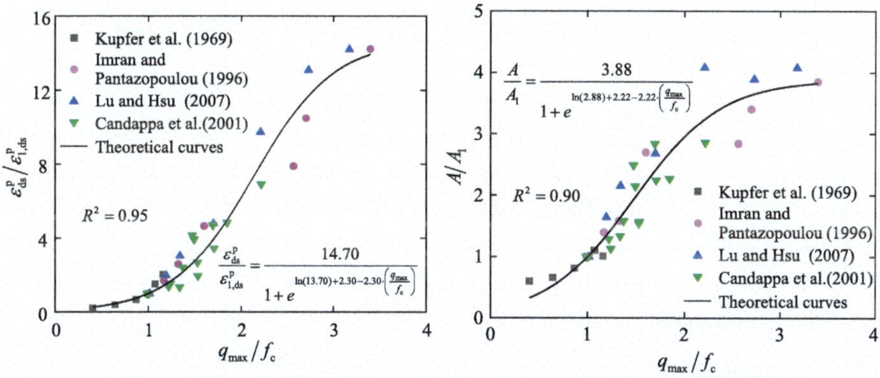

(a) ε^{p}_{ds} with the stress state

(b) A with the stress state

Fig. 6.3 Relationship of material parameters with the stress state

6.1.3 Flow Rule

A non-orthogonal flow rule (Lu et al., 2019, 2022) is employed, which directly determines $\mathbf{r}$ via the developed dynamic yield function in Eq. (6.1):

$$\mathbf{r} = \frac{\partial^{\mu} f}{\partial p^{\mu}} \frac{\partial p}{\partial \boldsymbol{\sigma}} + \frac{\partial^{\mu} f}{\partial q^{\mu}} \frac{\partial q}{\partial \boldsymbol{\sigma}} \tag{6.9}$$

where μ indicates the fractional order and $\partial^{\mu} f / \partial(\cdot)^{\mu}$ is the Riemann–Liouville fractional derivative operator, the details of which can be found in reference (Lu et al., 2022). $\partial p/\partial\boldsymbol{\sigma} = \boldsymbol{\delta}/3$ and $\partial q/\partial\boldsymbol{\sigma} = 1.5\mathbf{s}/q$ are used to achieve the conversion of coordinates from principal stress to general stress. Combining with Eq. (6.1), $\partial^{\mu} f/\partial p^{\mu}$ and $\partial^{\mu} f/\partial q^{\mu}$ can be obtained:

$$\begin{cases} \dfrac{\partial^{\mu} f}{\partial p^{\mu}} = \dfrac{a_0 p^{-\mu}}{\Gamma(1-\mu)} + \dfrac{a_1 \Gamma(2) p^{1-\mu}}{\Gamma(2-\mu)} + \dfrac{a_2 \Gamma(3) p^{2-\mu}}{\Gamma(3-\mu)} + \dfrac{a_3 \Gamma(4) p^{3-\mu}}{\Gamma(4-\mu)} + \dfrac{a_4 \Gamma(5) p^{4-\mu}}{\Gamma(5-\mu)} \\ \dfrac{\partial^{\mu} f}{\partial q^{\mu}} = \dfrac{b_0 q^{-\mu}}{\Gamma(1-\mu)} + \dfrac{b_1 \Gamma(2) q^{1-\mu}}{\Gamma(2-\mu)} + \dfrac{b_2 \Gamma(3) q^{2-\mu}}{\Gamma(3-\mu)} + \dfrac{b_3 \Gamma(4) q^{3-\mu}}{\Gamma(4-\mu)} + \dfrac{b_4 \Gamma(5) q^{4-\mu}}{\Gamma(5-\mu)} \end{cases} \tag{6.10}$$

where $\Gamma(\cdot)$ is the gamma function. Coefficients a_i and b_i $(i = 0,\ 1,\ 2,\ 3,\ 4)$ are given as follows respectively.

$$\begin{cases} a_0 = \dfrac{(1-H)^2 q^4}{81\left(f_c^d\right)^4} + \dfrac{2(1-H)q^3}{9\left(f_c^d\right)^3} + \dfrac{q^2}{\left(f_c^d\right)^2} + \dfrac{m_0^d H^2 r(\cos\theta) q}{3 f_c^d} - H^2 \\ a_1 = -\dfrac{4(1-H)^2 q^3}{27\left(f_c^d\right)^4} - \dfrac{4(1-H)q^2}{3\left(f_c^d\right)^3} - \dfrac{m_0^d H^2}{f_c^d} \\ a_2 = \dfrac{2(1-H)^2 q^2}{3\left(f_c^d\right)^4} + \dfrac{2(1-H)q}{\left(f_c^d\right)^3} \\ a_3 = -\dfrac{4(1-H)^2 q}{3\left(f_c^d\right)^4} \\ a_4 = \dfrac{(1-H)^2}{\left(f_c^d\right)^4} \end{cases} \tag{6.11}$$

$$\begin{cases} b_0 = \dfrac{(1-H)^2 p^4}{\left(f_c^d\right)^4} - \dfrac{m_0^d H^2 p}{f_c^d} - H^2 \\ b_1 = -\dfrac{4(1-H)^2 p^3}{3\left(f_c^d\right)^4} + \dfrac{2(1-H)p^2}{\left(f_c^d\right)^3} + \dfrac{m_0^d H^2 r(\cos\theta)}{3f_c^d} \\ b_2 = \dfrac{2(1-H)^2 p^2}{3\left(f_c^d\right)^4} - \dfrac{4(1-H)p}{3\left(f_c^d\right)^3} + \dfrac{1}{\left(f_c^d\right)^2} \\ b_3 = -\dfrac{4(1-H)^2 p}{27\left(f_c^d\right)^4} + \dfrac{2(1-H)}{9\left(f_c^d\right)^3} \\ b_4 = \dfrac{(1-H)^2}{81\left(f_c^d\right)^4} \end{cases} \tag{6.12}$$

The constitutive equation of the established model can be obtained by substituting Eqs. (6.1), (6.6) and (6.9) into the incremental stress–strain–environment relation.

6.1.4 *Loading/Unloading Criterion*

This section introduces the environmental variable u_d into the yield function as an independent constitutive variable. The loading–unloading criterion under the combined action of the strain and the strain rate can be presented in the $(\boldsymbol{\varepsilon}, u_d)$ coordinate space as follows:

$$\frac{\partial F}{\partial \boldsymbol{\varepsilon}} : \mathrm{d}\boldsymbol{\varepsilon} + \frac{\partial F}{\partial u_d}\mathrm{d}u_d = \begin{cases} \dfrac{\partial f}{\partial \boldsymbol{\sigma}} : \mathbb{D}^e : \mathrm{d}\boldsymbol{\varepsilon} + \dfrac{\partial f}{\partial u_d}\mathrm{d}u_d > 0 & \text{Loading} \\ \dfrac{\partial f}{\partial \boldsymbol{\sigma}} : \mathbb{D}^e : \mathrm{d}\boldsymbol{\varepsilon} + \dfrac{\partial f}{\partial u_d}\mathrm{d}u_d = 0 & \text{Neutral loading} \\ \dfrac{\partial f}{\partial \boldsymbol{\sigma}} : \mathbb{D}^e : \mathrm{d}\boldsymbol{\varepsilon} + \dfrac{\partial f}{\partial u_d}\mathrm{d}u_d < 0 & \text{Unloading} \end{cases} \tag{6.13}$$

where $(\partial F/\partial \boldsymbol{\varepsilon}, \partial F/\partial u_d)$ represents the dynamic closed yield surface's normal direction.

6.2 Test of Loading Under Constant- and Variable-Strain Rate

Two types of constant- and variable-strain rate loading experiments are conducted in the Rapid Triaxial Testing Systems of Beijing University of Technology. The variable strain rate loading tests are conducted under uniaxial compressive conditions.

The constant-strain rate tests are performed under conventional triaxial compressive conditions.

6.2.1 Testing Procedures and Apparatus

The loading process at a constant-strain rate is realized by setting the linear relationship of axis strain to time, including four strain rates, i.e., $\dot{\varepsilon}_1$ of 10^{-5} s^{-1}, 10^{-4} s^{-1}, 10^{-3} s^{-1}, and 10^{-2} s^{-1}, and three confining pressures, i.e., σ_3 of 0, 10 MPa, and 20 MPa. The concrete specimen is first subjected to the confining pressure at a rate of 5 MPa/min, during which the synchronous increase of axial stress puts the specimen in a hydrostatic state. The confining pressure remains constant after the target value is achieved. Then, utilizing displacement control, the axial loading is applied until the test is finished. For another set of dynamic tests with variable strain rate loading, the axial strain changes with time as a quadratic function, that is, the axial strain rate increases linearly with time. In the three sets of variable-strain rate tests, the strain rate increment is 10^{-7} s^{-2}, 10^{-5} s^{-2}, and 10^{-3} s^{-2}, respectively.

All dynamic loading tests were performed on the Rapid Triaxial Testing Systems (RTR-1500) shown in Fig. 6.4. It primarily comprises a confining pressure intensifier apparatus, a pore pressure intensifier apparatus, a digital-signal conditioning control unit, and high-pressure triaxial cells. Its compressive loading capacity is up to 1500 kN, and the confining pressure and pore pressure can be increased to 140 MPa. The axial strain is calculated by the gauge axial deformation and the axial gauge length of the specimen. The strain hinge measures the lateral strain. The length of the strain hinge can be adjusted according to the specimen, and the testing system automatically converts the lateral strain according to the change in the hinge circumference.

6.2.2 Preparation of Materials and Specimens

Table 6.1 lists the mixed proportion of the concrete, and the cementitious materials are 42.5 ordinary Portland cement. Crushed stone with a diameter of 5–20 mm makes up the coarse aggregate, while medium sand with a fineness modulus of 2.42 makes up the fine aggregate. First, materials for preparing mortar mixture are prepared according to the mix proportion. Then, the coarse and fine aggregates are mixed with the cement evenly. The tap water is gradually added to the mixture during the stirring. The well-mixed concrete is cast into a mold coated with a release agent and vibrated to a compact state on a vibrating table. All the specimens are demolded after curing for 24 hours at room temperature. After demolding, the specimens are further cured for 28 days in a standard curing room.

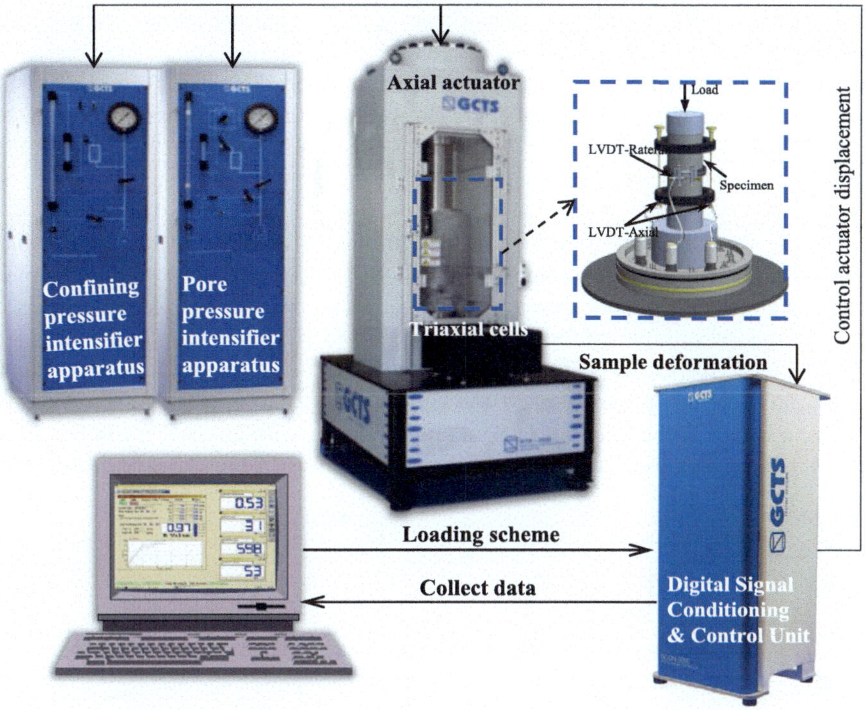

Fig. 6.4 A schematic of the Rapid Triaxial rock testing systems

Table 6.1 The mixture proportion of the concrete

Cement (kg/m^3)	Fine aggregate (kg/m^3)	Coarse aggregate (kg/m^3)	Water (kg/m^3)	Water-to-cement ratio
389	554	1196	190	0.49

6.2.3 Experimental Results

Figure 6.5 displays the stress–strain curves for each specimen at varied confining pressures and constant strain rates. The peak deviatoric stress $(\sigma_1 - \sigma_3)^{\mathrm{p}}$ can be used to characterize the shear strength of concrete under conventional triaxial compression conditions. It can be found that the $(\sigma_1 - \sigma_3)^{\mathrm{p}}$ rises as confining pressure and strain rate increase due to the restraint strengthening effect and strain rate effect of concrete. Numerous studies have delved into exploring the mechanism behind the increase in strength of concrete materials resulting from strain rate effects. According to consensus, the thermal activation mechanism, the Stefan effect, and inertial mechanisms are responsible for enhancing the overall load-bearing capacity of concrete at a macroscopic level (Qi et al., 2009; Zhang & Zhao, 2014). The thermal activation mechanism is caused by the thermal motion of atoms, which plays a role in the whole strain rate range. The latter two mechanisms only play within the range of medium to high strain rates. For the strain rates range tested, the strength of concrete increased due to the combined effects of thermal activation and the Stefan effect, with the former playing a more prominent role. The strain at the peak stress under conventional triaxial conditions significantly increases compared to the uniaxial condition at the same loading rate. The results show that the confining pressure improves the ductility and resistance of concrete to deformation.

For the variable-strain rate loading test, the actual loading process is obtained through the strain–time data points automatically collected by the testing instrument during the loading process (see Fig. 6.6b, d, and f). The strain rate increments of three groups of variable-strain rate tests are 1.26×10^{-7} s^{-2}, 1.26×10^{-5} s^{-2}, and 8.8×10^{-4} s^{-2} respectively. The $\sigma_1 - \varepsilon_1 - \dot{\varepsilon}_1$ and $\sigma_1 - \varepsilon_3 - \dot{\varepsilon}_1$ behavior is depicted in Fig. 6.6a, c, and e, where each working condition is repeated three times. As shown in Fig. 6.7, the strength of concrete increases with an increase in the strain rate increment. A larger strain rate increment means a faster increase in the strain

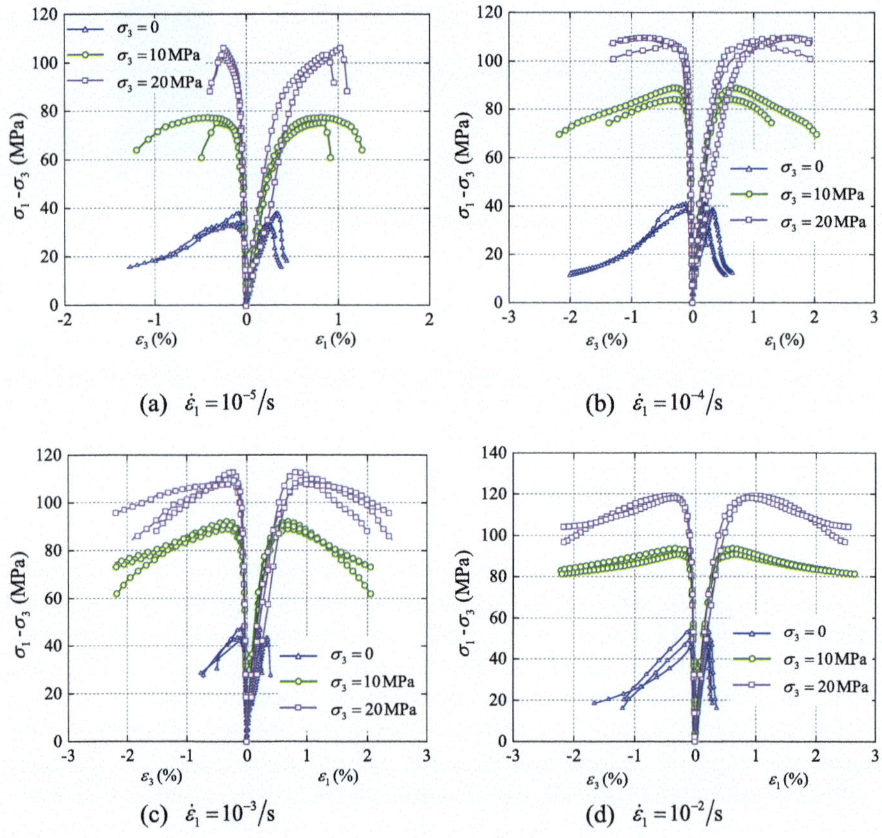

Fig. 6.5 Curves of stress–strain at different constant strain rates

rate. When the strain rate changes more rapidly before reaching its peak strength, the performance of the specimen resistant to external loads will be exerted more fully.

6.3 Comparison of Model Predictions and Test Results

The established model has eleven parameters, including two elastic parameters (E,ν), five strength parameters (f_c, $\text{DIF}_{c,\max}$, ξ_c, $\text{DIF}_{t,\max}$, ξ_t), three parameters controlling deformation ($\varepsilon^p_{1,ds}$, A_1, D) in the hardening–softening function, and a fractional order (μ) that captures the direction of plastic flow. E, ν and f_c can be obtained by uniaxial compression tests, and $f_t = 0.1f_c$ (Gebbeken & Krauthammer, 2013). $\varepsilon^p_{1,ds}$ is the value of ε^p_{ds} at f_c. A_1 represents the ratio of the secant modulus at f_c to the elastic modulus. D reflects the residual strength of concrete. Reference (Wang et al., 2018b) gives expressions for determining A_1 and D. μ can be calibrated by the stress and strain at the phase transition point (Zhou et al., 2020). Reference (Lu et al., 2017) gives the determination method of the dynamic strength parameters ($\text{DIF}_{c,\max}$, ξ_c, $\text{DIF}_{t,\max}$, ξ_t) and also provides the statistical values of these four parameters based on the analysis of numerous compression and tension tests.

The model's performance is evaluated using experimental data under both constant- and variable-strain rate conditions. The constant-strain rate tests include the conventional triaxial compression tests conducted by the authors and the uniaxial and biaxial compression tests obtained from published papers. The variable-strain rate conditions tests are performed by authors. In particular, a designed experiment with a sudden increase and decrease in strain rate is designed to supplement the analysis of variable-strain rate cases. Table 6.2 presents the model parameters adopted by the tests.

6.3.1 Constant-Strain Rate

First, the constant strain rate test data performed by the authors were adopted, including four strain rates, i.e., $\dot{\varepsilon}_1$ of 10^{-5} s^{-1}, 10^{-4} s^{-1}, 10^{-3} s^{-1}, and 10^{-2} s^{-1}, and three confining pressures, i.e., σ_3 of 0, 10, and 20 MPa. The model parameters are calibrated using test data at $\sigma_3 = 0$ MPa. These calibrated values are then used to predict test results at σ_3 of 10 MPa and 20 MPa.

Figure 6.8 shows the model's prediction curve and test results. According to test results, as confining pressure increases, concrete's ability to resist plastic deformation is enhanced, and the phenomena of softening weaken. The proposed model can well capture these experimental laws because it establishes two semi-empirical formulas for the increase of ε^p_{ds} and A with increasing stress (Eqs. 6.7 and 6.8) to characterize the impact of confining pressure on the deformation properties. In addition, as shown in Fig. 6.9, the test results under different confining pressures also show that the

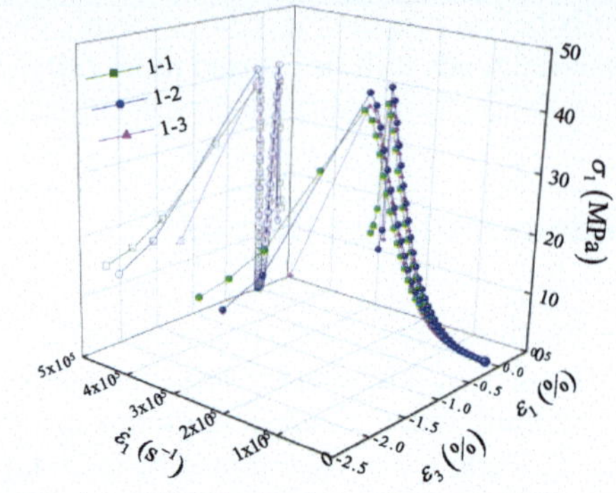

(a) $\sigma_1 - \varepsilon_1 - \dot{\varepsilon}_1$ and $\sigma_1 - \varepsilon_3 - \dot{\varepsilon}_1$ curves of group one

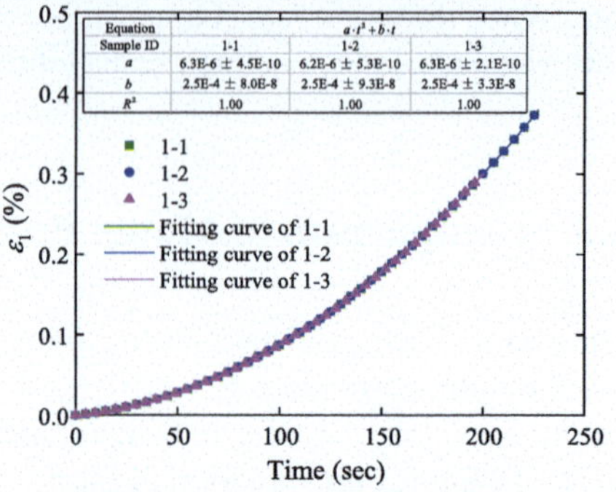

(b) Strain–time curves of group one

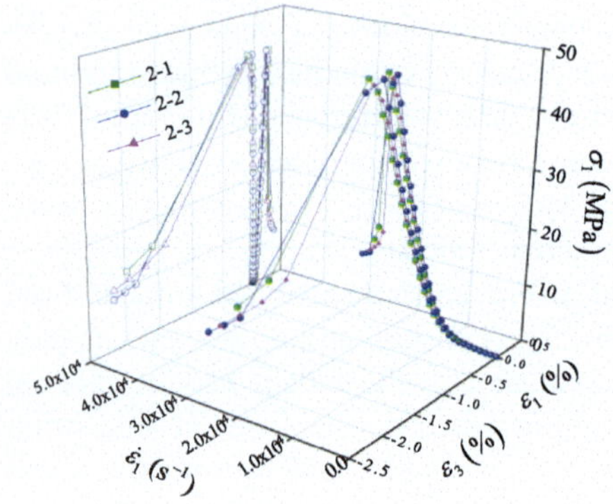

(c) $\sigma_1 - \varepsilon_1 - \dot{\varepsilon}_1$ and $\sigma_1 - \varepsilon_3 - \dot{\varepsilon}_1$ curves of group two

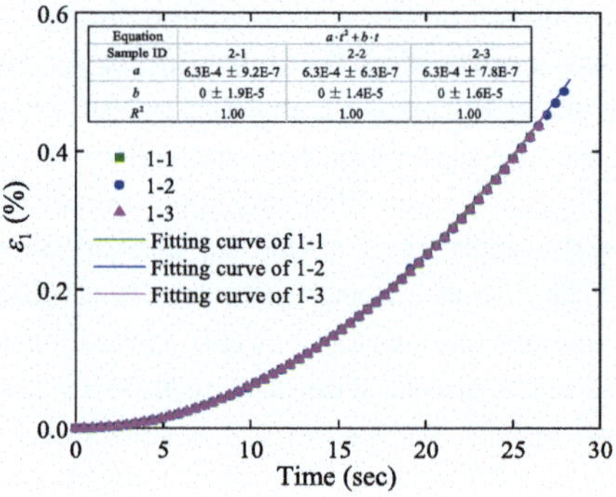

(d) Strain–time curves of group two

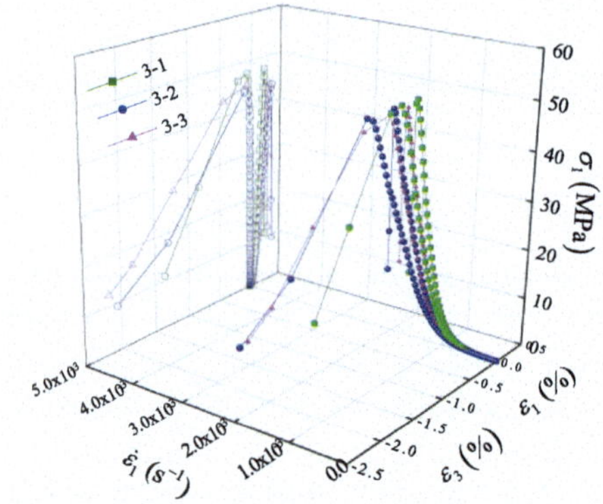

(e) $\sigma_1 - \varepsilon_1 - \dot{\varepsilon}_1$ and $\sigma_1 - \varepsilon_3 - \dot{\varepsilon}_1$ curves of group three

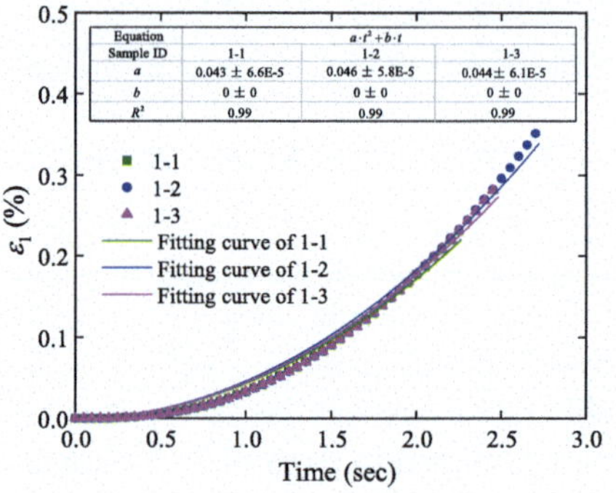

(f) Strain–time curves of group three

Fig. 6.6 Stress–strain–strain rate curves and strain–time curves during loading

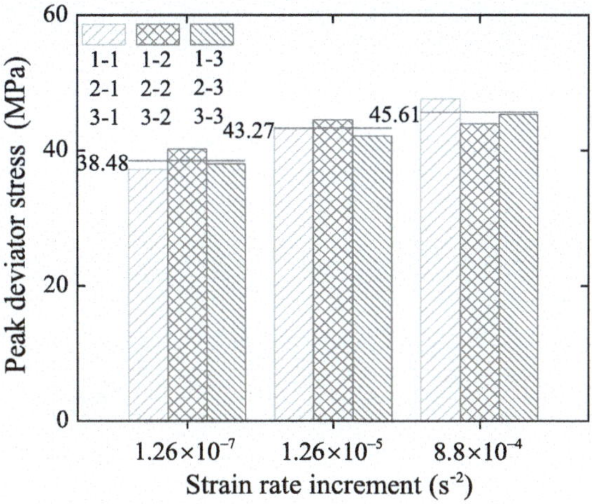

Fig. 6.7 Strength value in different strain rate increments

Table 6.2 Model parameters of concrete under dynamic loading

Data	E (MPa)	ν	f_c (MPa)	$DIF_{c,max}$	ξ_c	$DIF_{t,max}$	ξ_t	$\varepsilon^p_{1,ds}$	A_1	D	μ
Constant-strain rate	22,000	0.2	33.97	1.4	1.4	16	1.4	0.0005	6	0.2	1.9
Yan and Lin (2007)	14,000	0.2	9.84	1.38	1.3	16	1.4	0.0004	8	0.8	1.9
Grote et al. (2001)	42,500	0.2	46.00	5.60	6.0	16	1.4	0.004	14	1.0	1.0
Variable-strain rate	22,000	0.2	33.97	1.42	1.9	16	1.4	0.0005	8	0.4	1.0

strength of concrete rises as the strain rate does. The developed dynamic strength criterion in Eq. (6.1) is proven to be rational by comparing the model's strength curve with the test results.

The following evaluates the ability of the model using biaxial compressive–compressive loading tests with σ_3 of zero reported by Yan and Lin (2007). The ratio of σ_2 to σ_1 is set at 0:1.0, 0.25:1.0, 0.50:1.0, 0.75:1.0, and 1.0:1.0, and the values of $\log_{10}\dot{\varepsilon}_1$ equal −5, −4, and −3. Figure 6.10a–e illustrates the prediction results of stress versus strain at different strain rates and various stress ratios. It can be seen that the proposed model is capable of portraying the stress–strain performance of concrete when subjected to varying loading conditions. Based on the comparison of strength values between the model and experimental results in Fig. 6.10f, it can be concluded that the developed yield function can accurately characterize the multiaxial dynamic strength of concrete.

To confirm the model's ability at high strain rates, four split Hopkinson pressure bar tests performed by Grote et al. (2001) are adopted, where $\dot{\varepsilon}_1$ equals 290 s^{-1}, 620 s^{-1}, 1050 s^{-1}, 1500 s^{-1}. When $\dot{\varepsilon}_1 = 1500$ s^{-1}, dynamic strength f_c^d, nearly 3.5 times

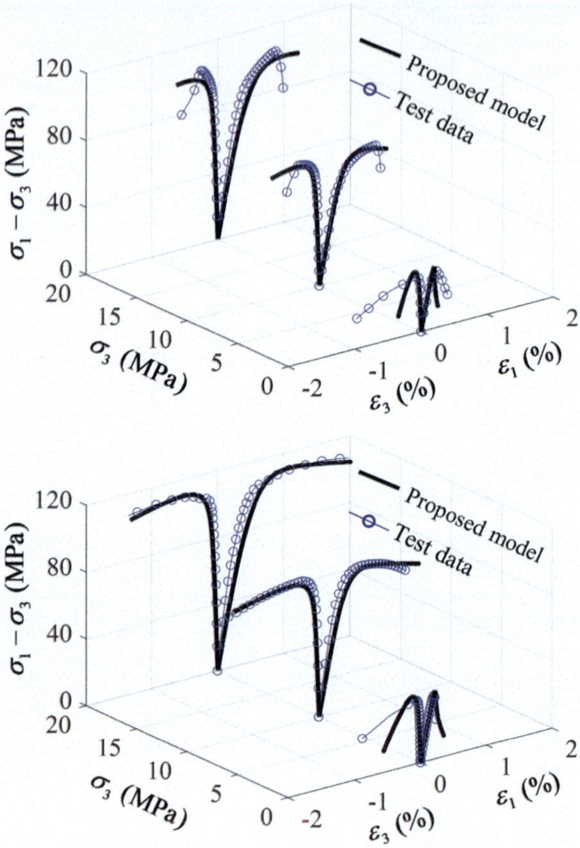

Fig. 6.8 Comparison results between the model and test under conventional triaxial compression conditions

the static strength f_c, is 160 MPa. This indicates that there is a significant increase in compressive strength with an increase in $\dot{\varepsilon}_1$. A comparison is shown in Fig. 6.11 between the $\sigma_1 - \varepsilon_1$ curve of the predicted and tested results. The prediction results show that the proposed model can describe concrete behavior at high strain rates.

6.3.2 *Variable-Strain Rate*

To further evaluate the capabilities of the developed model, variable strain rate tests with axial strain as a quadratic function of time are performed. In the three sets of variable-strain rate tests, the strain rate increment is 10^{-7} s^{-2}, 10^{-5} s^{-2}, and 10^{-3} s^{-2}, respectively. Figure 6.12 illustrates the comparison results between model predictions and test data in the stress–strain–strain rate coordinate space. The changes in strain rate during loading are evident from the curve depicted on the strain and strain rate plane in Fig. 6.12, which is different from a constant strain rate. At this

Fig. 6.8 (continued)

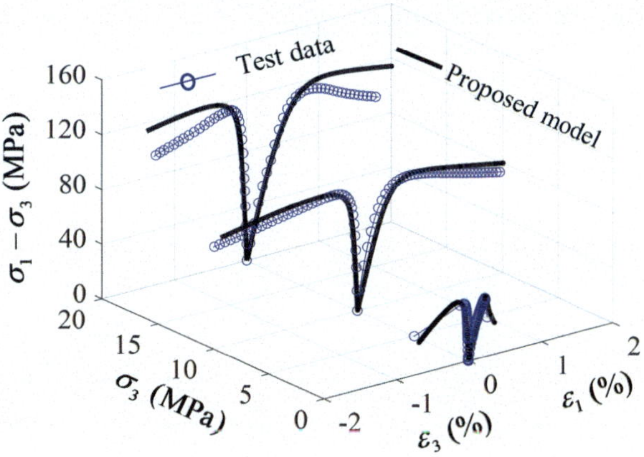

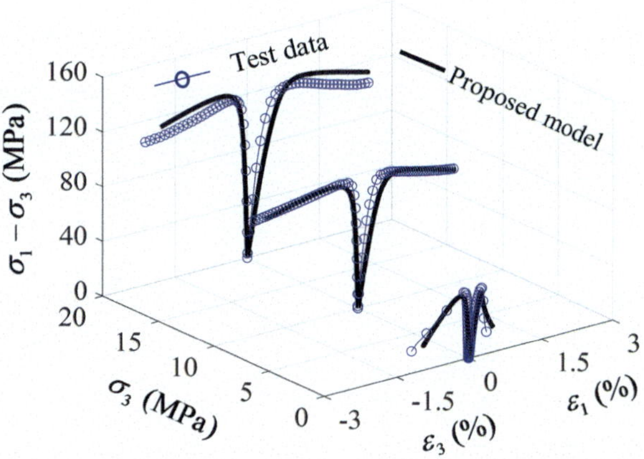

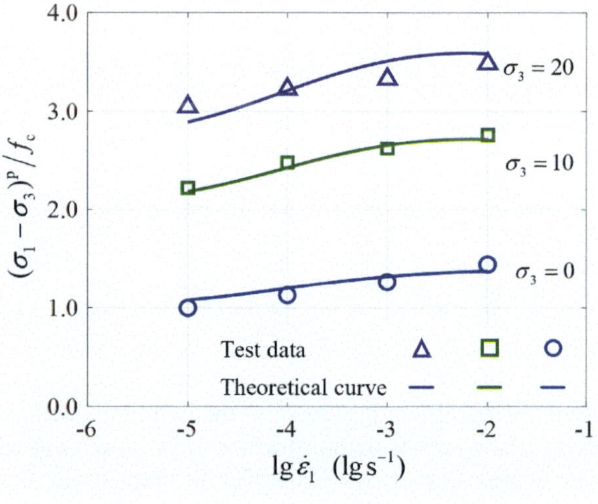

Fig. 6.9 Comparison of strength results between model prediction and test data

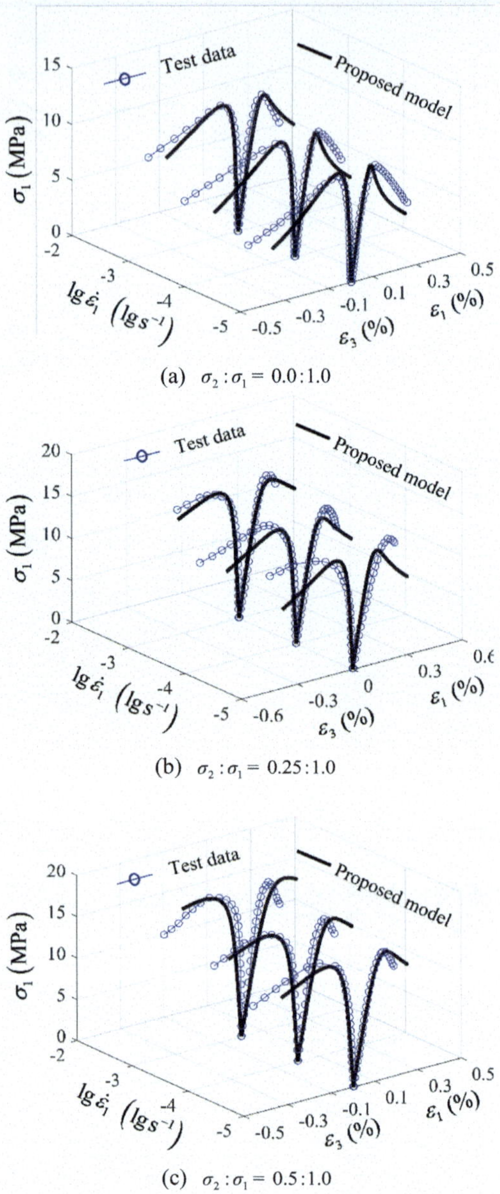

Fig. 6.10 Model predictions versus the test data for biaxial compression conducted by Yan and Lin (2007)

time, the specimen is subject to the combined action of strain and strain rate. The comparison results demonstrate that the model can predict the stress–strain–strain rate relationship of concrete under different strain rate increments conditions.

Fig. 6.10 (continued)

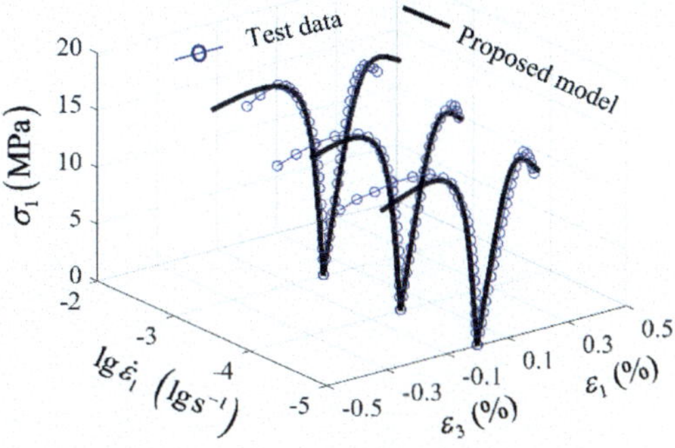

(d) $\sigma_2:\sigma_1 = 0.75:1.0$

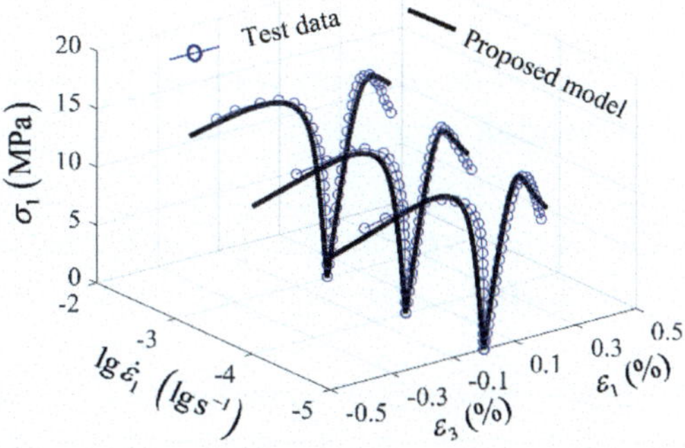

(e) $\sigma_2:\sigma_1 = 1.0:1.0$

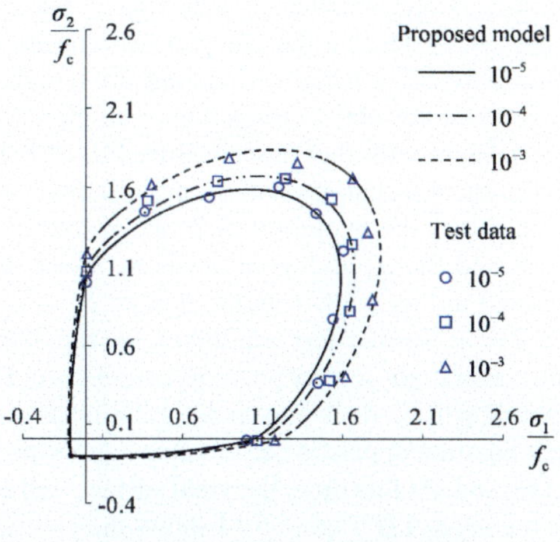

(f) value of the dynamic multiaxial strength

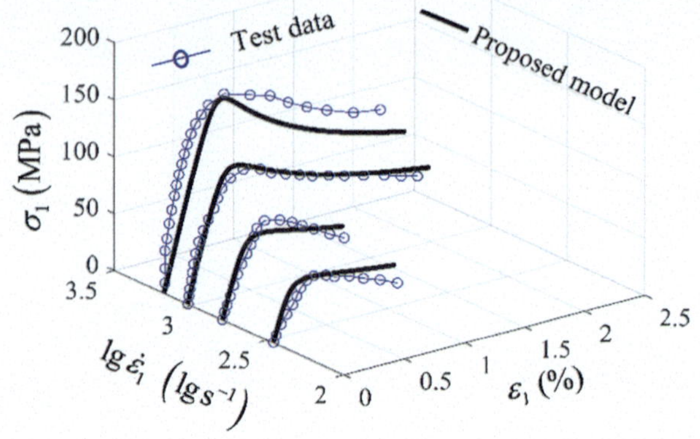

Fig. 6.11 Comparing the model predictions with the test results at high strain rates

6.4 Analysis of Incremental Stress–Strain–Strain Rate Model

6.4.1 *Loading Paths with Independent Variation in Strain Rate*

To analyze the effect of strain rate as a constitutive variable, a loading path with reduced strain rate and constant stress similar to the creep process is discussed as an example for model analysis. Table 6.3 shows the model parameters used in the model analysis, where the dynamic strength parameters' statistical values and other parameters are obtained by referring to the concrete test (Kupfer et al., 1969; Lu et al., 2017). During the creep evolution, the stress remains constant and the total strain increases gradually. Constant stress makes the elastic strain increment zero, so the increase in total strain is due to an increase in plastic strain. This phenomenon is consistent with that described by Eq. (2.11), that is, the plastic strain caused by environmental changes will occur even under constant stress conditions. The plastic strain drives the evolution of H as an internal variable. The relationship between the H and the u_d and axial stress σ_1 under the uniaxial compression condition is depicted in Fig. 6.13. The initial point N_1 is loaded to $H = 0.3$ under the condition of $u_\mathrm{d} = 4$. Similar test conditions for the discussed loading path N_1N_2 have been studied for high-density polyethylene (Reis et al., 2014). According to Eq. (2.11), the material will produce plastic deformation as a result of the change in strain rate. The accumulation of plastic deformation causes the hardening parameter to increase from 0.3 to 0.84. At this time, the yield surface will expand outward correspondingly with the increase of H. Figure 6.14 shows the yield surfaces at the two state points N_1 and N_2, and the four-dimensional hypersurface is still displayed by two three-dimensional surfaces. As seen in Fig. 6.14, the yield surface's expansion makes the new state point N_2 still on the subsequent yield surface, which means that the consistency condition

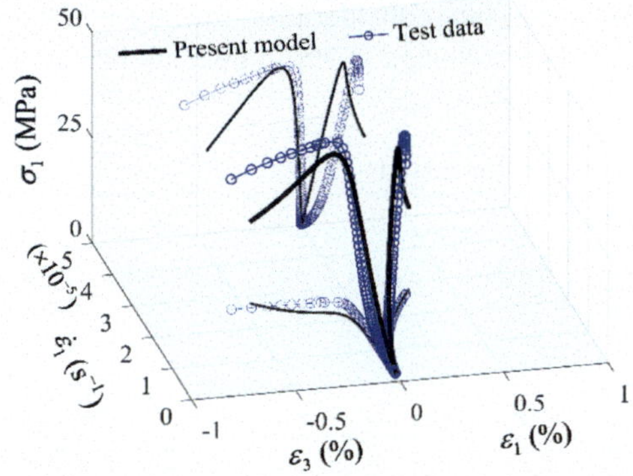

(a) Stress–strain–strain rate curve with strain rate increments of 1.26×10^{-7} s^{-2}

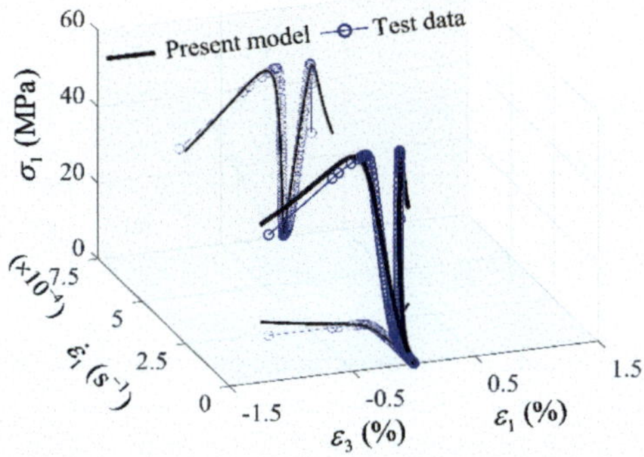

(b) Stress–strain–strain rate curve with strain rate increments of 1.26×10^{-5} s^{-2}

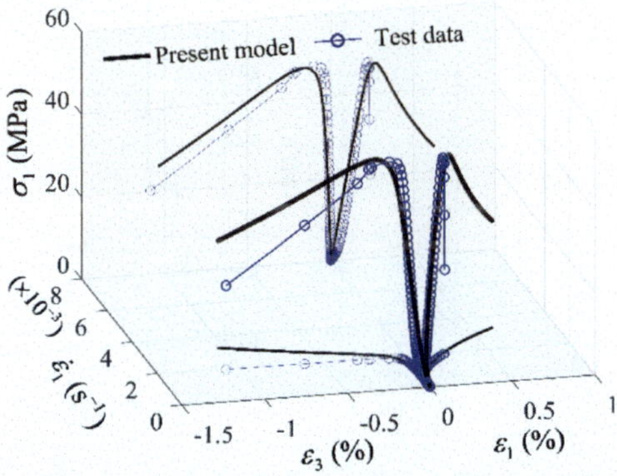

(c) Stress–strain–strain rate curve with strain rate increments of 8.8×10^{-4} s^{-2}

Fig. 6.12 Comparing the model predictions with the test results under different strain rate increment conditions is strictly satisfied. By taking into consideration the strain rate dimension, the yield surface expansion caused by a change in strain rate can be captured.

Table 6.3 The model parameters for analysis

E (MPa)	ν	f_c (MPa)	$\mathrm{DIF}_{c,\max}$	ξ_c	$\mathrm{DIF}_{t,\max}$	ξ_t	$\varepsilon^{p}_{1,ds}$	A_1	D	μ
30,000	0.2	32.8	4.72	1.2	16	1.4	0.003	10	0.8	1.90

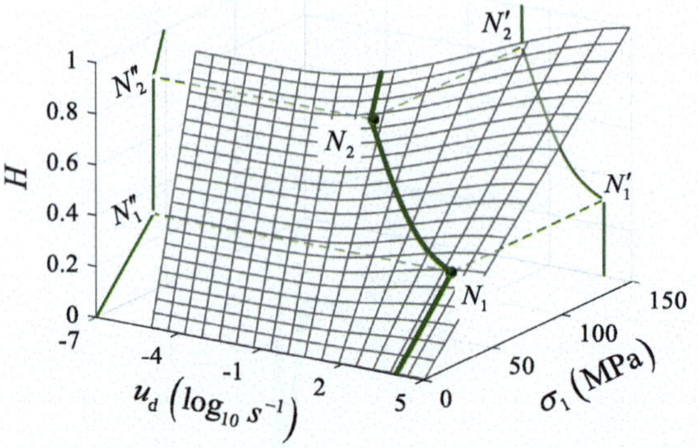

Fig. 6.13 The influence of strain rate and stress on the evolution of hardening parameter

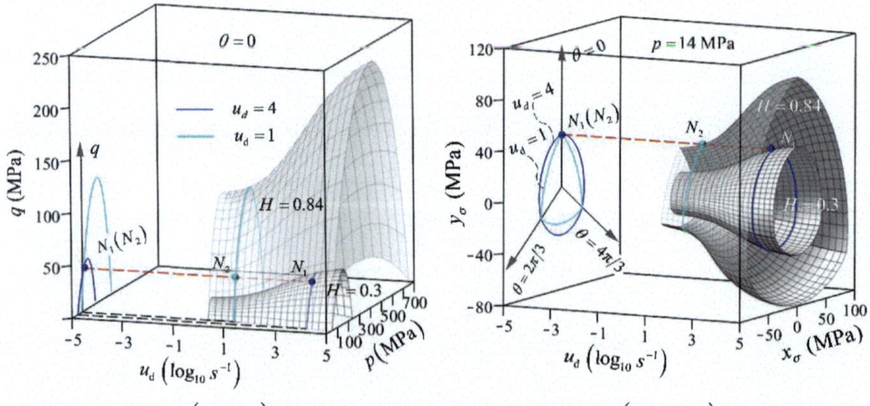

(a) Yield surface in (u_d, p, q) coordinate space (b) Yield surface in $(u_d, x_\sigma, y_\sigma)$ coordinate space

Fig. 6.14 Influence of strain rate change on yield surface evolution

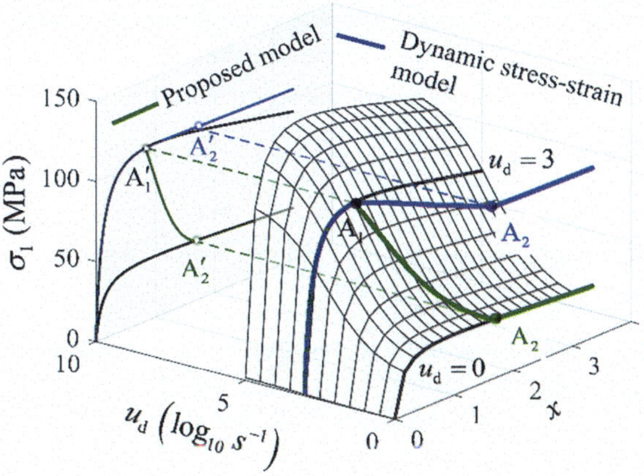

Fig. 6.15 Comparison of models between stress–strain–strain rate and dynamic stress–strain

6.4.2 Designed Variable-Strain Rate Loading Test

The variable strain rate test is conducted only with a slowly increasing strain rate due to limitations in test conditions. To highlight the performance of the proposed model, a designed dynamic loading test is analyzed, in which the strain rate sharply decreases from 1000 s^{-1} to 1 s^{-1} during the A_1A_2 loading path (see Fig. 6.15). The variable strain rate test results of composite materials have shown that a reduction in strain rate will result in a corresponding reduction in stress during the loading process (Khan & Liu, 2012; Reis et al., 2014). A dynamic stress–strain model ignoring strain rate increments is also employed as a comparison in which strain rate is only the influencing factor of material parameters rather than a constitutive variable. f, H, and the flow rule of the dynamic stress–strain model are the same as those used in the proposed models.

Figure 6.15 shows the stress–plastic internal variable–strain rate ($\sigma_1 - x - u_d$) surface. Any point on the surface satisfies the consistency condition. Accordingly, the state points inside the surface are in the elastic range. State points above the surface violate the consistency condition, which means an overestimation of the yield stress of the concrete. The calculation results of the $\sigma_1 - x - u_d$ curve indicate that the proposed model incorporates the reduction in stress caused by variations in strain rate during the A_1A_2 stage. The loading–unloading criterion can be used to judge whether changes in strain rate cause the plastic flow of concrete materials. For the plastic loading state with a reduced strain rate, u_d is considered as a constitutive variable, so the consistency condition is strictly met throughout the entire loading process. When the yield surface shrinks owing to the reduction of u_d, the stress is simultaneously reduced to ensure that the stress point is always on the yield surface, as shown by the green curve in Fig. 6.15. u_d is not considered a constitutive variable but rather a constant in the dynamic stress–strain model. This violates the consistency condition when u_d changes during loading. The stress evolves along the curve above the u_d equal to 1 s^{-1} as shown by the blue curve in Fig. 6.15. This will cause the dynamic stress–strain model to overestimate the concrete strength values, as shown in the curve of the $\sigma_1 - x$ plane in Fig. 6.15. As a result, the proposed model can be used to forecast the mechanical response of concrete under actual dynamic loading with variable strain rates.

6.5 Summary

The classical dynamic stress–strain relationship is developed into a stress–strain–strain rate relationship by introducing the environmental variables into the yield function as constitutive variables. The established model can describe the mechanical response of materials caused by strain rate changes. A loading–unloading criterion was established in the combined coordinate space of the strain and the strain rate. The state information and load increment of the material could be presented geometrically to evaluate their position relationship with the current yield surface. The conditions under which the material undergoes plastic deformation can be reasonably determined.

A dynamic closed yield function is proposed that considers strain rate in strength parameters to determine the elastic domain for various rates. The plastic deformation produced by the independent change in strain rate can be captured since strain rate is treated as a constitutive variable when performing consistency conditions. The model's capability is evaluated under different stress states and strain rate ranges by contrasting model simulations with experimental findings obtained by the author and published papers. In particular, the simulation results for an experiment with a rapidly changing strain rate show that the established model is capable of capturing stress state changes caused by strain rate changes, which is beyond the depth of the dynamic stress–strain model. The good performance of the established model further confirms the feasibility of the proposed modeling method.

References

Aráoz, G., & Luccioni, B. (2015). Modeling concrete like materials under sever dynamic pressures. *International Journal of Impact Engineering, 76*, 139–154.

Bai, Z., Liu, Y. X., Yang, J. T., & He, S. Q. (2020). A constitutive model for concrete subjected to extreme dynamic loadings. *International Journal of Impact Engineering, 138*, 103483.

Borrvall, T., & Riedel, W. (2011). The RHT concrete model in LS-DYNA. In: *Proceedings of the 8th European LS-DYNA users conference* (pp. 5–7). The Alyotech group, Strasbourg, France.

Bridgman, P. W. (1964). Studies in large plastic flow and fracture: with special emphasis on the effects of hydrostatic pressure. Harvard University Press.

Candappa, D. C., Sanjayan, J. G., & Setunge, S. (2001). Complete triaxial stress-strain curves of high-strength concrete. *Journal of Materials in Civil Engineering, 13*(3), 209–215.

Etse, G., & Willam, K. (1994). Fracture energy formulation for inelastic behavior of plain concrete. *Journal of Engineering Mechanics, 120*(9), 1983–2011.

Gebbeken, N., & Krauthammer, T. (2013). Understanding the dynamic response of concrete to loading: Practical examples. In J. Weerheijm (Ed.), *In understanding the tensile properties of concrete* (pp. 338–369). Woodhead Publishing.

Grassl, P., & Jirásek, M. (2006). Damage-plastic model for concrete failure. *International Journal of Solids and Structures, 43*(22–23), 7166–7196.

Grote, D. L., Park, S. W., & Zhou, M. (2001). Dynamic behavior of concrete at high strain rates and pressures: I. Experimental characterization. *International Journal of Impact Engineering, 25*(9), 869–886.

Heeres, O. M., Suiker, A. S. J., & de Borst, R. (2002). A comparison between the Perzyna viscoplastic model and the consistency viscoplastic model. *European Journal of Mechanics - A/Solids, 21*(1), 1–12.

Holmquist, T. J., & Johnson, G. R. (2011). A computational constitutive model for glass subjected to large strains, high strain rates and high pressures. *Journal of Applied Mechanics, 78*(5), 051003.

Hossain, A. B., & Weiss, J. (2004). Assessing residual stress development and stress relaxation in restrained concrete ring specimens. *Cement and Concrete Composites, 26*(5), 531–540.

Imran, I., & Pantazopoulou, S. J. (1996). Experimental study of plain concrete under triaxial stress. *ACI Materials Journal, 93*(6), 589–601.

Kang, H. D., & Willam, K. J. (2000). Performance evaluation of elastoviscoplastic concrete model. *Journal of Engineering Mechanics, 126*(9), 995–1000.

Khan, A. S., & Liu, H. (2012). Variable strain rate sensitivity in an aluminum alloy: Response and constitutive modeling. *International Journal of Plasticity, 36*, 1–14.

Kong, X. Z., Fang, Q., Li, Q. M., Wu, H., & Crawford, J. E. (2017). Modified K&C model for cratering and scabbing of concrete slabs under projectile impact. *International Journal of Impact Engineering, 108*, 217–228.

Kong, X. Z., Fang, Q., Chen, L., & Wu, H. (2018). A new material model for concrete subjected to intense dynamic loadings. *International Journal of Impact Engineering, 120*, 60–78.

Kupfer, H., Hilsdorf, H. K., & Rusch, H. (1969). Behavior of concrete under biaxial stresses. *Journal Proceedings, 66*(8), 656– 666.

Lu, X. B., & Hsu, C. T. (2007). Stress-strain relations of high-strength concrete under triaxial compression. *Journal of Materials in Civil Engineering, 19*(3), 261–268.

Lu, D. C., Wang, G. S., Du, X. L., & Wang, Y. (2017). A nonlinear dynamic uniaxial strength criterion that considers the ultimate dynamic strength of concrete. *International Journal of Impact Engineering, 103*, 124–137.

Lu, D. C., Zhou, X., Du, X. L., & Wang, G. S. (2019). A 3D fractional elastoplastic constitutive model for concrete material. *International Journal of Solids and Structures, 165*, 160–175.

Lu, D. C., Zhou, X., Du, X. L., & Wang, G. S. (2020). 3D dynamic elastoplastic constitutive model of concrete within the framework of rate-dependent consistency condition. *Journal of Engineering Mechanics, 146*(11), 04020124.

Lu, D. C., Meng, F. P., Zhou, X., Wang, G. S., & Li, D. X. (2022). Double scalar variables plastic-damage model for concrete. *Journal of Engineering Mechanics, 148*(2), 04021143.

Naghdi, P. M., & Murch, S. A. (1963). On the mechanical behavior of viscoelastic/plastic solids. *Journal of Applied Mechanics, 30*(3), 321–328.

Qi, C. Z., Wang, M. Y., & Qian, Q. H. (2009). Strain-rate effects on the strength and fragmentation size of rocks. *International Journal of Impact Engineering, 36*(12), 1355–1364.

Qiao, Y. F., Ferrari, A., Laloui, L., & Ding, W. Q. (2016). Nonstationary flow surface theory for modeling the viscoplastic behaviors of soils. *Computers and Geotechnics, 76*, 105–119.

Rahnavard, R., Craveiro, H. D., Simões, R. A., Laím, L., & Santiago, A. (2022). Fire resistance of concrete-filled cold-formed steel (CF-CFS) built-up short columns. *Journal of Building Engineering, 48*, 103854.

Reis, J. M. L., Pacheco, L. J., & Da Costa Mattos, H. S. (2014). Temperature and variable strain rate sensitivity in recycled HDPE. *Polymer Testing, 39*, 30–35.

Shkolnik, I. E. (2008). Influence of high strain rates on stress–strain relationship, strength and elastic modulus of concrete. *Cement and Concrete Composites, 30*(10), 1000–1012.

Torelli, G., Mandal, P., Gillie, M., & Tran, V. (2018). A confinement-dependent load-induced thermal strain constitutive model for concrete subjected to temperatures up to 500 °C. *International Journal of Mechanical Sciences, 144*, 887–896.

Wang, G. S., Lu, D. C., Du, X. L., & Zhou, X. (2018a). Dynamic multiaxial strength criterion for concrete based on strain rate–dependent strength parameters. *Journal of Engineering Mechanics, 144*(5), 04018018.

Wang, G. S., Lu, D. C., Du, X. L., Zhou, X., & Cao, S. T. (2018b). A true 3D frictional hardening elastoplastic constitutive model of concrete based on a unified hardening/softening function. *Journal of the Mechanics and Physics of Solids, 119*, 250–273.

Yan, D., & Lin, G. (2007). Dynamic behaviour of concrete in biaxial compression. *Magazine of Concrete Research, 59*(1), 45–52.

Yu, S. S., Lu, Y. B., & Cai, Y. (2013). The strain-rate effect of engineering materials and its unified model. *Latin American Journal of Solids and Structures, 10*(4), 833–844.

Zhang, Q. B., & Zhao, J. (2014). A review of dynamic experimental techniques and mechanical behaviour of rock materials. *Rock Mechanics and Rock Engineering, 47*(4), 1411–1478.

Zhou, X., Lu, D. C., Du, X. L., Wang, G. S., & Meng, F. P. (2020). A 3D non-orthogonal plastic damage model for concrete. *Computer Methods in Applied Mechanics and Engineering, 360*, 112716.

Chapter 7
Unconstrained Stress Update Algorithm for Elastoplastic Models

The stress update problem of elastoplastic models is an initial value problem of the ODEs constrained by inequalities. The ODEs need to be discretized into nonlinear algebraic equations over the time domain to solve based on the explicit (Gao et al., 2022; Zhao & Cai, 2010) or implicit integral schemes (Borja & Lee, 1990; De Borst & Heeres, 2002). Many researchers prefer an implicit integral strategy in the numerical implementation of elastoplastic models due to its unconditional stability. There are two computational difficulties for solving implicit integral equations for the elastoplastic model, i.e., the non-smoothness from inequality constraints and the convergence of the solution of nonlinear equations. The most popular implicit stress updating algorithm may be the return-mapping algorithm (Simo & Taylor, 1986) where the inequality constraints are addressed by the loading/unloading estimation and the nonlinear equations are solved by the Newton method. This computational paradigm is also followed by the cutting plane algorithm (Starman et al., 2014), the semi-implicit algorithm (Moran et al., 1990), and some other advanced stress update algorithms (Petalas & Dafalias, 2019). However, it is found that the iterations may not converge for the Newton method when the initial value is far from the final solution or the problem is highly nonlinear due to complex model formulations. The loading/unloading estimation of the return mapping stress strategy also makes the stress update procedure more cumbersome. A more concise and robust computational paradigm is eagerly desired for thriving elastoplastic models of geomaterials.

This chapter mainly focuses on addressing the non-smoothness in elastoplastic computations. More robust numerical algorithms for solving nonlinear equations will be studied in Chap. 8. First, this chapter presents the standard implicit stress integration format for elastoplastic models. Then, the classic return mapping stress update strategy is reviewed, which comprises elastic prediction and plastic correction. In Sect. 7.3, we introduce a new technique for dealing with non-smoothness in elastoplastic problems, namely the unconstrained stress update strategy, where the loading/unloading inequality is equivalently replaced by a smooth function. Subsequently, a comparative analysis of two stress update strategies is conducted through

D. Lu et al., *Multi-physical and Non-orthogonal Geotechnical Plasticity Modeling and Algorithms*, https://doi.org/10.1007/978-981-96-2369-3_7

the examination of a 1D elastoplastic problem. In the remaining content, the NOEP model in Sect. 3.2 was implemented in ABAQUS based on the unconstrained stress update algorithm, and the performance of the algorithm was evaluated based on several boundary value problems.

7.1 Implicit Stress Integral Scheme for Elastoplastic Models

For a classical rate-independent elastoplastic constitutive model, the mathematical equations describing the stress–strain relationship are generally defined by a set of ODEs with constraints as follows:

$$\begin{cases} \mathrm{d}\boldsymbol{\sigma} = \mathbb{D} : \mathrm{d}\boldsymbol{\varepsilon}^{\mathrm{e}} = \mathbb{D} : (\mathrm{d}\boldsymbol{\varepsilon} - \mathrm{d}\boldsymbol{\varepsilon}^{\mathrm{p}}) \\ \mathrm{d}\boldsymbol{\varepsilon}^{\mathrm{p}} = \mathrm{d}\phi\mathbf{r} \\ \mathrm{d}H = \mathrm{d}\phi h \\ \mathrm{d}\phi \geq 0, f \leq 0, \mathrm{d}\phi f = 0 \end{cases} \tag{7.1}$$

where the four parts of Eq. (7.1) are known as Hooke's Law, flow rule, hardening law, and *KKT* conditions, respectively. $\mathbb{D}$ is the elastic stiffness and $\mathbb{D} = \mathbb{I}^{\mathrm{vol}}(3K - 2G) + 2\mathbb{I}^{\mathrm{sym}}G$, where $\mathbb{I}^{\mathrm{vol}}$ and $\mathbb{I}^{\mathrm{sym}}$ are the volumetric and symmetric fourth-order unit tensors, respectively. K and G denote the elastic bulk modulus and shear modulus, respectively. $\mathrm{d}\boldsymbol{\sigma}$, $\mathrm{d}\boldsymbol{\varepsilon}$, $\mathrm{d}\boldsymbol{\varepsilon}^{\mathrm{e}}$, $\mathrm{d}\boldsymbol{\varepsilon}^{\mathrm{p}}$, and $\mathrm{d}H$ are the differential or infinitesimal increment of the stress tensor, total strain tensor, elastic strain tensor, plastic strain tensor, and plastic internal variable, respectively. $\mathbf{r}$ and h denote the plastic flow direction and the gradient of the plastic internal variable, respectively. f and $\mathrm{d}\phi$ are the yield function and plastic multiplier, respectively. Note that the *KKT* conditions constrain the allowable state variables, namely, $\mathrm{d}\phi \geq 0$ and $f(\boldsymbol{\sigma}, H) = 0$ for loading, and $\mathrm{d}\phi = 0$ and $f \leq 0$ for unloading.

The constitutive equations in Eq. (7.1) are defined in differential form. In the numerical implementation, it needs to be discretized into algebraic equations in time by a specific integral strategy. Based on the backward Euler integration formula, Eq. (7.1) can be written as follows:

$$\begin{cases} \boldsymbol{\sigma}_{n+1} = \boldsymbol{\sigma}_n + \mathbb{D} : \left(\Delta\boldsymbol{\varepsilon}_{\mathrm{n+1}} - \Delta\boldsymbol{\varepsilon}^{\mathrm{p}}_{\mathrm{n+1}}\right) \\ \Delta\boldsymbol{\varepsilon}^{\mathrm{p}}_{n+1} = \Delta\phi_{n+1}\mathbf{r}_{n+1} \\ \Delta H_{n+1} = \Delta\phi_{n+1}h_{n+1} \\ \Delta\phi_{n+1} \geq 0, f_{n+1} \leq 0, \Delta\phi_{n+1}f_{n+1} = 0 \end{cases} \tag{7.2}$$

where Δ represents the increment with the finite size. The aim of the stress updating algorithm is to solve Eq. (7.2) based on a given set of state variables $(\boldsymbol{\sigma}_n, H_n)$ at step n and the strain increment $\Delta\boldsymbol{\varepsilon}_{n+1}$ to obtain the state variables $(\boldsymbol{\sigma}_{n+1}, H_{n+1})$ at step

$n + 1$. Note that the difficulty of solving Eq. (7.2) lies mainly in the non-smoothness caused by inequality constraints and the nonlinearity of f, $\mathbf{r}$, h, and $\mathbb{D}$.

7.2 Return Mapping Stress Update Strategy

For the inequality constraints in Eq. (7.2), the return mapping stress update strategy appears to provide a good treatment. In this stress updating strategy, the inequality constraints are first activated by the elastic prediction, where the trial stress is computed by $\boldsymbol{\sigma}_{n+1}^{trial} = \boldsymbol{\sigma}_n + \mathbb{D}:\Delta\boldsymbol{\varepsilon}_{n+1}$. There are three loading/unloading cases for elastic prediction, as shown in Fig. 7.1.

Case ①: $f(\boldsymbol{\sigma}_n, H_n) \le 0$ and $f\left(\boldsymbol{\sigma}_{n+1}^{\mathrm{trial}}, H_n\right) < 0$, which represents that the stress path at the current step is inside of the initial yield surface. $\Delta\boldsymbol{\varepsilon}$ is pure elastic. $\boldsymbol{\sigma}_{n+1} = \boldsymbol{\sigma}_{n+1}^{\mathrm{trial}}$ and other variables related to plasticity remain unchanged.

Case ②: $f(\boldsymbol{\sigma}_n, H_n) < 0$ and $f\left(\boldsymbol{\sigma}_{n+1}^{\mathrm{trial}}, H_n\right) > 0$, which represents that the stress path at the current step passes through the initial yield surface. $\Delta\boldsymbol{\varepsilon}$ is consisted of the elastic part and the elastoplastic part.

Case ③: $f(\boldsymbol{\sigma}_n, H_n) = 0, f\left(\boldsymbol{\sigma}_{n+1}^{\mathrm{trial}}, H_n\right) > 0$, which represents that the stress path at the current step lies outside the initial yield surface. The current step is a plastic loading process.

In the implicit return mapping stress update algorithm, the trial stress in Case ② and Case ③ needs to be pulled back to the yield surface by the plastic correction.

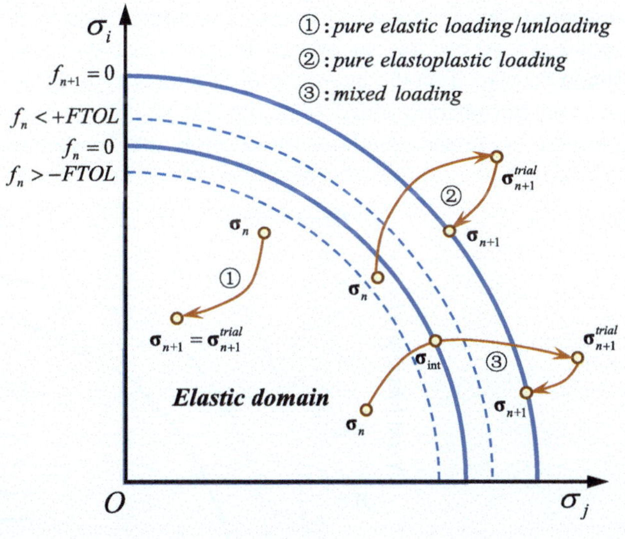

Fig. 7.1 Return mapping stress update strategy

7.3 Unconstrained Stress Update Strategy

The non-smoothness of the elastoplastic problem is inherited from the *KKT* conditions defined by Eq. (7.2). In numerical optimization, the non-smooth *KKT* conditions can be replaced equivalently by the smoothing function, and then Eq. (7.2) is transformed into the smooth version shown below:

$$\begin{cases} \boldsymbol{\sigma}_{n+1} = \boldsymbol{\sigma}_n + \mathbb{D} : \left(\Delta\boldsymbol{\varepsilon}_{\mathrm{n+1}} - \Delta\boldsymbol{\varepsilon}^{\mathrm{p}}_{\mathrm{n+1}}\right) \\ \Delta\boldsymbol{\varepsilon}^{\mathrm{p}}_{n+1} = \Delta\phi_{n+1}\mathbf{r}_{n+1} \\ \Delta H_{n+1} = \Delta\phi_{n+1}h_{n+1} \\ \sqrt{(c_{\mathrm{d}}\Delta\phi_{n+1})^2 + f_{n+1}^2 + 2\beta} - c_{\mathrm{d}}\Delta\phi_{n+1} + f_{n+1} = 0 \end{cases} \tag{7.3}$$

where the *Fischer–Burmeister* (*FB*) function, i.e., Eq. $(7.3)_4$, is employed (Fischer, 1992; Kanzow, 1996). c_{d} is a dimensional parameter. β is a small algorithm parameter. Fig. 7.2 shows the effect of β on the *FB* smoothing curve. The smoothing function will converge to *KKT* conditions when β trends to 0.

By using the *FB* smooth function instead of *KKT* conditions, Eq. (7.3) can give rise to an unconstrained stress updating strategy without the need for loading/unloading estimations, as shown in Fig. 7.3. The stress states on both the elastic domain and yield surfaces, as illustrated in Fig. 7.1, are projected onto a smooth curve. As a result, the stress–strain behavior under the pure elastic loading/unloading condition, elastoplastic loading, and mixed loading can be described uniformly by a set of smooth equations. Under this computational paradigm, one of the difficulties in solving the elastoplastic problems, i.e., the non-smoothness, can be bypassed. The only change required is to use the smoothing function instead of *KKT* conditions. More focus should be placed on the treatment of nonlinearity in Eq. (7.3). Furthermore, by combining the unconstrained stress update strategy with the numerical algorithm for solving nonlinear equations, for example, the Newton method, LSM,

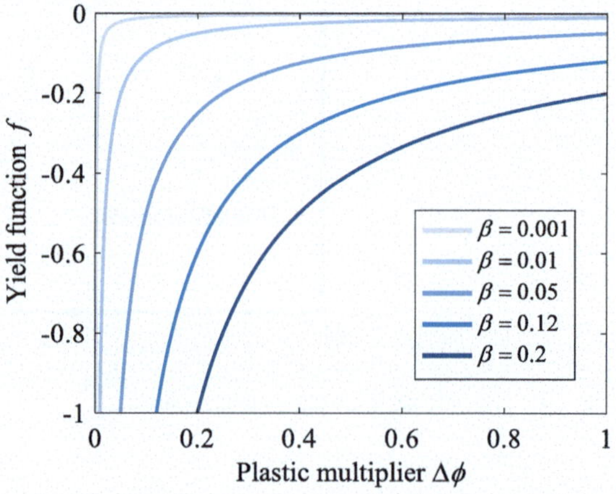

Fig. 7.2 Smoothing curves with the different values of β

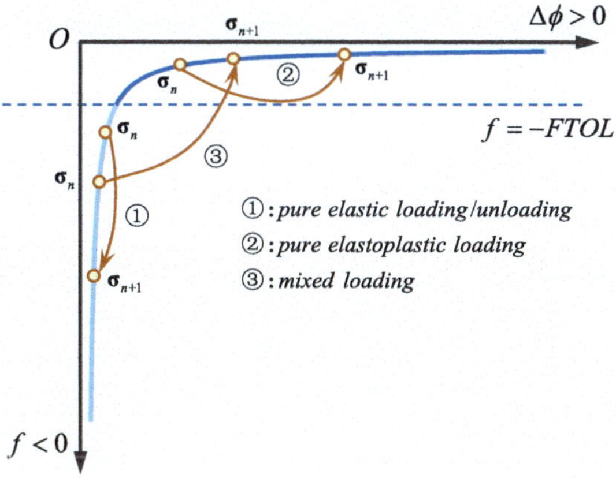

Fig. 7.3 Unconstrained stress updating strategy

and Trust region method, we can obtain corresponding unconstrained stress update algorithms.

7.4 Application to 1D Compression Problem

In what follows, a 1D spring-slider physical model shown in Fig. 7.4 is used as an example to further illustrate the computational framework of the two stress update strategies. Corresponding to the Eq. (7.2), the integral equations of the 1D physical model is given:

$$\begin{cases} \sigma_{n+1} = \sigma_n + E\left(\Delta\varepsilon_{n+1} - \Delta\varepsilon^{\mathrm{p}}_{n+1}\right) \\ \Delta\varepsilon^{\mathrm{p}}_{n+1} = \Delta\phi_{n+1}\mathrm{sign}(\sigma_{n+1}) \\ \Delta H_{n+1} = \left|\Delta\varepsilon^{\mathrm{p}}_{n+1}\right| \\ \Delta\phi_{n+1} \geq 0,\ f(\sigma_{n+1}, H_{n+1}) \leq 0, \text{ and } \Delta\phi_{n+1} f(\sigma_{n+1}, H_{n+1}) = 0 \end{cases} \tag{7.4}$$

where $f(\sigma, \alpha) = |\sigma| - \sigma_Y - K^{\mathrm{p}}H$ and the simple linear hardening law is assumed. E and K^{p} are the spring stiffness and the hardening modulus of the slider. σ_Y is the initial sliding stress. sign(·) is the sign function and Eq. $(7.4)_2$ denotes that the direction of $\Delta\varepsilon^{\mathrm{p}}_{n+1}$ is the same as that of stress in the 1D condition. When only the

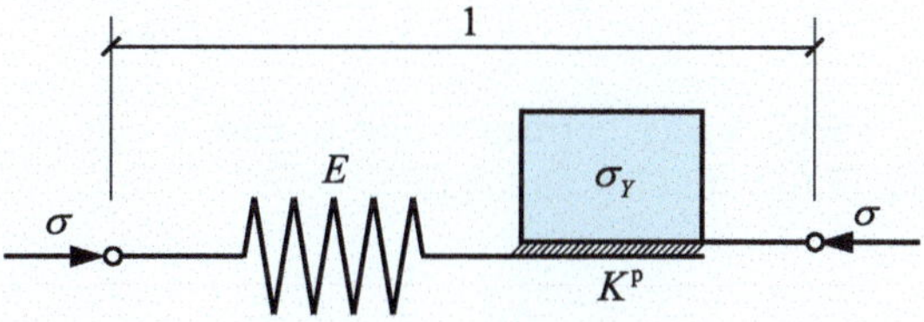

Fig. 7.4 1D spring-slider physical model

compression situation is considered, the sign and absolute value functions can be omitted.

7.4.1 Return Mapping Stress Update Strategy Solution

The key to solving the elastoplastic problem defined by Eq. (7.4) is to determine $\Delta\phi_{n+1}$ under the action of an external load. In the return mapping stress update strategy, three loading/unloading cases may need to be estimated before calculating $\Delta\phi_{n+1}$ due to the presence of *KKT* conditions, as demonstrated in Fig. 7.5. First, the trial stress $\sigma_{n+1}^{trial} = \sigma_n + E\Delta\varepsilon_{n+1}$ is determined based on the elastic Hooke's Law. Then, the loading and unloading cases under the current increment step can be estimated by the values of $f(\sigma_n, H_n)$ and $f\left(\sigma_{n+1}^{trial}, H_n\right)$ as follows:

$$\begin{cases} \text{case } a\text{: pure elastic loading/unloading} & \text{if } f(\sigma_n, H_n) \le 0 \text{ and } f\left(\sigma_{n+1}^{trial}, H_n\right) \le 0 \\ \text{case } b\text{: pure elastoplastic loading} & \text{if } f(\sigma_n, H_n) = 0 \text{ and } f\left(\sigma_{n+1}^{trial}, H_n\right) > 0 \\ \text{case } c\text{: mixed loading} & \text{if } f(\sigma_n, H_n) < 0 \text{ and } f\left(\sigma_{n+1}^{trial}, H_n\right) > 0 \end{cases} \tag{7.5}$$

For case *a*, the current increment step is purely elastic. $\Delta\phi_{n+1} = 0$ and the σ_{n+1}^{trial} is accepted as the σ_{n+1}. For case *b*, the plastic deformation mechanism is triggered under the current increment step. The plastic correction needs to be used to pull the trial stress back toward the real stress σ_{n+1}, as shown in Fig. 7.6. $\Delta\phi_{n+1}$ can be obtained by substituting Eq. $(7.4)_{1\sim3}$ and $\sigma_n = \sigma_Y + K^p H_n$ into the equation $f(\sigma_{n+1}, \alpha_{n+1}) = 0$ as follows:

$$\Delta\phi_{n+1} = \frac{E\Delta\varepsilon_{n+1}}{E + K^p} \tag{7.6}$$

Case *c* indicates that the stress point undergoes a mechanical behaviour transition from elasticity to plasticity at the current increment step. The partial strain increment $\xi\,\Delta\varepsilon_{n+1}$ corresponding to the pure elastic loading in Fig. 7.5c needs to be determined in advance. The scale factor $\xi\,(0 < \xi < 1)$ is determined by solving the equation $f(\sigma_{\text{int}}, H_n) = 0$ and considering the condition $\sigma_n < \sigma_Y + K^p H_n$, where σ_{int} is the largest historical yield stress:

$$\xi = \frac{\sigma_Y + K^p H_n - \sigma_n}{E\Delta\varepsilon_{n+1}} \tag{7.7}$$

Then, $\Delta\phi_{n+1}$ corresponding to case c can be obtained by replacing $\Delta\varepsilon_{n+1}$ in Eq. (7.6) with $(1 - \xi)\Delta\varepsilon_{n+1}$:

$$\Delta\phi_{n+1} = \frac{E\Delta\varepsilon_{n+1} - (\sigma_Y + H_n K^p - \sigma_n)}{E + K^p} \tag{7.8}$$

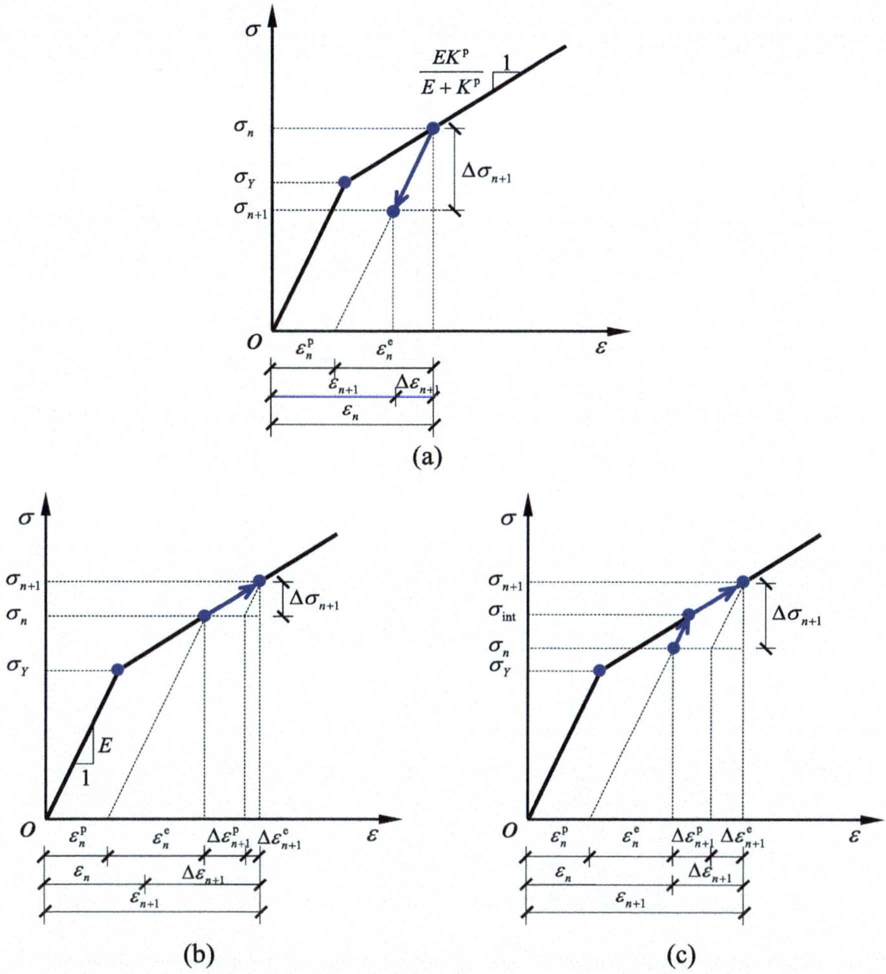

Fig. 7.5 Different loading/unloading cases: **a** Pure elastic loading/unloading; **b** Pure elastoplastic loading; **c** Mixed elastic and elastoplastic loading

Finally, the updating of state variables for the different loading and unloading cases can be obtained by substituting $\Delta\phi_{n+1} = 0$, Eqs. (7.6) and (7.8) into Eq. (7.4) as follows:

For case *a*:

$$\begin{cases} \sigma_{n+1} = \sigma_n + E\Delta\varepsilon_{n+1} \\ H_{n+1} = H_n \\ \Delta\phi_{n+1} = 0 \end{cases} \tag{7.9}$$

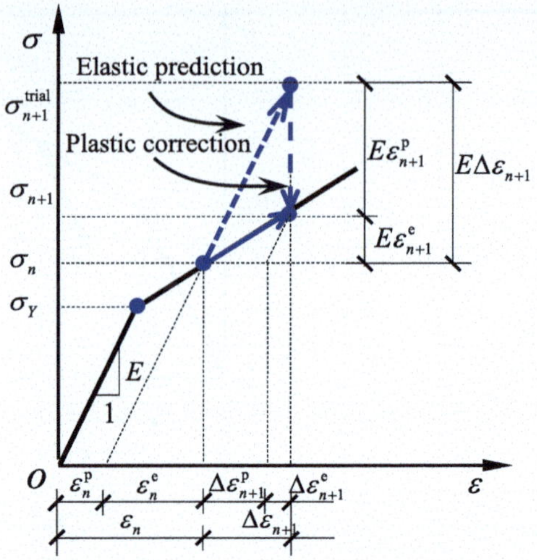

Fig. 7.6 Stress update procedure for the purely elastoplastic loading case

For case *b*:

$$\begin{cases} \sigma_{n+1} = \sigma_n + \frac{EK^{\text{p}}\Delta\varepsilon_{n+1}}{E+K^{\text{p}}} \\ H_{n+1} = H_n + \frac{E\Delta\varepsilon_{n+1}}{E+K^{\text{p}}} \\ \Delta\phi_{n+1} = \frac{E\Delta\varepsilon_{n+1}}{E+K^{\text{p}}} \end{cases} \tag{7.10}$$

For case *c*:

$$\begin{cases} \sigma_{n+1} = \sigma_n + \frac{EK^{\text{p}}\Delta\varepsilon_{n+1}}{E+K^{\text{p}}} + \frac{E(\sigma_{\text{Y}}+H_nK^{\text{p}}-\sigma_n)}{E+K^{\text{p}}} \\ H_{n+1} = H_n + \frac{E\Delta\varepsilon_{n+1}-(\sigma_{\text{Y}}+H_nK^{\text{p}}-\sigma_n)}{E+K^{\text{p}}} \\ \Delta\phi_{n+1} = \frac{E\Delta\varepsilon_{n+1}-(\sigma_{\text{Y}}+H_nK^{\text{p}}-\sigma_n)}{E+K^{\text{p}}} \end{cases} \tag{7.11}$$

In the return mapping stress update strategy, the loading/unloading judgment is achieved by the elastic predictor. Although this method provides an excellent distinction between the different loading/unloading cases, it leads to non-smoothness in the calculation. Additional attention needs to be paid to case *c* (if the explicit integration scheme is adopted). It can be found that the root of the non-smoothness in the *KTT* conditions arises from the inequality constraints. Replacing the inequality constraints with a smoothing function to eliminate the non-smoothness in the calculation becomes the biggest research motivation of this Chapter.

7.4.2 Unconstrained Stress Update Strategy Solution

Replacing Eq. (7.4)$_4$ with the *FB* smooth function, the elastoplastic problem constrained by the loading/unloading inequalities can be transformed into the following unconstrained form:

$$\begin{cases} \sigma_{n+1} = \sigma_n + E\left(\Delta\varepsilon_{n+1} - \Delta\varepsilon^{\mathrm{p}}_{n+1}\right) \\ \Delta\varepsilon^{\mathrm{p}}_{n+1} = \Delta\phi_{n+1} \\ \Delta H_{n+1} = \Delta\varepsilon^{\mathrm{p}}_{n+1} \\ \sqrt{(c_{\mathrm{d}}\Delta\phi)^2 + f^2(\sigma_{n+1}, H_{n+1}) + 2\beta} - c_{\mathrm{d}}\Delta\phi_{n+1} + f(\sigma_{n+1}, H_{n+1}) = 0 \end{cases} \tag{7.12}$$

Now, the elastoplastic problem depicted by Fig. 7.4 can be solved directly by Eq. (7.12) without dealing with inequality constraints. For Eq. (7.12)$_4$, we have:

$$\beta = -c_{\mathrm{d}}\Delta\phi_{n+1}f(\sigma_{n+1}, H_{n+1}) \tag{7.13}$$

Substituting $f(\sigma_{n+1}, H_{n+1}) = \sigma_n + E(\Delta\varepsilon_{n+1} - \Delta\phi_{n+1}) - \sigma_Y - K^{\mathrm{p}}(H_n + \Delta\phi_{n+1})$ into Eq. (7.13), we can obtain:

$$c_{\mathrm{d}}\Delta\phi^2_{n+1}(E + K^{\mathrm{p}}) - c_{\mathrm{d}}\Delta\phi_{n+1}(\sigma_n + E\Delta\varepsilon_{n+1} - \sigma_Y - H_nK^{\mathrm{p}}) - \beta = 0 \tag{7.14}$$

Utilizing the root of the quadratic equation, the nonnegative solutions of $\Delta\phi_{n+1}$ can be obtained as follows:

$$\Delta\phi_{n+1} = \frac{c_{\mathrm{d}}\left(\sigma_n + E\Delta\varepsilon_{n+1} - \sigma_{\mathrm{Y}} - H_nK^{\mathrm{p}}\right) + \sqrt{c_{\mathrm{d}}^2\left(\sigma_n + E\Delta\varepsilon_{n+1} - \sigma_{\mathrm{Y}} - H_nK^{\mathrm{p}}\right)^2 + 4c_{\mathrm{d}}(E + K^{\mathrm{p}})\beta}}{2c_{\mathrm{d}}(E + K^{\mathrm{p}})} \tag{7.15}$$

When β is close to 0, the following solutions for $\Delta\phi_{n+1}$ can be automatically obtained for a given $\Delta\varepsilon_{n+1}$ and σ_n:

$$\lim_{\beta\to 0}\Delta\phi_{n+1} = \begin{cases} 0, & \sigma_n \le \sigma_Y + H_nK^{\mathrm{p}} \text{ and } \sigma_n + E\Delta\varepsilon_{n+1} \le \sigma_Y + H_nK^{\mathrm{p}} \\ \frac{E\Delta\varepsilon_{n+1}}{E+K^{\mathrm{p}}}, & \sigma_n = \sigma_Y + H_nK^{\mathrm{p}} \text{ and } \sigma_n + E\Delta\varepsilon_{n+1} > \sigma_Y + H_nK^{\mathrm{p}} \\ \frac{E\Delta\varepsilon_{n+1} - (\sigma_Y + H_nK^{\mathrm{p}} - \sigma_n)}{E+K^{\mathrm{p}}}, & \sigma_n < \sigma_Y + H_nK^{\mathrm{p}} \text{ and } \sigma_n + E\Delta\varepsilon_{n+1} > \sigma_Y + H_nK^{\mathrm{p}} \end{cases} \tag{7.16}$$

where there is no need to perform the loading/unloading judgment in advance or to calculate the intersection point σ_{int} in Fig. 7.5c. Substituting Eq. (7.16) into Eq. (7.12), the updating formulas of the same state variable as Eqs. (7.9), (7.10), and (7.11) can be obtained correctly for the given $\Delta\varepsilon_{n+1}$. The non-smoothness of the elastoplastic problem is bypassed by the *FB* smoothing function.

It is noted that the constitutive equations for the elastoplastic problems in this chapter are solved manually due to the simplicity of the 1D physical model. However, the mechanical behaviors of most materials, especially for geomaterials, are often complex and exhibit highly nonlinear characteristics. The nonlinear stress integrations need to be solved by iteration.

7.5 Application to Non-orthogonal Elastoplastic Model for Clay

In this section, the unconstrained stress update algorithm will be used to solve the NOEP model in Chap. 3. We first introduce the implicit stress integration scheme of the model, in which the *KKT* conditions are equivalently replaced by the *FB* smooth function. Then, the smooth nonlinear stress integration equations are solved through the standard Newton method. In addition, this section includes the derivation of the consistent tangent stiffness matrix corresponding to the algorithm, which plays a crucial role in ensuring the quadratic convergence speed of the global equilibrium iteration.

7.5.1 Implicit Stress Update Implementation

In the model implementation, the ODEs need to be discretized into the algebraic equations for the time interval $\left[t_n,\ t_{n+1}\right]$. Based on the Backward Euler method, the governing equations of the NOEP clay model are given by:

$$\begin{cases} \boldsymbol{\sigma}_{n+1} = \boldsymbol{\sigma}_n + \overline{\mathbb{D}}:\Delta\boldsymbol{\varepsilon}_{n+1} - \Delta\phi_{n+1}\overline{\mathbb{D}}:\left(\dfrac{\partial^{\mu}\mathrm{f}_{n+1}}{\partial \mathrm{p}^{\mu}_{n+1}}\dfrac{\partial \mathrm{p}_{n+1}}{\partial \boldsymbol{\sigma}_{n+1}} + \dfrac{\partial^{\mu}\mathrm{f}_{n+1}}{\partial \mathrm{q}^{\mu}_{n+1}}\dfrac{\partial \mathrm{q}_{n+1}}{\partial \boldsymbol{\sigma}_{n+1}}\right) \\ p_{\mathrm{c},\,n+1} = p_{\mathrm{c},\,n}\exp\left(c_{\mathrm{p}}\Delta\varepsilon^{\mathrm{p}}_{\mathrm{v},\,n+1}\right) \\ \Delta\phi_{n+1} \geq 0,\ f_{n+1} \leq 0, \Delta\phi_{n+1}f_{n+1} = 0 \end{cases} \tag{7.17}$$

where $c_{\mathrm{p}} = (1 + e_0)/(\lambda - \kappa)$. $\overline{\mathbb{D}} = \mathbb{D}\left(\overline{K}, \overline{G}\right)$ is the secant elastic stiffness tensor. The secant bulk modulus $\overline{K}$ is:

$$\overline{K} = \frac{p_n}{\Delta\varepsilon^{\mathrm{e}}_{\mathrm{v},\,n+1}}\left[\exp\left(c_{\kappa}\Delta\varepsilon^{\mathrm{e}}_{\mathrm{v},\,n+1}\right) - 1\right] \tag{7.18}$$

where $\Delta\varepsilon^{\mathrm{e}}_{\mathrm{v},\,n+1}$ represents elastic volume strain increment and $\Delta\varepsilon^{\mathrm{e}}_{\mathrm{v},\,n+1} = \Delta\varepsilon_{\mathrm{v},\,n+1} - \Delta\varepsilon^{\mathrm{p}}_{\mathrm{v},\,n+1}$. Equation (7.17) contains 8 equality equations and 2 inequality constraints. The number of stress integral equations can be simplified to reduce the difficulty of the solution. The stress tensor $\boldsymbol{\sigma}_{n+1}$ can be decomposed into its isotropic part and deviatoric part:

$$\boldsymbol{\sigma}_{n+1} = p_{n+1}\mathbf{1} + \mathbf{s}_{n+1} \tag{7.19}$$

where $\mathbf{s}_{n+1}$ is the deviatoric stress tensor. p_{n+1} and $\mathbf{s}_{n+1}$ can be expressed as follows:

$$p_{n+1} = \frac{1}{3}\boldsymbol{\sigma}_{n+1}:\mathbf{1} = p_n + \overline{K}\left(\Delta\varepsilon_{\mathrm{v},\,n+1} - \Delta\varepsilon^{\mathrm{p}}_{\mathrm{v},\,n+1}\right) \tag{7.20}$$

$$\mathbf{s}_{n+1} = \mathbf{s}_n + 2\overline{G}(\Delta\boldsymbol{\gamma}_{n+1} - \Delta\boldsymbol{\gamma}^{\mathrm{p}}_{n+1}) \tag{7.21}$$

where $\overline{G} = r\overline{K}$ represents the secant shear modulus, $\Delta\boldsymbol{\gamma}_{n+1}$ and $\Delta\boldsymbol{\gamma}^{\mathrm{p}}_{n+1}$ denote the total deviatoric strain increment and its plastic part. $\Delta\varepsilon^{\mathrm{p}}_{\mathrm{v},\,n+1}$ and $\Delta\boldsymbol{\gamma}^{\mathrm{p}}_{n+1}$ are expressed by:

$$\begin{aligned}\Delta\varepsilon^{\mathrm{p}}_{\mathrm{v},\,n+1} &= \Delta\boldsymbol{\varepsilon}^{\mathrm{p}}_{n+1}:\mathbf{1}\\ &= \Delta\phi_{n+1}\left[\frac{q^2_{n+1}}{p^{\mu}_{n+1}N^2\Gamma(1-\mu)} + \frac{2p^{2-\mu}_{n+1}}{\Gamma(3-\mu)} - \frac{p_{\mathrm{c},\,n+1}p^{1-\mu}_{n+1}}{\Gamma(2-\mu)}\right]\end{aligned} \tag{7.22}$$

$$\begin{aligned}\Delta\boldsymbol{\gamma}^{\mathrm{p}}_{n+1} &= \mathbb{P}:\Delta\boldsymbol{\varepsilon}^{\mathrm{p}}_{\mathrm{n+1}}\\ &= 3\Delta\phi_{n+1}\mathbf{s}_{n+1}\left[\frac{q^{1-\mu}_{n+1}}{N^2\Gamma(3-\mu)} + \frac{p_{n+1}\left(p_{n+1} - p_{c,\,n+1}\right)}{2q^{1+\mu}_{n+1}\Gamma(1-\mu)}\right]\end{aligned} \tag{7.23}$$

where the fourth-order projection tensor $\mathbb{P}$ is determined by $\mathbb{P} = \mathbb{I} - \mathbf{1}\otimes\mathbf{1}/3$. Substituting Eq. (7.23) into Eq. (7.21), one can get the expression for $\mathbf{s}_{n+1}$:

$$\mathbf{s}_{n+1} = \frac{\mathbf{s_n} + 2\overline{G}\Delta\boldsymbol{\gamma}_{\mathbf{n+1}}}{1+c} \tag{7.24}$$

where

$$c = 6\overline{G}\Delta\phi_{n+1}\left[\frac{q^{1-\mu}_{n+1}}{N^2\Gamma(3-\mu)} + \frac{p_{n+1}\left(p_{n+1} - p_{\mathrm{c,\,n+1}}\right)}{2q^{1+\mu}\Gamma(1-\mu)}\right] \tag{7.25}$$

Substituting Eq. (7.22) into Eq. (7.20) and considering Eq. (7.24), the update formulas of p_{n+1} and q_{n+1} can be obtained to replace that of $\boldsymbol{\sigma}_{n+1}$ in Eq. (7.17).

$$\begin{aligned}p_{n+1} = p_n\exp\Bigg\{&c_\kappa\Delta\varepsilon_{\mathrm{v},\,n+1} - c_\kappa\Delta\phi_{n+1}\Bigg[\frac{q^2_{n+1}}{p^{\mu}_{n+1}N^2\Gamma(1-\mu)}\\ &+\frac{2p^{2-\mu}_{n+1}}{\Gamma(3-\mu)} - \frac{p_{\mathrm{c},\,n+1}p^{1-\mu}_{n+1}}{\Gamma(2-\mu)}\Bigg]\Bigg\}\end{aligned} \tag{7.26}$$

$$q_{n+1} = \sqrt{\frac{3}{2}}\frac{\left\|\mathbf{s}_n + 2\overline{G}\Delta\boldsymbol{\gamma}_{n+1}\right\|}{1+c} \tag{7.27}$$

Finally, the implicit stress integral equations of the NOEP clay model can be simplified in the following form:

$$\begin{Bmatrix} f_1 \\ f_2 \\ f_3 \\ f_4 \end{Bmatrix} = \begin{Bmatrix} p_{n+1} - p_n \exp\left(c_\kappa \Delta\varepsilon^{\mathrm{e}}_{\mathrm{v},\,n+1}\right) = 0 \\ q_{n+1} - \sqrt{\frac{3}{2}} \frac{\left\| \mathbf{s}_n + 2\overline{G}\Delta\boldsymbol{\gamma}_{n+1} \right\|}{1+c} = 0 \\ p_{\mathrm{c},\,n+1} - p_{\mathrm{c},\,n} \exp\left(c_{\mathrm{p}} \Delta\varepsilon^{\mathrm{p}}_{\mathrm{v},\,n+1}\right) = 0 \\ \sqrt{(c_{\mathrm{d}}\Delta\phi_{n+1})^2 + f_{n+1}^2 + 2\beta} - c_{\mathrm{d}}\Delta\phi_{n+1} + f_{n+1} = 0 \end{Bmatrix} \tag{7.28}$$

where c_{d} is recommended equal to $\left\| \boldsymbol{\sigma}_n + \overline{\mathbb{D}}\left(\overline{K}, \overline{G}\right) : \Delta\boldsymbol{\varepsilon}_{n+1} \right\|^3$ to balance the magnitude and dimension difference between $\Delta\phi_{n+1}$ and f_{n+1}. In this Chapter, the elastic trial stress point $\boldsymbol{\sigma}^{trial}_{n+1} = \boldsymbol{\sigma}_n + \overline{\mathbb{D}}\left(\overline{K}, \overline{G}\right) : \Delta\boldsymbol{\varepsilon}_{n+1}$ is used as the initial iteration point. Selecting $\left\| \boldsymbol{\sigma}_n + \overline{\mathbb{D}}\left(\overline{K}, \overline{G}\right) : \Delta\boldsymbol{\varepsilon}_{n+1} \right\|^3$ as the value of c_{d} will save the computational cost and reduce the influence of the step size on the numerical stability to some extent. Equation (7.28) can be solved iteratively by the Newton method with the quadratic convergence speed.

$$\{\mathbf{x}\}^{k+1}_{n+1} = \{\mathbf{x}\}^{k}_{n+1} + \{\mathbf{d}\}^k \tag{7.29}$$

where $\{\mathbf{d}\}^k$ is the search direction of Newton's method and can be determined by

$$\{\mathbf{d}\}^k = -[\nabla\mathbf{f}(\mathbf{x})]^{-1}_k \{\mathbf{f}(\mathbf{x})\}^k \tag{7.30}$$

where $\{\mathbf{f}\} = \left\{ f_1\ f_2\ f_3\ f_4 \right\}^T$. $[\nabla\mathbf{f}(\mathbf{x})]_{n+1}$ is the derivative of nonlinear equations $\{\mathbf{f}\}_{n+1}$ with respect to variables $\{\mathbf{x}\}_{n+1}$, also known as the Jacobian matrix. The implicit stress update procedure of the NOEP model is shown in Fig. 7.7.

7.5.2 Smoothing Consistent Tangent Operator

In the local calculation of nonlinear finite element analysis, two things need to be done. One is to update the model's state variables using the stress update algorithm, and the other is to provide the CTO $\partial\boldsymbol{\sigma}_{n+1}/\partial\boldsymbol{\varepsilon}_{n+1}$ that is consistent with the integral equations of the model. In the global calculation, the updated state variables are used to determine the structural internal forces, while the CTO is used to generate the global stiffness of the structure. When the Newton method is employed to solve the equilibrium equations in the global calculation, the CTO can preserve the global solution's quadratic convergence speed. Based on the operator splitting stress updating

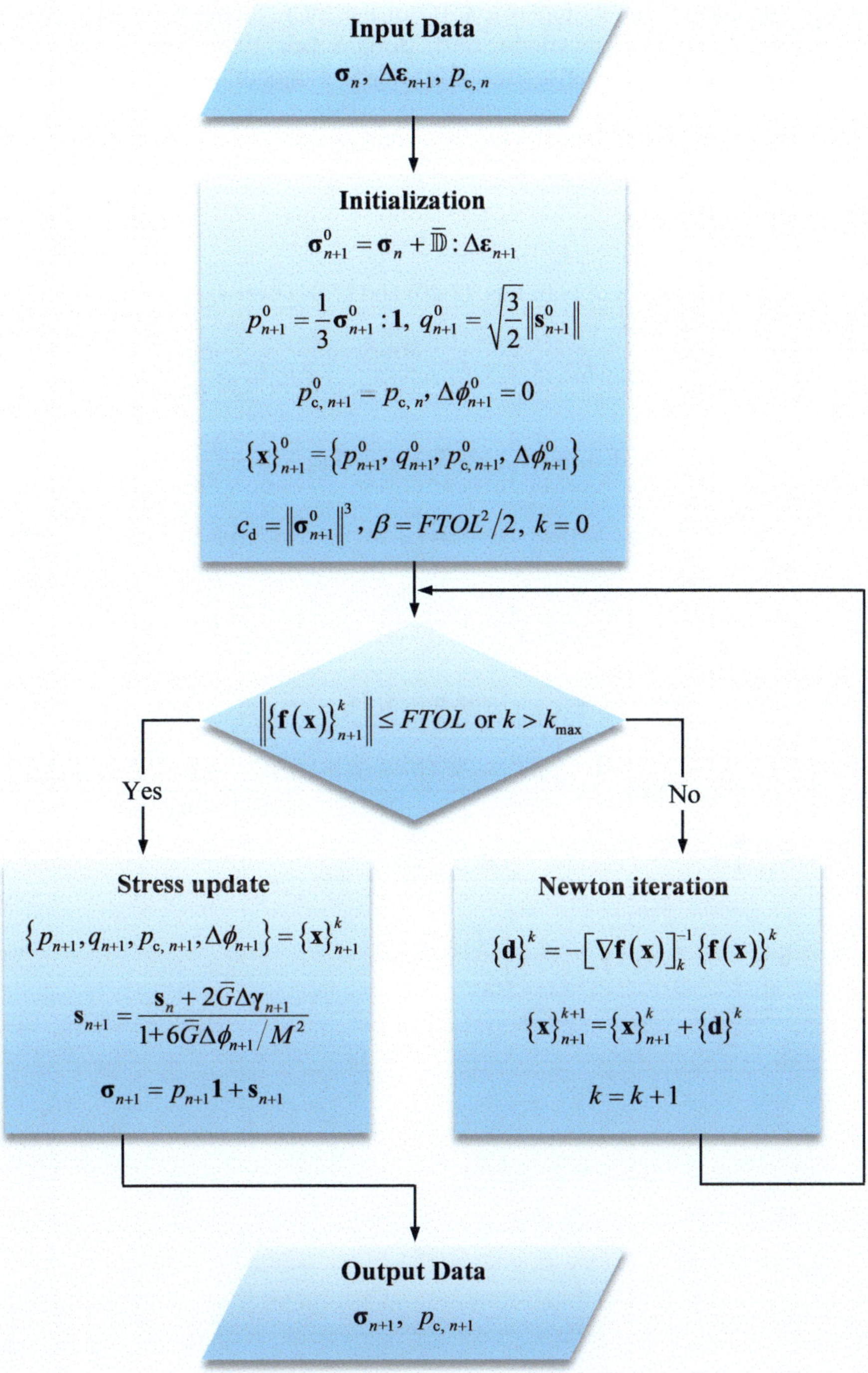

Fig. 7.7 Flow chart of the proposed algorithm

strategy, the elastic and elastoplastic CTO of the MCC model have been derived by Borja and his co-workers (Borja, 1991; Borja & Lee, 1990). In this section, the unconstrained stress updating strategy gives rise to a smoothing CTO with a unified form.

The consistent tangent operator can be expressed by:

$$\frac{\partial \boldsymbol{\sigma}_{n+1}}{\partial \boldsymbol{\varepsilon}_{n+1}} = \mathbf{1} \otimes \frac{\partial p_{n+1}}{\partial \boldsymbol{\varepsilon}_{n+1}} + \frac{\partial \mathbf{s}_{n+1}}{\partial \boldsymbol{\varepsilon}_{n+1}} \tag{7.31}$$

Taking the derivatives of both Eqs. (7.20) and (7.24) over $\boldsymbol{\varepsilon}_{n+1}$, one can obtain:

$$\frac{\partial p_{n+1}}{\partial \boldsymbol{\varepsilon}_{n+1}} = a_1 K^*_{n+1} \mathbf{1} + a_2 K^*_{n+1} \frac{\partial \Delta\phi_{n+1}}{\partial \boldsymbol{\varepsilon}_{n+1}} + a_3 K^*_{n+1} \frac{\partial q_{n+1}}{\partial \boldsymbol{\varepsilon}_{n+1}} \tag{7.32}$$

$$\begin{aligned}\frac{\partial \mathbf{s}_{n+1}}{\partial \boldsymbol{\varepsilon}_{n+1}} &= 2a_8 \mathbb{R}{:}\left(\bar{G}\mathbb{P} + a_1 a_9 \boldsymbol{\Delta\gamma}_{n+1} \otimes \mathbf{1} + a_2 a_9 \boldsymbol{\Delta\gamma}_{n+1} \otimes \frac{\partial \Delta\phi_{n+1}}{\partial \boldsymbol{\varepsilon}_{n+1}}\right) \\ &\quad - 2\sqrt{6} a_8 q_{n+1} \mathbb{R}{:}\mathbf{n} \otimes \left[a_1 a_7 a_9 \Delta\phi_{n+1} \mathbf{1} + a_7 \left(a_2 a_9 \boldsymbol{\Delta}\phi_{n+1} + \bar{G}\right) \frac{\partial \Delta\phi_{n+1}}{\partial \boldsymbol{\varepsilon}_{n+1}} \right. \\ &\quad \left. + \frac{\bar{G} \Delta\phi_{n+1} K^*_{n+1}}{2 q_{n+1}^{1+\mu} \Gamma(1-\mu)} \left(c_1 \mathbf{1} + c_2 \frac{\partial \Delta\phi_{n+1}}{\partial \boldsymbol{\varepsilon}_{n+1}}\right)\right]\end{aligned} \tag{7.33}$$

where $K^*_{n+1} = c_\kappa p_n \exp\left(c_\kappa \Delta\varepsilon^{\mathrm{e}}_{\mathrm{v},\, n+1}\right)$ and $\mathbf{n} = \mathbf{s}/\|\mathbf{s}\|_2$.

$$\begin{aligned}\mathbb{R} = \Bigg\{ &\mathbb{I} + 6 a_8 q_{n+1} \boldsymbol{\Delta}\varphi_{n+1} \left[\bar{G} \frac{2 q_{n+1}^{-\mu} + N^2\left(\mu^2 - \mu - 2\right)\left(p_{n+1}^2 - p_{n+1} p_{\mathrm{c},\, n+1}\right) q_{n+1}^{-2-\mu}}{2N^2 (2-\mu)\Gamma(1-\mu)} \right. \\ &\left. + \frac{\bar{G} K^*_{n+1} (2 p_{n+1} - p_{\mathrm{c},n+1}) a_3 - \bar{G} K^*_{n+1} p_{n+1} a_6}{2 q_{n+1}^{1+\mu} \Gamma(1-\mu)} + a_3 a_7 a_9 \right] \\ &\mathbf{n} \otimes \mathbf{n} - \sqrt{6} a_3 a_8 a_9 \boldsymbol{\Delta\gamma}_{\mathbf{n+1}} \otimes \mathbf{n} \Bigg\}^{-1}\end{aligned} \tag{7.34}$$

$$
\begin{cases}
a = 1 + K_{n+1}^{*}\Delta\phi_{n+1}\left(\dfrac{2p_{n+1}^{1-\mu}}{\Gamma(2-\mu)} - \dfrac{\mu q_{n+1}^{2}}{p_{n+1}^{1+\mu}N^{2}\Gamma(1-\mu)} - \dfrac{p_{\text{c},\,n+1}}{p_{n+1}^{\mu}\Gamma(1-\mu)}\right) \\
\quad + \dfrac{p_{\text{c},\,n+1}c_{\text{p}}\Delta\phi_{n+1}p_{n+1}^{1-\mu}}{\Gamma(2-\mu)} \\
a_1 = \left(1 + \dfrac{p_{\text{c},\,n+1}c_{\text{p}}\Delta\phi_{n+1}p_{n+1}^{1-\mu}}{\Gamma(2-\mu)}\right)\Big/ a \\
a_2 = \left(\dfrac{p_{\text{c},\,n+1}p_{n+1}^{1-\mu}}{\Gamma(2-\mu)} - \dfrac{q_{n+1}^{2}}{p_{n+1}^{\mu}N^{2}\Gamma(1-\mu)} - \dfrac{2p_{n+1}^{2-\mu}}{\Gamma(3-\mu)}\right)\Big/ a \\
a_3 = \dfrac{-2\Delta\phi_{n+1}q_{n+1}}{ap_{n+1}^{\mu}N^{2}\Gamma(1-\mu)} \\
a_4 = p_{\text{c},\,n+1}c_{\text{p}}\Delta\phi_{n+1}\left(\dfrac{2p_{n+1}^{1-\mu}}{\Gamma(2-\mu)} - \dfrac{\mu q_{n+1}^{2}}{p_{n+1}^{1+\mu}N^{2}\Gamma(1-\mu)} - \dfrac{p_{\text{c, n + 1}}}{p_{n+1}^{\mu}\Gamma(1-\mu)}\right)\Big/ a \\
a_5 = p_{\text{c},\,n+1}c_{\text{p}}\left(\dfrac{q_{n+1}^{2}}{p_{n+1}^{\mu}N^{2}\Gamma(1-\mu)} + \dfrac{2p_{n+1}^{2-\mu}}{\Gamma(3-\mu)} - \dfrac{p_{\text{c},\,n+1}p_{n+1}^{1-\mu}}{\Gamma(2-\mu)}\right)\Big/ \left(aK_{n+1}^{*}\right) \\
a_6 = \left(2p_{\text{c},\,n+1}c_{\text{p}}\Delta\phi_{n+1}q_{n+1}\right)\big/\left[p_{n+1}^{\mu}N^{2}K_{n+1}^{*}a\Gamma(1-\mu)\right] \\
a_7 = \dfrac{q_{n+1}^{1-\mu}}{N^{2}\Gamma(3-\mu)} + \dfrac{p_{n+1}^{2} - p_{n+1}p_{\text{c},\,n+1}}{2q_{n+1}^{1+\mu}\Gamma(1-\mu)} \\
a_8 = \left(1 + 6\overline{G}\Delta\phi_{n+1}a_7\right)^{-1} \\
a_9 = r\left(K^{*} - \overline{K}\right)\big/\Delta\varepsilon_{\text{v},\,n+1}^{\text{e}} \\
c_1 = a_1(2p_{n+1} - p_{\text{c},\,n+1}) - p_{n+1}a_4 \\
c_2 = a_2(2p_{n+1} - p_{\text{c},\,n+1}) - p_{n+1}a_5
\end{cases}
\tag{7.35}
$$

$\frac{\partial p_{\text{c},\,n+1}}{\partial \boldsymbol{\varepsilon}_{n+1}}$ can be derived by taking Eq. (7.17)$_3$ with respect to $\boldsymbol{\varepsilon}_{n+1}$:

$$
\frac{\partial p_{\text{c},\,n+1}}{\partial \boldsymbol{\varepsilon}_{n+1}} = a_4K_{n+1}^{*}\mathbf{1} + a_5K_{n+1}^{*}\frac{\partial \Delta\phi_{n+1}}{\partial \boldsymbol{\varepsilon}_{n+1}} + a_6K_{n+1}^{*}\frac{\partial q_{n+1}}{\partial \boldsymbol{\varepsilon}_{n+1}} \tag{7.36}
$$

There are four unknowns (i.e., $\partial p_{n+1}/\partial \boldsymbol{\varepsilon}_{n+1}$, $\partial \mathbf{s}_{n+1}/\partial \boldsymbol{\varepsilon}_{n+1}$, $\partial p_{\text{c},\,n+1}/\partial \boldsymbol{\varepsilon}_{n+1}$, and.$\partial \Delta\phi_{n+1}/\partial \boldsymbol{\varepsilon}_{n+1}$) in Eqs. (7.32), (7.33), and (7.36). An additional constraint is needed to close the equations involving unknowns. The total differential of Eq. (7.28)$_4$ can yield:

$$
\begin{aligned}
&\frac{\partial f_4}{\partial f}\left[\frac{3\mathbf{s}_{n+1}}{M^2}:\frac{\partial \mathbf{s}_{n+1}}{\partial \boldsymbol{\varepsilon}_{n+1}} + \left(2p_{n+1} - p_{\text{c},\,n+1}\right)\frac{\partial p_{n+1}}{\partial \boldsymbol{\varepsilon}_{n+1}} - p_{n+1}\frac{\partial p_{\text{c},\,n+1}}{\partial \boldsymbol{\varepsilon}_{n+1}}\right] \\
&+ \frac{\partial f_4}{\partial \Delta\phi_{n+1}}\frac{\partial \Delta\phi_{n+1}}{\partial \boldsymbol{\varepsilon}_{n+1}} = 0
\end{aligned}
\tag{7.37}
$$

From Eqs. (7.32), (7.33), (7.36), and (7.37), $\partial p_{n+1}/\partial \boldsymbol{\varepsilon}_{n+1}$, $\partial \mathbf{s}_{n+1}/\partial \boldsymbol{\varepsilon}_{n+1}$, $\partial p_{c,\,n+1}/\partial \boldsymbol{\varepsilon}_{n+1}$, and $\partial \Delta\phi_{n+1}/\partial \varepsilon_{n+1}$ can be derived as follows:

$$\begin{aligned}\frac{\partial p_{n+1}}{\partial \varepsilon_{n+1}} &= (a_1 + a_2 b_1) K^*_{n+1} \mathbf{1} + \left(a_2 b_2 + \sqrt{6} a_3 a_8 \overline{G}\right) K^*_{n+1} \mathbf{n}{:}\mathbb{R}{:}\mathbb{P} \\ &+ \sqrt{6} a_3 a_8 a_9 K^*_{n+1} (a_1 + a_2 b_1) \mathbf{n}{:}\mathbb{R}{:}\Delta\boldsymbol{\gamma}_{n+1} \otimes \mathbf{1} \\ &+ \sqrt{6} a_2 a_3 a_8 a_9 b_2 K^*_{n+1} \mathbf{n}{:}\mathbb{R}{:}\boldsymbol{\Delta\gamma}_{n+1} \otimes \mathbf{n}{:}\mathbb{R}{:}\mathbb{P} \\ &- 6 a_3 a_8 b_4 q_{n+1} K^*_{n+1} \mathbf{n}{:}\mathbb{R}{:}\mathbf{n} \otimes \mathbf{1} \\ &- 6 a_3 a_8 b_2 b_5 q_{n+1} K^*_{n+1} \mathbf{n}{:}\mathbb{R}{:}\mathbf{n} \otimes \mathbf{n}{:}\mathbb{R}{:}\mathbb{P}\end{aligned} \tag{7.38}$$

$$\begin{aligned}\frac{\partial \mathbf{s}_{n+1}}{\partial \boldsymbol{\varepsilon}_{n+1}} &= 2 a_8 \overline{G} \mathbb{R}{:}\mathbb{P} + 2 a_8 a_9 (a_1 + a_2 b_1) \mathbb{R}{:}\boldsymbol{\Delta\gamma}_{n+1} \otimes \mathbf{1} \\ &+ 2 a_2 a_8 a_9 b_2 \mathbb{R}{:}\Delta\boldsymbol{\gamma}_{n+1} \otimes \mathbf{n}{:}\mathbb{R}{:}\mathbb{P} \\ &- 2\sqrt{6} a_8 b_4 q_{n+1} \mathbb{R}{:}\mathbf{n} \otimes \mathbf{1} - 2\sqrt{6} a_8 b_2 b_5 q_{n+1} \mathbb{R}{:}\mathbf{n} \otimes \mathbf{n}{:}\mathbb{R}{:}\mathbb{P}\end{aligned} \tag{7.39}$$

$$\frac{\partial p_{c,\,n+1}}{\partial \boldsymbol{\varepsilon}_{n+1}} = a_4 K^*_{n+1} \mathbf{1} + a_5 K^*_{n+1} \frac{\partial \Delta\phi_{n+1}}{\partial \boldsymbol{\varepsilon}_{n+1}} + a_6 K^*_{n+1} \frac{\partial q_{n+1}}{\partial \boldsymbol{\varepsilon}_{n+1}} \tag{7.40}$$

$$\frac{\partial \phi_{n+1}}{\partial \boldsymbol{\varepsilon}_{n+1}} = b_1 \mathbf{1} + b_2 \mathbf{n}{:}\mathbb{R}{:}\mathbb{P} \tag{7.41}$$

where the coefficients

$$\begin{cases} b = 2 a_2 a_8 a_9 b_3 \mathbf{n}{:}\mathbb{R}{:}\boldsymbol{\Delta\gamma}_{\mathbf{n+1}} \dfrac{\partial f_4}{\partial f} - 2\sqrt{6} a_8 b_3 q_{n+1} \\ \quad \left\{ a_2 a_7 a_9 \Delta\phi_{n+1} + a_7 \overline{G} + \dfrac{\overline{G} \Delta\phi_{n+1} K^*_{n+1} c_2}{2 q_{n+1}^{1+\mu} \Gamma(1-\mu)} \right\} \mathbf{n}{:}\mathbb{R}{:}\mathbf{n} \dfrac{\partial f_4}{\partial f} \\ \quad + K^*_{n+1} c_2 \dfrac{\partial f_4}{\partial f} + \dfrac{\partial f}{\partial \Delta\phi_{n+1}} \\ b_1 = -\dfrac{\partial f_4}{\partial f} \Big[2 a_1 a_8 a_9 b_3 \mathbf{n}{:}\mathbb{R}{:}\boldsymbol{\Delta\gamma}_{\mathbf{n+1}} - 2\sqrt{6} a_8 b_3 q_{n+1} \mathbf{n}{:}\mathbb{R}{:}\mathbf{n} \\ \quad \left(a_1 a_7 a_9 \Delta\phi_{n+1} + \dfrac{\overline{G} \Delta\phi_{n+1} K^*_{n+1} c_1}{2 q_{n+1}^{1+\mu} \Gamma(1-\mu)} \right) + K^*_{n+1} c_1 \Big] \Big/ b \\ b_2 = -\left(2 a_8 b_3 \overline{G} \dfrac{\partial f_4}{\partial f} \right) \Big/ b \\ b_3 = \sqrt{6} \dfrac{q}{N^2} + \sqrt{\dfrac{3}{2}} K^*_{n+1} [(2 p_{n+1} - p_{c,\,n+1}) a_3 - p_{n+1} a_6] \\ b_4 = a_1 a_7 a_9 \Delta\phi_{n+1} + a_2 a_7 a_9 b_1 \Delta\phi_{n+1} + a_7 b_1 \overline{G} \\ \quad + \dfrac{\overline{G} \Delta\phi_{n+1} K^*_{n+1} (c_1 + b_1 c_2)}{2 q_{n+1}^{1+\mu} \Gamma(1-\mu)} \\ b_5 = a_2 a_7 a_9 \Delta\phi_{n+1} + a_7 \overline{G} + \dfrac{\overline{G} \Delta\phi_{n+1} K^*_{n+1} c_2}{2 q_{n+1}^{1+\mu} \Gamma(1-\mu)} \end{cases} \tag{7.42}$$

Substituting Eqs. (7.38) and (7.39) into Eq. (7.31), the consistent tangent operator is obtained analytically as follows:

$$
\begin{aligned}
\frac{\partial \boldsymbol{\sigma}_{n+1}}{\partial \boldsymbol{\varepsilon}_{n+1}} =&(a_1 + a_2 b_1) K_{n+1}^{*} \mathbf{1} \otimes \mathbf{1} + \left(a_2 b_2 + \sqrt{6} a_3 a_8 \overline{G}\right) K_{n+1}^{*} \mathbf{1} \otimes \mathbf{n}:\mathbb{R}:\mathbb{P} \\
&+ \sqrt{6} a_3 a_8 a_9 (a_1 + a_2 b_1) K_{n+1}^{*} \mathbf{1} \otimes \mathbf{n}:\mathbb{R}:\Delta \boldsymbol{\gamma}_{n+1} \otimes \mathbf{1} \\
&+ \sqrt{6} a_2 a_3 a_8 a_9 b_2 K_{n+1}^{*} \mathbf{1} \otimes \mathbf{n}:\mathbb{R}:\Delta \boldsymbol{\gamma}_{n+1} \otimes \tilde{\mathbf{n}}:\mathbb{R}:\mathbb{P} \\
&- 6 a_3 a_8 b_4 q_{n+1} K_{n+1}^{*} \mathbf{1} \otimes \mathbf{n}:\mathbb{R}:\mathbf{n} \otimes \mathbf{1} \\
&- 6 a_3 a_8 b_2 b_5 q_{n+1} K_{n+1}^{*} \mathbf{1} \otimes \mathbf{n}:\mathbb{R}:\mathbf{n} \otimes \mathbf{n}:\mathbb{R}:\mathbb{P} \\
&+ 2 \overline{G} a_8 \mathbb{R}:\mathbb{P} + 2 a_8 a_9 (a_1 + a_2 b_1) \mathbb{R}:\Delta \boldsymbol{\gamma}_{n+1} \otimes \mathbf{1} \\
&+ 2 a_2 a_8 a_9 b_2 \mathbb{R}:\Delta \boldsymbol{\gamma}_{n+1} \otimes \mathbf{n}:\mathbb{R}:\mathbb{P} \\
&- 6\sqrt{\frac{2}{3}} a_8 b_4 q_{n+1} \mathbb{R}:\mathbf{n} \otimes \mathbf{1} - 6\sqrt{\frac{2}{3}} a_8 b_2 b_5 q_{n+1} \mathbb{R}:\mathbf{n} \otimes \mathbf{n}:\mathbb{R}:\mathbb{P}
\end{aligned}
\tag{7.43}
$$

where $\partial \boldsymbol{\sigma}_{n+1} / \partial \boldsymbol{\varepsilon}_{n+1}$ will degenerate into that of the MCC model in the case of $\mu = 1$.

7.6 Numerical Examples

Based on the unconstrained stress update algorithm, the NOEP model has been implemented in ABAQUS finite element software via the external subroutine UMAT. In this section, the correctness of the model implementation is first evaluated based on the problem of cylindrical cavity expansion with analytical solutions. Then, the performance of the algorithm is analyzed and discussed in depth by comparing it with the ABAQUS/Standard default algorithm (Simo & Hughes, 2006) in the case of $\mu = 1$. Table 7.1 lists the material parameters of the NOEP model used for boundary value problems, where e_1 denotes the void ratio of the soil at $p = 1$ kPa.

Table 7.1 Material parameters of NOEP clay model used for boundary value problems

Boundary value problems	M	λ	κ	ν	e_1	μ
Cylindrical cavity expansion problem	1.2	0.15	0.03	0.278	1.823	0.5–1.5
Cylinder subjected to cyclic shear loads	1.0	0.25	0.05	0.3	1.83	1.0
Strip footing	0.898	0.25	0.05	0.3	1.6	1.0

7.6.1 Cylindrical Cavity Expansion Problems

The proposed algorithm is compared with the analytical solution of the cylindrical cavity expansion problem provided by the literature (Chen & Abousleiman, 2012, 2013) for both undrained and drained conditions. As shown in Fig. 7.8a, a cylindrical cavity with the initial radius r_0 exists in cylindrical soil with infinite height and radius, which is subjected to initial total vertical stress σ_{v0}, horizontal stress σ_{h0}, and internal pressure σ_0. Then, under the action of internal pressure, the radius of the cylindrical cavity gradually expands from r_0 to r_1. As can be noticed, the cylindrical cavity expansion is an axisymmetric plane strain problem, which can be analyzed by the simplified model shown in Fig. 7.8b. The geometric simplification can be implemented by using the eight nodes axisymmetric pore pressure element (CAX8P) for the undrained case and the eight nodes axisymmetric element (CAX8) for the drained, respectively. The initial stress state and the pre-consolidation pressure are set to $\sigma_r = \sigma_\theta = 100$ kPa, $\sigma_z = 160$ kPa, and $p_{\mathrm{c}} = 145$ kPa.

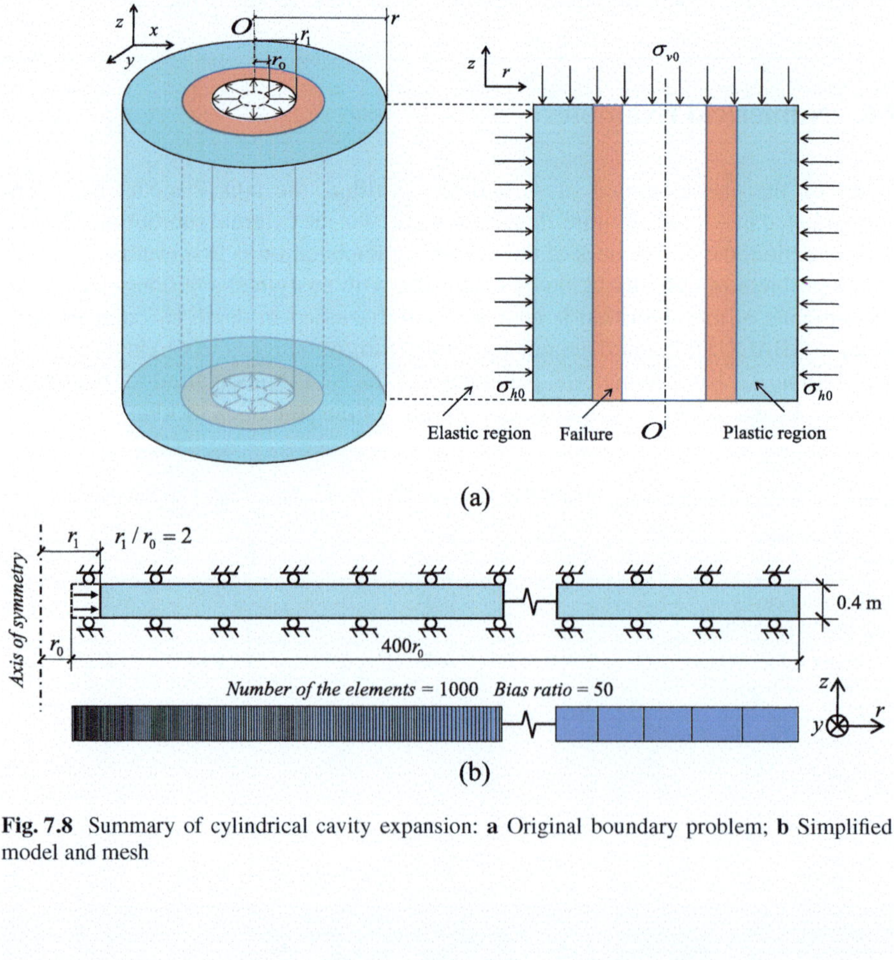

Fig. 7.8 Summary of cylindrical cavity expansion: **a** Original boundary problem; **b** Simplified model and mesh

The simulation process includes three analysis steps. In the initial step, the translational degree of freedom of the top and bottom edges in the 3-direction and the translational degree of freedom of the left edge in the 1-direction are fixed. The translational degree of freedom of the right edge in the 1-direction is fixed only for the drained case. In the geostatic step, the pressure load is applied on the right edge of the model to balance the in-situ stress. In the third analysis step, the displacement load in the 1-direction is applied on the left edge of the model to simulate the cavity expansion process. It is worth noting that, for the undrained case, the permeability coefficient and the total time of the third analysis step are set to 2.3×10^{-3} and 0.001 s (Liu et al., 2019). It can be considered that there is almost no dissipation of pore water pressure in the cavity expansion process due to the very short drainage time. The initial pore water pressure is set to $u_p = 100$ kPa. Figures 7.9 and 7.10 show the comparison results for the undrained and drained conditions, respectively. For different values of μ, the changes law of the hydrostatic pressure p, generalized shear stress q, and the excess pore water pressure Δu_p from the proposed algorithm and the analytical solution are in good agreement, which proves the correctness of the model implementation.

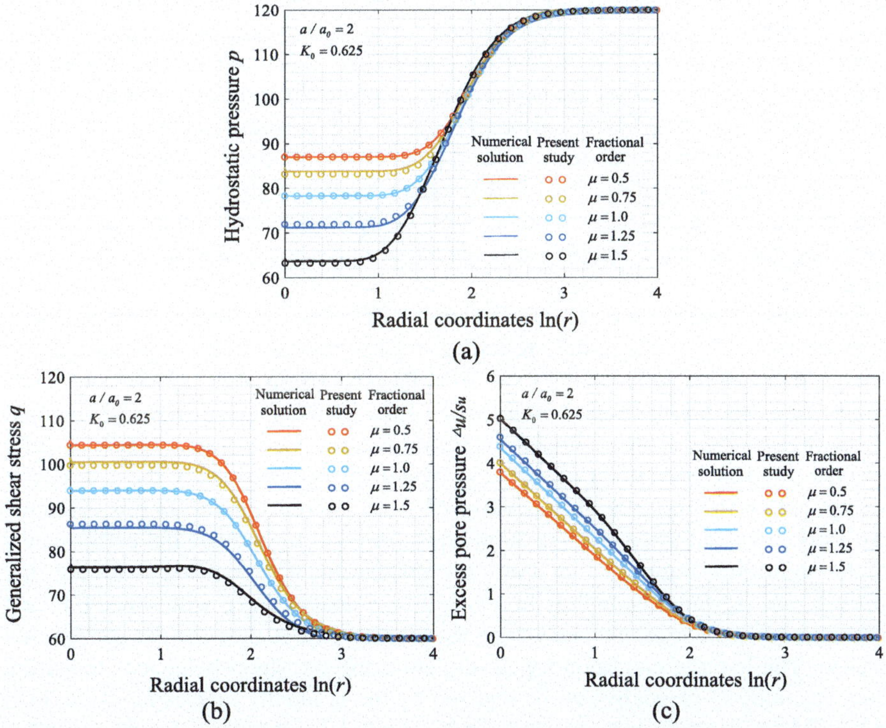

Fig. 7.9 Comparison between the proposed algorithm and analytical solution under undrained conditions: **a** Hydrostatic pressure; **b** Generalized shear stress; **c** Excess pore water pressure

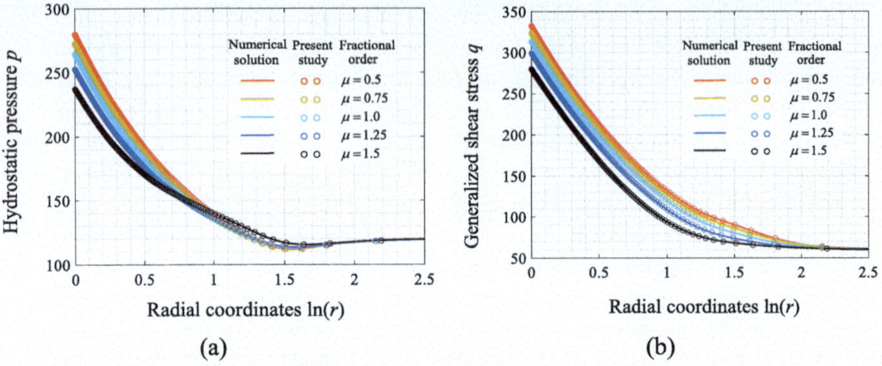

Fig. 7.10 Comparison between the proposed algorithm and analytical solution under drained conditions: **a** Hydrostatic pressure; **b** Generalized shear stress

7.6.2 Cylinder under Cyclic Shear Loads

The effectiveness of the unconstrained stress update algorithm for cyclic loading and unloading situations is further studied, where a problem involving a cylinder sample subjected to cyclic shear loads is considered. This problem has been used to evaluate the performance of the implicit stress update algorithm in Shterenlikht and Alexander's (2012) and Lester and Scherzinger's (2017) works. One major difference between the question in this section and the one in the literature (Scherzinger, 2017; Shterenlikht & Alexander, 2012) is that cyclic loading with more demand is used instead of monotonic loading to obtain more complex stress conditions. In addition, the ABAQUS/Standard default stress updating algorithm is used as a comparison object.

The cylinder in the problem is 0.2 m long with a cross-section of diameter 0.1 m, as shown in Fig. 7.11. The initial stress state and the pre-consolidation pressure are set to $\sigma_1 = \sigma_2 = \sigma_3 = 200$ kPa and $p_c = 200$ kPa. The origin of the global coordinate system is at the centre of the bottom face of the cylinder. All degrees of freedom of the bottom face of the cylinder are fixed. The top face of the cylinder is fully coupled to a reference point (0, 0, 0.25). The displacement load in the 1-direction is imposed on the top face of the cylinder by the reference point. In this example, two load cases and three discretization types are used to evaluate the performance of the algorithm. Figure 7.12a shows the displacement time-history curve in the loading case *a*, which is applied in a single load analysis step with 40 equidistant time steps. For the example analysis in the loading case *a*, only the discretization *a* is used. The loading condition repeatedly suffering the stress behaviour transition from elasticity to plasticity can be realized by the gradually increasing cyclic load in Fig. 7.12a. The time-history curves of the reaction force of the reference point for the two algorithms are demonstrated in Fig. 7.12b. The results obtained from the two algorithms are in good agreement, which also shows that the smoothing function is effective in replacing the *KKT* conditions.

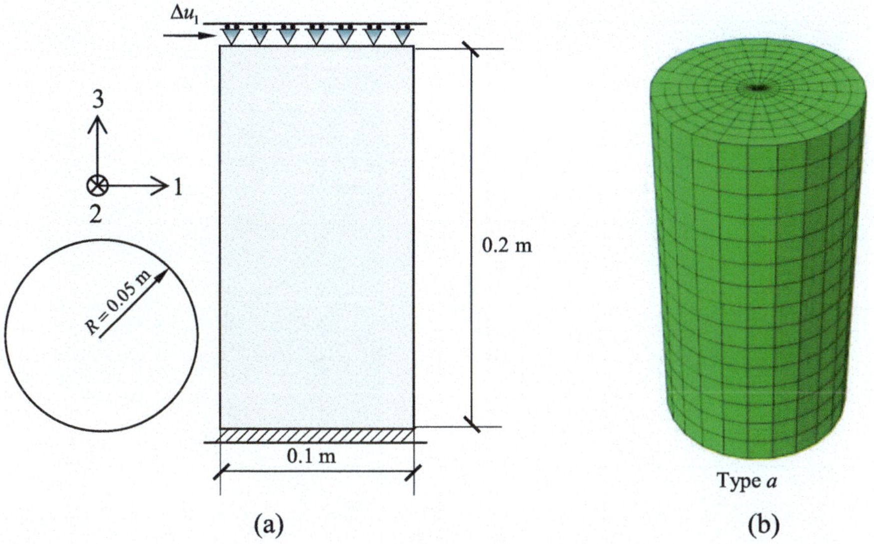

Fig. 7.11 Geometry and discretization of the cylinder sample: **a** Geometry; **b** Discretization

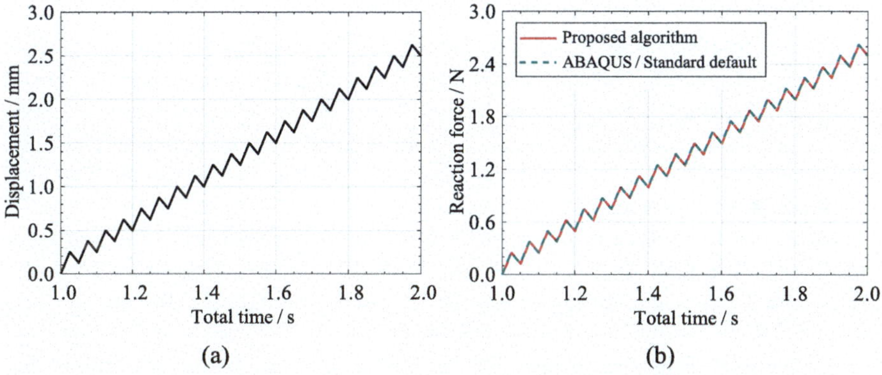

Fig. 7.12 Load and response: **a** Displacement load; **b** Reaction force response in the 1-direction

7.6.3 Foundation Bearing Capacity Test

The following numerical example is a bearing capacity test of rigid strip footing, which has been widely used to assess the numerical implementation of critical state models due to the strong rotation of principal stress and the singularity at the edge of footing (Scherzinger, 2017; Shterenlikht & Alexander, 2012). The geometric information, boundary conditions, and mesh of the foundation are shown in Fig. 7.13, where the element type is set to 8 nodes brick element (C3D8). The elements 4 and 2260 at the edge of the footing are indicated additionally for subsequent analysis.

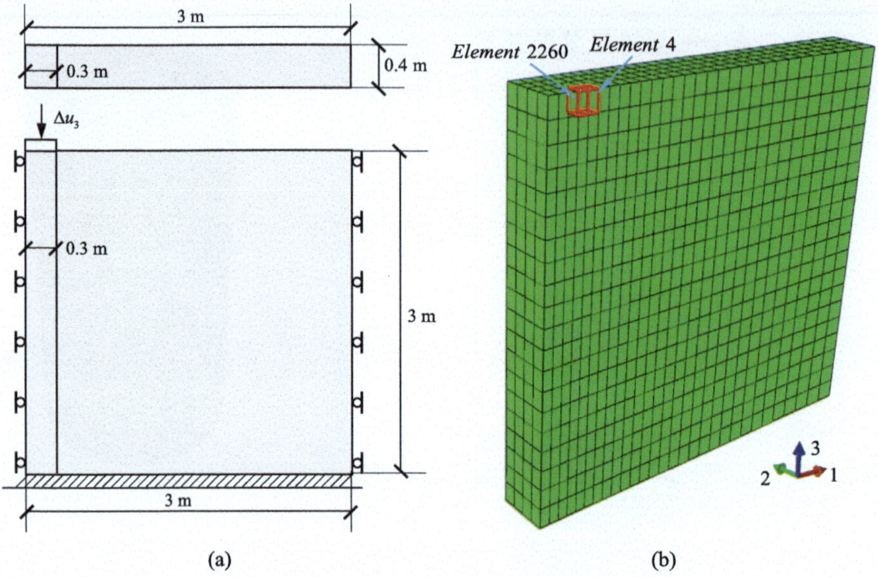

Fig. 7.13 String footing: **a** Geometry and boundary conditions; **b** Mesh

The unit weight of soil is set to $\gamma_w = 6\,\text{kN/m}^3$. The initial stress field and initial void ratio e_0 are generated by the submerged weight of soil and pre-load 50 kPa imposed on the ground surface in the geostatic stress balance analysis. In the unload analysis step, the pre-load is completely removed with 4 equal load increments. Then, the displacement with $U_3 = -0.15$ m is applied on the footing in the load analysis step with 16 equal load increments.

Figure 7.14a presents the generalized shear stress distribution of the soil layer at the end of the load. In Fig. 7.14b, the load–displacement responses from the unconstrained stress update algorithm and ABAQUS/Standard DIM are in good agreement. Note that the continuum tangent operator is also considered to highlight the effectiveness of the smoothing CTO for the convergence behavior of the global solution.

The convergence behavior at the critical node for different algorithms is depicted in Fig. 7.15. In comparison to the continuum tangent operator, the CTO shows excellent convergence. The reason is that the CTO can guarantee the quadratic convergence speed of the Newton method used in the global equilibrium problem. Using the continuum tangent operator instead of the CTO will disturb the search direction of the Newton method and further destroy the convergence speed and radius of the Newton method.

Similar results are also observed in Fig. 7.16. For the unconstrained stress update algorithm with CTO and ABAQUS/Standard DIM, the overall number of iterations and the CPU time consumed are similar. This high similarity of convergence speed also verifies the validity of smoothing CTO derived by the unconstrained stress

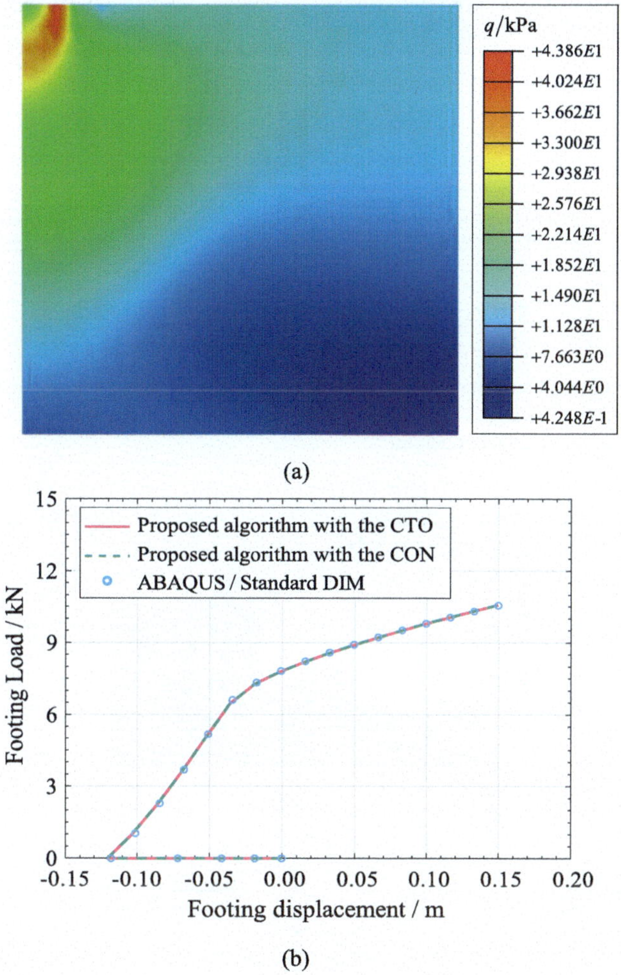

Fig. 7.14 Simulation results of bearing capacity test of foundation: **a** Generalized shear stress distribution; **b** Footing load versus footing displacement

updating strategy. For the unconstrained stress update algorithm with the continuum tangent operator, three to four times the iteration number and computation time are required, significantly increasing the computational cost of numerical analysis. Undoubtedly, the use of CTO is key to ensuring efficient calculation in elastoplastic analysis.

In what follows, the evolution of the stress state and yield surface corresponding to elements 4 and 2260 in Fig. 7.13 are analyzed, which will aid knowledge of the unconstrained stress updating strategy. Note that, in both elements, only integral point 8 is employed. In the unloading analysis step, i.e., the path from stress point 1 to stress point 5, the stress points of the two elements break away from the yield

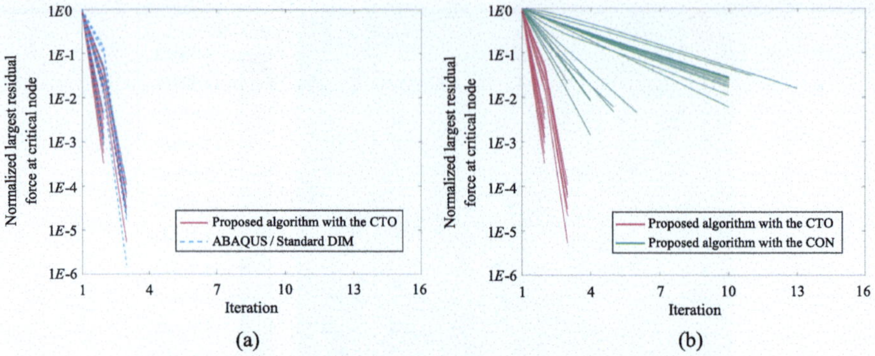

Fig. 7.15 Convergence behaviour at the critical node: **a** Unconstrained stress update algorithm with the CTO and the ABAQUS/Standard DIM; **b** Unconstrained stress update algorithm with the CTO and the continuum tangent operator

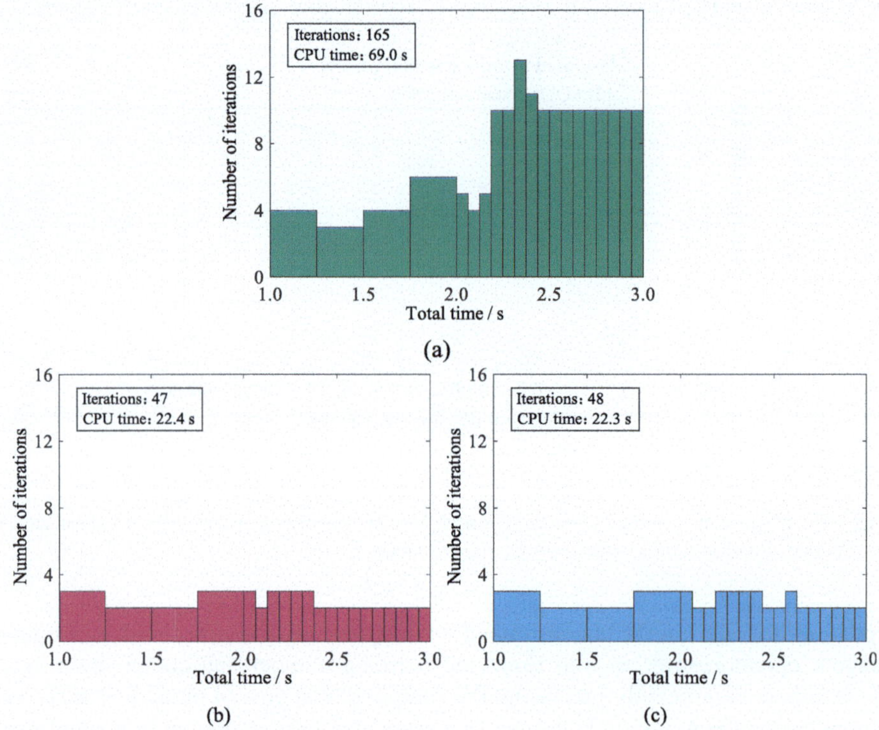

Fig. 7.16 Number of iterations at each load increment: **a** LSM with the continuum tangent operator; **b** LSM with the CTO; **c** ABAQUS/Standard DIM

surface and move in the elastic region, as shown in Figs. 7.17a and 7.18a. The yield surface remains unchanged because there is no plastic deformation. In the loading analysis step, i.e., the path from stress point 5 to stress point 21, the stress point moves to the yield surface again. The plastic deformation starts to occur after the stress point reaches the yield surface.

For element 4, as shown in Fig. 7.17b, the stress point first lies on the 'dry' side of the critical state line, then temporarily travels to the 'wet' side, and eventually returns to the 'dry' side. The change of stress path indicates that the soil around the integral point undergoes a transformation from strain-softening to strain-hardening and back again. Correspondingly, the yield surface shrinks first, then expands, and eventually shrinks again. The stress state of the integral point for element 2260 is always on the 'wet' side of the critical state line, as shown in Fig. 7.18b. The strain-hardening behavior is the only thing that is seen. The yield surface is always expanding. It is worth emphasizing that due to the use of the smoothing function, the stress point is always on or inside the yield surface. From the perspective of the unconstrained stress updating strategy, the allowable stress region in Figs. 7.17a and 7.18a, including the elastic region and the yield surface, is projected onto a smoothing curve. Whether loading or unloading cases, the stress point is always on the smoothing curve, as shown in Figs. 7.17c and 7.18c. The *KKT* conditions are always satisfied due to the fact that the smoothing function is an equivalent approximation of *KKT* conditions.

7.7 Summary

The non-smoothness has always been a challenge that needs to be overcome for the numerical implementation of elastoplastic models. The classic return mapping stress update algorithm requires loading/unloading judgment when dealing with the non-smoothness of elastoplastic models, which leads to the complexity of elastoplastic calculations. This chapter proposes an unconstrained stress update algorithm without the need for the loading/unloading judgment by utilizing the *FB* smooth function to reformulate the governing equations of the elastoplastic model. The performance of the algorithm is evaluated based on the NOEP model and a series of boundary value problems. The main contributions of this chapter can be listed below:

(1) The root of non-smoothness for elastoplastic models lies in loading/unloading inequality constraints i.e., *KKT* conditions in the field of numerical optimization. The stress integration equations of elastoplastic are transformed into a smooth form by replacing the *KKT* conditions with the *FB* smooth function, developing an unconstrained stress updating strategy without the need for the loading/unloading judgment.

(2) Taking the NOEP as an example, we derive a smooth consistent tangent operator corresponding to the unconstrained stress update strategy. The stress integration equations and the smoothing CTO corresponding to this stress updating strategy have a unified form regardless of the loading and unloading cases. The benefit

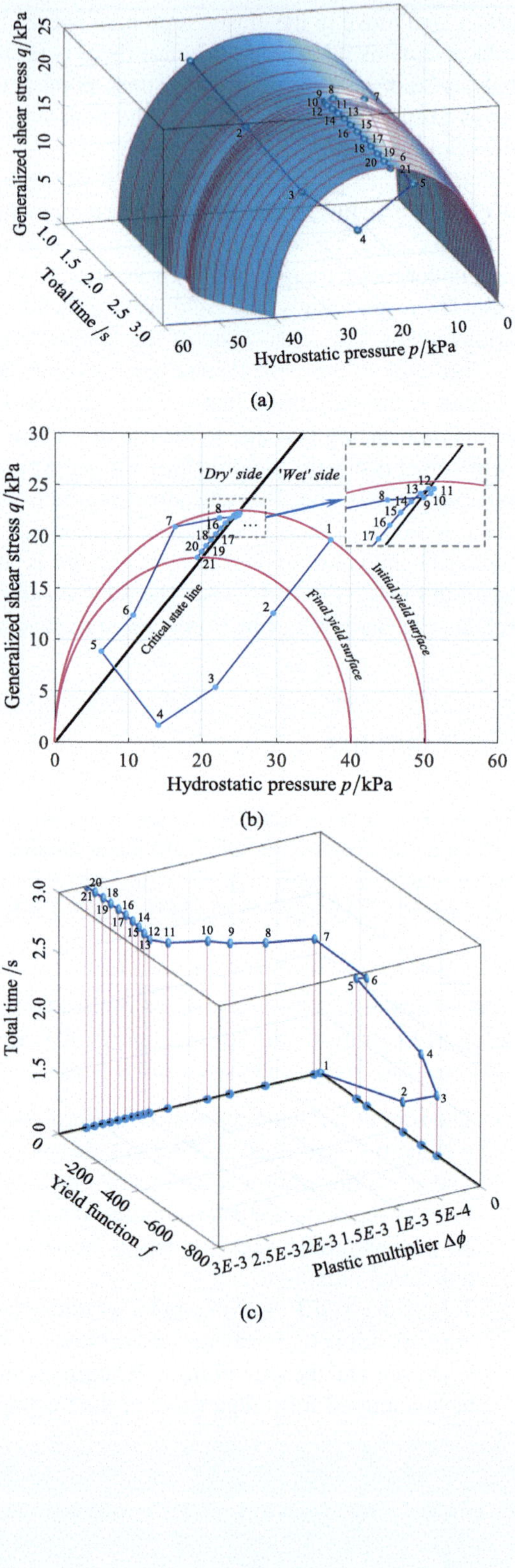

Fig. 7.17 Stress path at integral point 8 of element 4: **a** The result in the *p-q-time* coordinate system; **b** The result in the *p-q* coordinate system; **c** The result in the f-$\Delta\phi$-*time* coordinate system

Fig. 7.18 Stress path at integral point 8 of element 2260: **a** The result in the *p-q-time* coordinate system; **b** The result in the *p-q* coordinate system; **c** The result in the *f*-$\Delta\phi$-*time* coordinate system

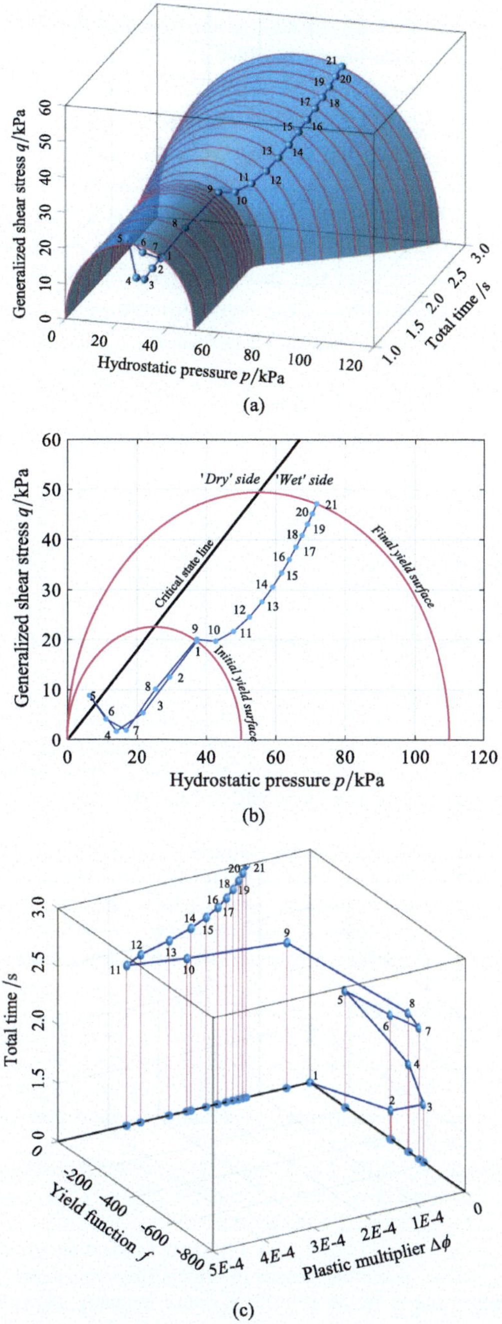

is that it provides a concise computational framework for the numerical implementation of the elastoplastic model, avoiding the complexity of elastoplastic calculations.

(3) Combined with the unconstrained stress update algorithm, the NOEP has been implemented in ABAQUS software based on the UMAT subroutine. The correctness of the model implementation has been verified by comparing them with the analytical solutions of the NOEP model in the cylindrical cavity expansion problem. The numerical analysis of the remaining boundary value problems shows that the unconstrained stress update algorithm has the same calculation results as the return mapping algorithm, even under conditions involving repeated loading and unloading. In addition, the smooth consistent tangent operator exhibits a quadratic convergence rate in the global equilibrium iteration, once again verifying the effectiveness of the proposed algorithm for numerical calculations of elastoplastic models.

References

Borja, R. I. (1991). Cam-clay plasticity, part II: Implicit integration of constitutive equation based on a nonlinear elastic stress predictor. *Computer Methods in Applied Mechanics and Engineering, 88*, 225–240.

Borja, R. I., & Lee, S. R. (1990). Cam-clay plasticity, part I: Implicit integration of elasto-plastic constitutive relations. *Computer Methods in Applied Mechanics and Engineering, 78*, 49–72.

Chen, S. L., & Abousleiman, Y. N. (2012). Exact undrained elasto-plastic solution for cylindrical cavity expansion in modified Cam Clay soil. *Géotechnique, 62*(5), 447–456.

Chen, S. L., & Abousleiman, Y. N. (2013). Exact drained solution for cylindrical cavity expansion in modified Cam Clay soil. *Géotechnique, 63*(6), 510–517.

De Borst, R., & Heeres, O. M. (2002). A unified approach to the implicit integration of standard, non-standard and viscous plasticity models. *International Journal for Numerical and Analytical Methods in Geomechanics, 26*(11), 1059–1070.

Fischer, A. (1992). A special newton-type optimization method. *Optimization, 24*, 269–284.

Gao, Z. W., Li, X., & Lu, D. C. (2022). Nonlocal regularization of an anisotropic critical state model for sand. *Acta Geotechnica, 17*(2), 427–439.

Kanzow, C. (1996). Some noninterior continuation methods for linear complementarity problems. *Siam Journal on Matrix Analysis and Applications, 17*(4), 851–868.

Liu, K., Chen, S. L., & Voyiadjis, G. Z. (2019). Integration of anisotropic modified Cam Clay model in finite element analysis: Formulation, validation, and application. *Computers and Geotechnics, 116*, 103198.

Moran, B., Ortiz, M., & Shih, C. F. (1990). Formulation of implicit finite element methods for multiplicative finite deformation plasticity. *International Journal for Numerical Methods in Engineering, 29*(3), 483–514.

Petalas, A. L., & Dafalias, Y. F. (2019). Implicit integration of incrementally non-linear, zero- elastic range, bounding surface plasticity. *Computers and Geotechnics, 112*, 386–402.

Scherzinger, W. M. (2017). A return mapping algorithm for isotropic and anisotropic plasticity models using a line search method. *Computer Methods in Applied Mechanics and Engineering, 317*, 526–553.

Sheng, D. C., Sloan, S. W., & Yu, H. S. (2000). Aspects of finite element implementation of critical state models. *Computational Mechanics, 26*(2), 185–196.

Shterenlikht, A., & Alexander, N. A. (2012). Levenberg-Marquardt vs Powell's dogleg method for Gurson-Tvergaard-Needleman plasticity model. *Computer Methods in Applied Mechanics and Engineering, 237–240*, 1–9.

Simo, J. C., & Hughes, T. J. (2006). *Computational inelasticity*. Springer Science & Business Media.

Simo, J. C., & Taylor, R. L. (1986). A return mapping algorithm for plane stress elastoplasticity. *International Journal for Numerical Methods in Engineering, 22*(3), 649–670.

Starman, B., Halilovič, M., Vrh, M., & Atok, B. (2014). Consistent tangent operator for cutting-plane algorithm of elasto-plasticity. *Computer Methods in Applied Mechanics and Engineering, 272*, 214–232.

Zhao, X. G., & Cai, M. (2010). Influence of plastic shear strain and confinement-dependent rock dilation on rock failure and displacement near an excavation boundary. *International Journal of Rock Mechanics and Mining Sciences, 47*(5), 723–738.

Chapter 8
Stress Update Algorithm Based on Numerical Optimization Methods

In Chap. 7, we use the smooth function instead of *KKT* conditions to reformulate the non-smooth elastoplastic problem into a system of smooth nonlinear equations. The Newton method is often used to solve nonlinear equations due to its quadratic convergence speed. However, its effectiveness will be challenged under large load step size and strong nonlinear cases. In the numerical optimization, the problem of nonlinear equations is equivalent to an unconstrained minimization problem of a merit function that is constructed by nonlinear equations. The solution of nonlinear equations corresponds to the minimum point of a merit function in multidimensional space. Numerical optimization algorithms such as LSM and trust region method can solve unconstrained minimization problems with strong nonlinear characteristics more effectively. In the LSM, for a given search direction, the step size is optimized to make the merit function get more decline along this direction. In the trust region method, the size of the search step is limited to the radius of the trust region. Then seek a direction and step that attain the best improvement possible subject to this distance constraint. In general, both methods can obtain better convergence properties than the standard Newton method by optimizing the step size or search direction.

This chapter develops two robust unconstrained stress update algorithms by combining numerical optimization methods with unconstrained stress update strategies. The MCC model is used as an application object for these two algorithms. The unconstrained minimization problem corresponding to the nonlinear stress integral equations is solved separately through inexact LSM and NMTR method, ensuring the convergence of the iterative solution under large load step size and strong nonlinear cases. Based on the proposed algorithms, the MCC model is numerically implemented in ABAQUS software by the UMAT subroutine. The performance of the algorithms is further evaluated through numerical analysis at the integration point level and structural level.

D. Lu et al., *Multi-physical and Non-orthogonal Geotechnical Plasticity Modeling and Algorithms*, https://doi.org/10.1007/978-981-96-2369-3_8

8.1 Unconstrained Minimization Problem: Modified Cam-Clay Model Case

Based on the backward Euler integral scheme and the *Fischer–Burmeister* (*FB*) function, the stress integral equations of the elastoplastic model can be written in the following general form:

$$\begin{aligned}
&\boldsymbol{\sigma}_{n+1} - \boldsymbol{\sigma}_n - \mathbb{D} : \left(\Delta\boldsymbol{\varepsilon}_{\mathrm{n+1}} - \Delta\boldsymbol{\varepsilon}^{\mathrm{p}}_{\mathrm{n+1}}\right) = \mathbf{0} \\
&\boldsymbol{\varepsilon}^{\mathrm{p}}_{n+1} - \boldsymbol{\varepsilon}^{\mathrm{p}}_{n} - \Delta\phi_{n+1}\mathbf{r}_{n+1} = \mathbf{0} \\
&H_{n+1} - H_n - \Delta\phi_{n+1}h_{n+1} = 0 \\
&\sqrt{(c_{\mathrm{d}}\Delta\phi_{n+1})^2 + f_{n+1}^2 + 2\beta} - c_{\mathrm{d}}\Delta\phi_{n+1} + f_{n+1} = 0
\end{aligned} \tag{8.1}$$

where the stress $\boldsymbol{\sigma}_{n+1}$, the plastic strain $\boldsymbol{\varepsilon}^{\mathrm{p}}_{n+1}$, the plastic internal variable H_{n+1}, and the plastic multiplier $\Delta\phi_{n+1}$ at step $n+1$ are the unknown variables to be solved. The plastic flow direction $\mathbf{r}_{n+1}$, the gradient of plastic internal variable h_{n+1}, the yield function f_{n+1} are functions of these unknown variables. Substituting Eq. (7.3)$_2$ into Eq. (7.3)$_1$, the number of nonlinear equations and unknowns in Eq. (7.3) can be further reduced to save the cost of iterative calculation:

$$\begin{Bmatrix} \mathbf{f}_\sigma \\ f_H \\ f_{\Delta\phi} \end{Bmatrix} = \begin{Bmatrix} \boldsymbol{\sigma}_{n+1} - \boldsymbol{\sigma}_n - \mathbb{D} : (\Delta\boldsymbol{\varepsilon}_{\mathrm{n+1}} - \Delta\phi_{\mathrm{n+1}}\mathbf{r}_{\mathrm{n+1}}) \\ H_{n+1} - H_n - \Delta\phi_{n+1}h_{n+1} \\ \sqrt{(c_{\mathrm{d}}\Delta\phi_{n+1})^2 + f_{n+1}^2 + 2\beta} - c_{\mathrm{d}}\Delta\phi_{n+1} + f_{n+1} \end{Bmatrix} = \begin{Bmatrix} \mathbf{0} \\ 0 \\ 0 \end{Bmatrix} \tag{8.2}$$

For any smooth nonlinear system of equations, such as Eq. (8.2), we can transform it into the unconstrained minimization problem of a merit function as follows:

$$\min\ \psi(\{\mathbf{x}_{n+1}\}) = \frac{1}{2}\{\mathbf{f}(\mathbf{x}_{n+1})\}^T\{\mathbf{f}(\mathbf{x}_{n+1})\} \tag{8.3}$$

where $\mathbf{x}_{n+1}$ denotes unknown variables. The merit function ψ is a simple quadratic function of the nonlinear equations defined by Eq. (8.2).

$$\{\mathbf{x}_{n+1}\} = \{\boldsymbol{\sigma}_{n+1}, H_{n+1}, \Delta\phi_{n+1}\}^T \tag{8.4}$$

$$\{\mathbf{f}(\mathbf{x})\} = \left\{\mathbf{f}_\sigma, f_H, f_{\Delta\phi}\right\}^T \tag{8.5}$$

It was obvious that the local minimum point of the merit function $\psi(\{\mathbf{x}_{n+1}\}) = 0$ is also the solution of nonlinear equations $\{\mathbf{f}(\mathbf{x}_{n+1})\} = \{\mathbf{0}\}$. The solution of Eq. (8.2) will be discussed based on the LSM and the trust region method in Sects. 8.2 and 8.3. Now, taking the the MCC as an example, we will give the details of Eq. (8.2). It is worth emphasizing again that the MCC model is a special case of the NOEP model. By setting fractional order $\mu = 1$, the stress integral equations of the NOEP model will degenerate into those of the MCC model. Therefore, the following is a brief introduction to the stress integral equations of the MCC model. Interested readers can refer to literature (Zhou et al., 2022) for derivation details.

In the MCC model, the yield function in elliptical form is employed as follows:

$$f_{n+1} = \frac{q_{n+1}^2}{M^2} + p_{n+1}\left(p_{n+1} - p_{\mathrm{c},\,n+1}\right) \tag{8.6}$$

where M denotes the slope of the critical state line in the p-q space. p_{n+1} and q_{n+1} are the mean effective stress and generalized shear stress. $p_{\mathrm{c},\,n+1}$ is the pre-consolidation pressure and its integral formula is expressed by:

$$p_{\mathrm{c},\,n+1} = p_{\mathrm{c},\,n} \exp\left[c_{\mathrm{p}} \Delta\phi_{n+1}\left(2p_{n+1} - p_{\mathrm{c},\,n+1}\right)\right] \tag{8.7}$$

where $c_{\mathrm{p}} = (1 + e_0)/(\lambda - \kappa)$. λ, κ, and e_0 denote the compression index, swell index, and the initial void ratio, respectively.

For the MCC model, the updating formula of the stress tensor can also be replaced by the mean effective stress p_{n+1} and the generalized shear stress q_{n+1} to reduce the number of integration equations:

$$p_{n+1} = \frac{1}{3}\boldsymbol{\sigma}_{n+1} : \mathbf{1} = p_n \exp\left\{c_\kappa\left[\Delta\varepsilon_{\mathrm{v},\,n+1} - \Delta\phi_{n+1}\left(2p_{n+1} - p_{\mathrm{c},\,n+1}\right)\right]\right\} \tag{8.8}$$

$$q_{n+1} = \sqrt{\frac{3}{2}}\left\|\mathbf{s}_{n+1}\right\| = \sqrt{\frac{3}{2}} \frac{\left\|\mathbf{s}_n + 2\overline{G}\Delta\boldsymbol{\gamma}_{n+1}\right\|}{1 + 6\overline{G}\Delta\phi_{n+1}/M^2} \tag{8.9}$$

where $\Delta\varepsilon_{\mathrm{v},\,n+1}$ is the total volume strain increment. $\overline{K}$ is the secant bulk modulus (Borja, 1991) and is defined by:

$$\overline{K} = \frac{p_n}{\Delta\varepsilon_{\mathrm{v},\,n+1} - \Delta\phi_{n+1}\left(2p_{n+1} - p_{\mathrm{c},\,n+1}\right)} \left\{\exp\left[c_\kappa\left(\Delta\varepsilon_{\mathrm{v},\,n+1} - \Delta\phi_{n+1}\left(2p_{n+1} - p_{\mathrm{c},\,n+1}\right)\right)\right] - 1\right\} \tag{8.10}$$

where $c_\kappa = (1 + e_0)/\kappa$. $\overline{G} = \overline{K}r$ is the secant shear modulus. $r = 3(1 - 2\nu)/(2(1 + \nu))$ where ν is the Poisson's ratio. $\Delta\boldsymbol{\gamma}_{n+1}$ is the total deviatoric strain increment tensor.

The nonlinear stress integral equations of the MCC model can be obtained by combining Eqs. (8.8), (8.9), (8.7), and (8.6), and the *FB* smooth function as follows:

$$\begin{Bmatrix} f_1 \\ f_2 \\ f_3 \\ f_4 \end{Bmatrix} = \begin{Bmatrix} p_{n+1} - p_n \exp\left[c_\kappa \Delta\varepsilon_{\mathrm{v},\, n+1} - c_\kappa \Delta\phi_{n+1}\left(2p_{n+1} - p_{\mathrm{c},\, n+1}\right)\right] \\ q_{n+1} - \sqrt{\frac{3}{2}} \frac{\left\|\mathbf{s}_n + 2\overline{G}\Delta\boldsymbol{\gamma}_{n+1}\right\|}{1+6\overline{G}\Delta\phi_{n+1}/M^2} \\ p_{\mathrm{c},\, n+1} - p_{\mathrm{c},\, n} \exp\left[c_\mathrm{p}\Delta\phi_{n+1}\left(2p_{n+1} - p_{\mathrm{c},\, n+1}\right)\right] \\ \sqrt{\left(c_\mathrm{d}\Delta\phi_{n+1}\right)^2 + f_{n+1}^2 + 2\beta} - c_\mathrm{d}\Delta\phi_{n+1} + f_{n+1} \end{Bmatrix} = \begin{Bmatrix} 0 \\ 0 \\ 0 \\ 0 \end{Bmatrix} \tag{8.11}$$

where the dimension parameter c_d is recommended equal to $\left\|\boldsymbol{\sigma}_n + \overline{\mathbb{D}}\left(\overline{\mathrm{K}}, \overline{\mathrm{G}}\right) : \Delta\boldsymbol{\varepsilon}_{\mathrm{n+1}}\right\|^3$.

After p_{n+1} and q_{n+1} are updated, the stress tensor can be updated by:

$$\mathbf{s}_{n+1} = \frac{\mathbf{s}_n + 2\overline{G}\Delta\boldsymbol{\gamma}_{n+1}}{1+6\overline{G}\Delta\phi_{n+1}/6\overline{G}\Delta\phi_{n+1}} \tag{8.12}$$

$$\boldsymbol{\sigma}_{n+1} = p_{n+1}\mathbf{1} + \mathbf{s}_{n+1} \tag{8.13}$$

In this situation, unknown variables are defined and nonlinear equations in Eq. (8.3) need to be rewritten as:

$$\{\mathbf{x}_{n+1}\} = \left\{p_{n+1}, q_{n+1}, p_{\mathrm{c},\, n+1}, \Delta\phi_{n+1}\right\}^T \tag{8.14}$$

$$\{\mathbf{f}(\mathbf{x})\} = \{f_1, f_2, f_3, f_4\}^T \tag{8.15}$$

8.2 Stress Update Algorithm with Line Search Method

In the LSM, the solution of the unconstrained minimization problem can be obtained through the following iterative process:

$$\{\mathbf{x}\}^{k+1} = \{\mathbf{x}\}^k + \alpha^k\{\mathbf{d}\}^k \tag{8.16}$$

where α^k and $\{\mathbf{d}\}^k$ are the step size and search direction vector at the kth iteration, respectively. The Newton method is usually a good choice to determine search direction due to its asymptotically quadratic rate of convergence:

$$\{\mathbf{d}\}^k = -[\nabla\mathbf{f}(\mathbf{x})]_k^{-1}\{\mathbf{f}(\mathbf{x})\}^k \tag{8.17}$$

The task of the LSM is to optimize the step length α^k for a given search direction to achieve more reduction of the merit function. This gives rise to a new one-dimensional minimization problem about α^k:

$$\min\ \psi\left(\alpha^k\right)=\frac{1}{2}\{\mathbf{f}(\alpha)\}_k^T\{\mathbf{f}(\alpha)\}^k \tag{8.18}$$

where $\{\mathbf{f}(\alpha)\}^k=\{\mathbf{f}(\mathbf{x}+\alpha\mathbf{d})\}^k$. The LSM will degenerate into the standard Newton method when α^k equals to 1. However, it is not easy to determine the optimal value of α^k by minimizing $\psi\left(\alpha^k\right)$ directly due to the fact that it will involve the computation of the Jacobian matrix. A more practical strategy is to minimize the approximation of $\psi\left(\alpha^k\right)$ to obtain an acceptable value of α^k that provides an adequate reduction in the merit function. Herein, a simple quadratic curve is used to fit $\psi\left(\alpha^k\right)$ as follows:

$$\hat{\psi}\left(\alpha^k\right)=A+B\alpha^k+C\alpha_k^2 \tag{8.19}$$

where the coefficient A, B, and C can be determined by three known points, i.e., $(0,\ \psi(0))$, $\left(\alpha_j^k,\ \psi\left(\alpha_j^k\right)\right)$, and $\left(0,\ \psi'(0)\right)$. The counter "$j$" represents the iterations when solving Eq. (8.19). α_j^k is a known step size at j th iteration. $\psi'(0)$ can be obtained by:

$$\left.\frac{\partial\psi}{\partial\alpha_k}\right|_{\alpha=0}=\mathbf{f}\left(\mathbf{x}_{n+1}^k+\alpha^k\mathbf{d}^k\right)^T\nabla\mathbf{f}\left(\mathbf{x}_{n+1}^k+\alpha^k\mathbf{d}^k\right)\mathbf{d}^k \tag{8.20}$$

Substituting Eq. (8.17) into Eq. (8.20), one can obtain:

$$\left.\frac{\partial\psi}{\partial\alpha_k}\right|_{\alpha=0}=-\left\{\mathbf{f}\left(\mathbf{x}_{n+1}^k\right)^T\right\}\left\{\mathbf{f}\left(\mathbf{x}_{n+1}^k\right)\right\}=-2\psi(0) \tag{8.21}$$

Then, A, B, and C can be determined by these three points into Eq. (8.19) as follows:

$$\begin{cases} A=\psi(0) \\ B=-2\psi(0) \\ C=\psi\left(\alpha_j^k\right)+\psi(0) \end{cases} \tag{8.22}$$

Substituting Eq. (8.22) into Eq. (8.19), one can obtain:

$$\hat{\psi}\left(\alpha^k\right)=\left(1-2\alpha^k+\alpha_k^2\right)\psi(0)+\alpha_k^2\psi\left(\alpha_j^k\right) \tag{8.23}$$

Minimizing Eq. (8.23), the following iterative formula is obtained to update α^k when the reduction of $\psi\left(\alpha^k\right)$ does not satisfy the requirements of the LSM:

$$\alpha_{j+1}^k=\frac{\psi(0)}{\psi(0)+\psi\left(\alpha_j^k\right)} \tag{8.24}$$

The same treatment can also be found in the literature (Scherzinger, 2017; Yoon et al., 2020). Then, the upper limit of α^k is determined by *Goldstein's condition* (Nocedal & Wright, 2006; Yoon et al., 2020) herein as follows:

$$\psi\left(\alpha_j^k\right) < \left(1 - 2\rho\alpha_j^k\right)\psi(0) \tag{8.25}$$

The lower limit of α^k is determined by the following expression to avoid having too small a step size:

$$\alpha_{j+1}^k = \max\left\{\varsigma\alpha_j^k, \frac{\psi(0)}{\psi(0) + \psi\left(\alpha_j^k\right)}\right\} \tag{8.26}$$

where Pérez-Foguet and Armero (2002) proposed to use $\rho = 10^{-4}$ and $\varsigma = 0.1$. Equations (8.25) and (8.26) specify an interval for the acceptable values of α^k.

The initial estimate of α_j^k is set to $\alpha_{j=0}^k = 1$. Once the *Goldstein's condition* is defined by Eq. (8.25) is satisfied, the iteration of α_j^k is terminated, otherwise, α_j^k will be updated by Eq. (8.26). It should be noted that, if α^k is fixed at 1, the LSM will degenerate into the classical Newton method. Finally, the iterative formula of α^k can be summarised as follows:

$$\begin{cases} \textit{Accept}\ \alpha_j^k\ \textit{and exit} & \textit{IF}\ \Psi\left(\alpha_j^k\right) < \left(1 - 2\rho\alpha_j^k\right)\Psi(0) \\ \alpha_{j+1}^k = \max\left\{\varsigma\alpha_j^k, \frac{\psi(0)}{\psi(0)+\psi\left(\alpha_j^k\right)}\right\} & \textit{ELSE} \end{cases} \tag{8.27}$$

Figure 8.1 shows that the calculation process of stress update algorithm with LSM for the MCC model. β is an input parameter. The smaller β, the closer the smooth curve is to the *KKT* condition. β cannot be too small to affect the numerical stability of floating-point calculation. Here, β is set to $FTOL^2/2$. This selection allows for a single step solution to be found. For example, the current step is the elastic loading, i.e., $\left\|\{\mathbf{f}\}_{n+1}^0\right\| = \sqrt{0^2 + 0^2 + 0^2 + f_4^2} = \sqrt{f_{n+1}^2 + FTOL^2} + f_{n+1}$. Then, the following inequality will hold: $\sqrt{f_{n+1}^2 + FTOL^2} + f_{n+1} \le FTOL \Rightarrow f_{n+1}^2 + FTOL^2 \le f_{n+1}^2 + FTOL^2 - 2f_{n+1}FTOL \Rightarrow 0 \le -2f_{n+1}FTOL$ due to $f_{n+1} < 0$ for the elastic step and $FTOL > 0$.

In addition, a necessary emphasis is needed for the meaning of symbols n, k, and j, in which n denotes the incremental load step, k denotes the local stress updating iteration, and j denotes the number of iterations required to obtain the optimal search step size of the LSM. Finally, we provide a synopsis of the LSM used in this chapter. First,$\alpha_0^k = 1$ is used as the initial value of step size. If the condition Eq. (8.25) is satisfied, then we set $\{\mathbf{x}\}^{k+1} = \{\mathbf{x}\}^k + \alpha^k\{\mathbf{d}\}^k$. If α_j^k exceeds the upper limit defined by Eq. (8.25), then update α_{j+1}^k using Eq. (8.26) to search the acceptable value of step size.

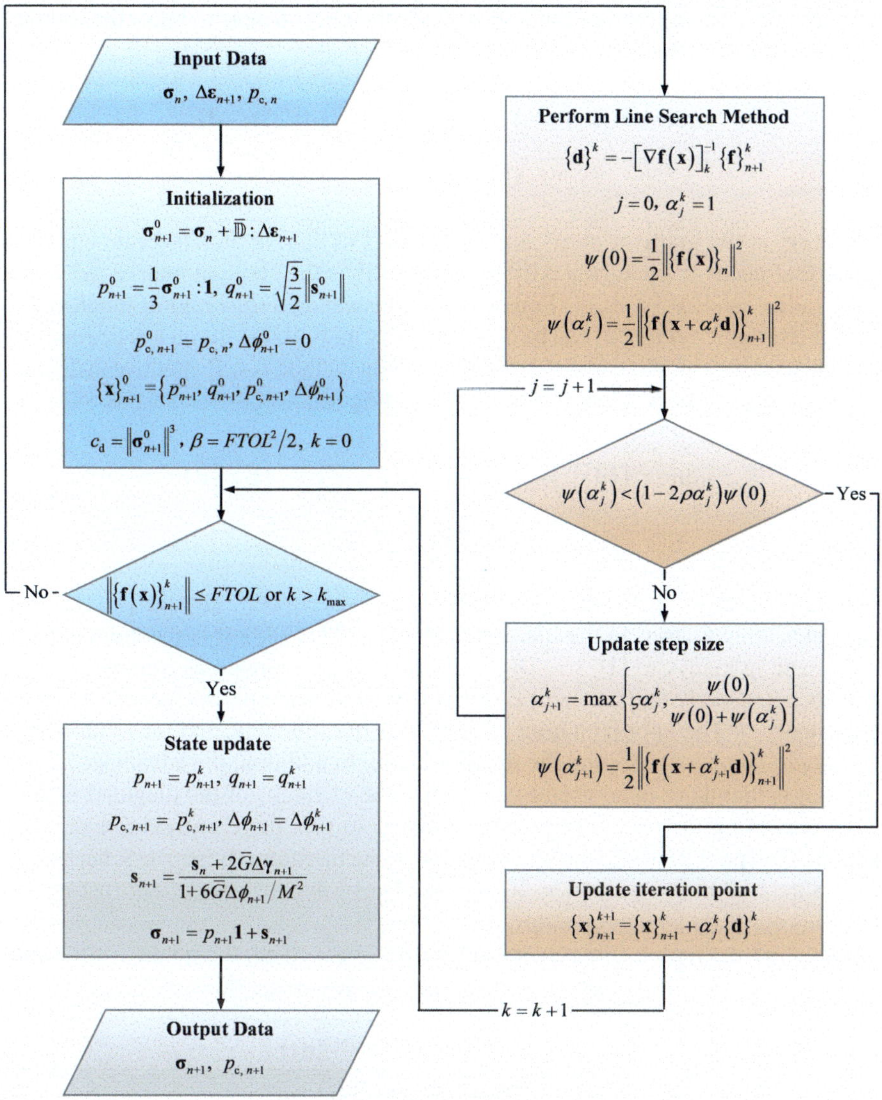

Fig. 8.1 Flow chart of unconstrained stress updating algorithm using the LSM

8.3 Stress Update Algorithm with Trust Region Method

Trust region method is another way to minimize ψ. In the trust region method, the step size is limited first by the trust region radius. Then, the search direction {d} in Eq. (8.16) is determined by the descent direction of a model function that is the approximation of the original merit function in a region around the current point.

Generally, the trust region method uses a quadratic model to approximate the original merit function as follows:

$$\min m^k(\{\mathbf{d}\}) = \psi^k + \{\mathbf{g}\}_k^T\{\mathbf{d}\} + \frac{1}{2}\{\mathbf{d}\}^T[\mathbf{B}]^k\{\mathbf{d}\}$$
$$s.t.\ \|\{\mathbf{d}\}\| \le \Delta^k \tag{8.28}$$

where $\{\mathbf{g}\}^k$ is the gradient of ψ^k and $\{\mathbf{g}\}^k = \{\mathbf{f}\}_k^T\{\nabla\mathbf{f}\}^k$. f is the algebraic equation system that needs to be solved. $[\mathbf{B}]^k = \{\nabla\mathbf{f}\}_k^T\{\nabla\mathbf{f}\}^k$ is the Hessian matrix. Δ^k is the region radius, and $\Delta^k \le \Delta_{\max}$. Equation (8.28) is called a trust region subproblem, where the variable vector $\{\mathbf{d}\}$ can be determined by minimizing the quadratic function m^k in the trust region. To ensure that the trust region method is still effective in dealing with an optimization problem that has poor scaling, Eq. (8.28) is recast as follows:

$$\min m^k\left(\{\tilde{\mathbf{d}}\}\right) = \psi^k + \{\tilde{\mathbf{g}}\}_k^T\{\tilde{\mathbf{d}}\} + \frac{1}{2}\{\tilde{\mathbf{d}}\}^T\left[\tilde{\mathbf{B}}\right]^k\{\tilde{\mathbf{d}}\}$$
$$s.t.\ \left\|\{\tilde{\mathbf{d}}\}\right\| \le \Delta^k \tag{8.29}$$

where $\{\tilde{\mathbf{g}}\}^k = [\mathbf{D}^s]^{-1}\{\tilde{\mathbf{g}}\}^k$, $\{\tilde{\mathbf{d}}\}^k = [\mathbf{D}^s]\{\mathbf{d}\}^k$, and $\left[\tilde{\mathbf{B}}\right]^k = [\mathbf{D}^s]^{-1}[\mathbf{B}]^k[\mathbf{D}^s]^{-1}$. $[\mathbf{D}^s]$ is a diagonal scaling matrix. In the literature (Moré, 1978, 1983), it is recommended to set the scaling matrix to $\left[\mathbf{D}_{ii}^s\right] = [\mathbf{B}_{ii}]^0$. This can be interpreted as follows: if B_{ii}^0 is large, then it is likely that the merit function ψ is more sensitive to changes in certain components x_i. The approximation of m to ψ is questionable, and thus, the search area should be restricted along the ith coordinate. In addition, the advantage of this approach is that there is no need to customize the elements of the diagonal scaling matrix according to specific physical problems, so it is more universal than other methods. The poor scaling problems reported in the literature (Lester & Scherzinger, 2017; Shterenlikht & Alexander, 2012) can also be avoided when the trust region method is applied to stress updating.

Once the trust region subproblem in Eq. (8.29) is solved, the quality of the trial step $\{\mathbf{x}^*\}^{k+1} = \{\mathbf{x}\}^k + \{\mathbf{d}\}$ can be evaluated by the following ratio:

$$\rho^k = \frac{\psi\left(\{\mathbf{x}\}^k\right) - \psi\left(\{\mathbf{x}\}^k + \{\mathbf{d}\}\right)}{m_k(\{\mathbf{0}\}) - m_k(\{\mathbf{d}\})} \tag{8.30}$$

where the numerator denotes the actual reduction of the merit function and the denominator denotes the predicted reduction of the model function. If ρ^k is close to 1, this shows that the model function and the merit function are very good approximations in the trust region. The trial point $\{\mathbf{x}^*\}^{k+1}$ can be accepted as the next iteration point $\{\mathbf{x}\}^{k+1}$, and the trust region radius Δ^k is amplified for the next iterative calculation. If ρ^k is close to 0 or even less than 0, then one should consider rejecting the trial step and reducing the trust region radius to resolve the trust region subproblem in Eq. (8.29). The trust region method that three areas need to be addressed for the successful application of the trust region method: (i) How can the trust region radius be determined? (ii) What are the criteria for accepting the trial point? (iii) How can the trust region subproblem be solved?

8.3.1 Trust Region Radius Determination

The reasonable choice of the trust region radius, including the initial trust region radius Δ^0 and the maximum trust region radius $\Delta_{\max}$, has a great influence on the convergence speed and stability of the trust region method (Walmag & Delhez, 2005; Walmag & Delhez, 2005). Utilizing too small a trust region radius reduces the convergence speed. If the trust region radius is too large, it will lose meaning. The reliability of the approximation of the model function to the merit function also decreases. In numerous studies regarding the trust region method that do not involve practical physical problems (Conn et al., 2000; Hei, 2003; Zhang et al., 2002), there is little evidence to support a particular setting of the trust region radius, or additional calculation efforts are required to provide a reasonable constraint range for the trust region radius. However, for the elastoplastic stress updating problem based on an incremental calculation, the maximum admissible stress increment can be determined by the difference between the converged stresses of step n and the elastic trial stress of step $n+1$. The upper limit of the allowable plastic strain increment is the total strain increment. Therefore, the maximum change among the four variables in Eq. (8.11) can be determined as follows:

$$\{\Delta\mathbf{x}\}_{\max} = \begin{Bmatrix} p_n \exp\left(c_{\mathrm{k}}\Delta\varepsilon_{\mathrm{v},\,n+1}\right) - p_n \\ \sqrt{\frac{3}{2}}\left\|\mathbf{s}_n + 2\overline{G}\Delta\boldsymbol{\gamma}_{n+1}\right\| - q_n \\ p_{\mathrm{c},\,n}\exp\left(c_{\mathrm{p}}\Delta\varepsilon_{\mathrm{v},\,n+1}\right) - p_{\mathrm{c},\,n} \\ \sqrt{\dfrac{\Delta\varepsilon_{\mathrm{v},\,n+1}^2 + 2\left\|\Delta\boldsymbol{\gamma}_{n+1}\right\|^2/3}{\left(2p_n - p_{\mathrm{c},\,n}\right)^2 + \left(2q_n/M^2\right)}} \end{Bmatrix} \tag{8.31}$$

and considering the influence of the diagonal scaling matrix on the trust region radius, the upper limit of the trust region radius can be determined by:

$$\Delta_{\max} = \left\|\left[\mathbf{D}^s\right]\{\Delta\mathbf{x}\}_{\max}\right\| \tag{8.32}$$

The initial trust region radius Δ^0 can be set to $\Delta_{\max}$ to consider the situation involving single-step iterative convergence.

8.3.2 Non-monotonic Trust Region Method

In the basic trust region method, the criteria for accepting the trial point is determined by considering only the reduction in the current iteration step. Trial points that cause increases in the merit function are not allowed to be accepted. The basic framework of the trust region method following this monotonic search strategy is detailed below:

Step 0. Initialization. The initial iteration point $\{\mathbf{x}\}^0$, the trust region radius Δ^0, and the small positive tolerance *FTOL* are given. $\Delta_{\max}$ is chosen as the upper limit of the trust region radius. Set $k = 0$.

Step 1. Convergence judgment. Compute the norm of the residual vector $\left\|\mathbf{f}\left(\{\mathbf{x}\}^k\right)\right\|$. If $\left\|\mathbf{f}\left(\{\mathbf{x}\}^k\right)\right\| \le FTOL$, the iterative process is terminated.

Step 2. Step calculation. Solve the trust region subproblem in Eq. (8.29) and obtain the trial point $\{\mathbf{x}^*\}^{k+1}$.

Step 3. Estimation of the trial point. Compute the ratio ρ^k by using Eq. (8.30).
If $\rho_k > 0.1$, accepted the trial point and set $\{\mathbf{x}\}^{k+1} = \{\mathbf{x}^*\}^{k+1}$; set $\{\mathbf{x}\}^{k+1} = \{\mathbf{x}\}^k$.

Step 4. Adjustment of the region radius. Adjust the trust region radius in the next iteration step as follows:

$$\Delta^{k+1} = \begin{cases} 0.5\Delta^k & \rho^k < 0.1 \\ \Delta^k & 0.1 \le \rho^k \le 0.75 \\ \min\left(2\Delta^k, \Delta_{\max}\right) & \rho^k > 0.75 \end{cases} \tag{8.33}$$

Set $k = k + 1$ and go to *Step 1.*

Based on the ratio ρ_k of predicted reduction and actual reduction, the quality of the trial step is divided into three kinds: 'bad', 'good', and 'very good'. If the trial step is 'bad', discard it and reduce the trust region radius. If the trial step is 'good' or 'very good', accept it. The difference is that the trust region radius will remain constant for the 'good' case and be enlarged for the 'very good' case. As seen from *Step 3*, however, in the monotonic trust region (MTR) method, the merit function is required to decrease continuously in each successful iteration because the trial point can be accepted only when ρ_k is greater than 0. However, this condition is too harsh for practical physical problems, and it may lead the search to fall into a local optimum and reduce the possibility of finding the optimal solution. Therefore, some researchers try to introduce a non-monotonic strategy in the acceptance criteria for trial points (Ahookhosh & Amini, 2012; Sun, 2004; Toint, 1997). For the non-monotonic strategy, the merit function in the search process is allowed a limited increase, but the overall trend needs to remain downward. This treatment can enhance the robustness of the trust region method with very little extra calculation cost. In this section, the non-monotonic strategy proposed by Toint (1997) is used to improve the MTR method above, where only *Step 3* needs to be adjusted.

Step 3 of the NMTR method. Estimation of the trial point. Compute the largest index such that:

$$\psi^{M(k)} = \max_{i=\max[l-M_{\max},0],\cdots,l} \psi^i \tag{8.34}$$

where $\psi^i = \psi\left(\{\mathbf{x}\}^i\right)$ is the sequence of values of the merit function at the ends of successful iterations. $M_{\max}$ is a positive integer indicating the maximum number of previously successful iterations that are considered when evaluating the trial point. It is recommended to set $M_{\max}$ to 10 (Toint, 1997). l denotes the number of successful iterations and is initialized to 0 in Step 0.

Then, compute $\psi\left(\{\mathbf{x}\}^k + \{\mathbf{d}\}\right)$ and the following ratio:

$$\rho_1^k = \begin{cases} \frac{\psi^{M(k)} - \psi\left(\{\mathbf{x}\}^k + \{\mathbf{d}\}\right)}{\sum_{i=M(k)}^{l} d^l} & \text{if } l > 0, \\ 0 & \text{otherwise} \end{cases} \tag{8.35}$$

Compute the ratio of the actual reduction to the predicted reduction at the current step:

$$\rho_2^k = \frac{\psi\left(\{\mathbf{x}\}^k\right) - \psi\left(\{\mathbf{x}\}^k + \{\mathbf{d}\}\right)}{m_k(\{\mathbf{0}\}) - m_k(\{\mathbf{d}\})} \tag{8.36}$$

Finally, the ratio used to evaluate the trial point is determined by:

$$\rho^k = \max\left(\rho_1^k, \rho_2^k\right) \tag{8.37}$$

If $\rho^k < 0.1$, then set $\{\mathbf{x}\}^{k+1} = \{\mathbf{x}\}^k$. Otherwise, set $\{\mathbf{x}\}^{k+1} = \{\mathbf{x}^*\}^{k+1}$, increment l by one and set:

$$\psi^l = \psi\left(\{\mathbf{x}\}^k + \{\mathbf{d}\}\right) \, and \, d^l = m_k(\{\mathbf{0}\}) - m_k(\{\mathbf{d}\}) \tag{8.38}$$

8.3.3 Double Dogleg Method for Subproblem Solving

Methods for solving trust region subproblems have been widely studied (Yuan, 2015), such as the singular dogleg method (Powell, 1970), double dogleg method (Dennis & Wei, 1979), two-dimensional subspace minimization method (Byrd et al., 1988), and conjugate gradient method (Steihaug, 1983). Herein, the double dogleg method is used to approximate the solution of the trust region subproblem. The double dogleg method is an improved version of the single dogleg method. In both methods, the dogleg step is determined by a combination of the Cauchy step and Newton step. According to the relative positions of the Cauchy point and the Gauss–Newton point to the trust region boundary, three cases for the determination of the dogleg step are schematically represented in Fig. 8.2.

First, the Cauchy point $\{\mathbf{x}_C\}^{k+1} = \{\mathbf{x}\}^k + \{\mathbf{d}_C\}$ is determined in the steepest descent direction, where $\{\mathbf{d}_C\} = -\{\tilde{\mathbf{g}}\}^k$. Then, the position of the Cauchy point relative to the trust region boundary needs to be estimated. If $\|\{\mathbf{d}_C\}\| \geq \Delta^k$, the intersection point between the Cauchy step and trust region boundary is used as the minimum point of

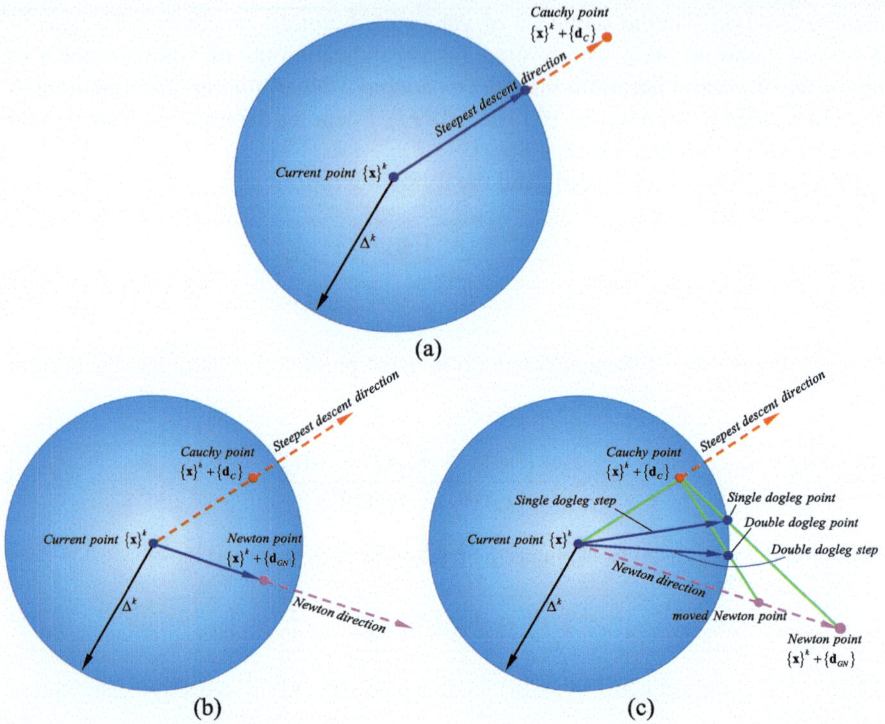

Fig. 8.2 Three cases of the dogleg method: **a** Cauchy step; **b** Newton step; **c** Single/double dogleg step

the quadratic model, as shown in Fig. 8.2a. If the Cauchy point is interior, we continue to calculate the Newton point $\{\mathbf{x}_{GN}\}^{k+1} = \{\mathbf{x}\}^k + \{\mathbf{d}_{GN}\}$, where $\{\mathbf{d}_{GN}\} = -\left[\tilde{\mathbf{B}}\right]_k^{-1}\{\tilde{\mathbf{g}}\}^k$. If the Newton point is interior, as shown in Fig. 8.2b, it is used as the minimum point of the quadratic model. If $\|\{\mathbf{d}_{GN}\}\| > \Delta^k$, the dogleg step is calculated, as shown in Fig. 8.2c. In the single dogleg method, the minimum point of the quadratic model is determined by the intersection point of between the trust region boundary and the dogleg segment, which is a segment connecting the Cauchy point and Newton point. In the double dogleg method, the position of the point connecting the Cauchy point to form the dogleg segment is moved towards the trust region boundary, as shown in Fig. 8.2c. The search direction generated by the double dogleg method is more inclined to the newton direction than that produced by the single dogleg step. Numerical experimental results have shown that the performance of the algorithm can be improved to a certain extent by this treatment (Dennis & Wei, 1979).

The double dogleg method is detailed below to complete *Step 2* in the NMTR method:

Step 2.0. Calculate the gradient and Hessian of the merit function at the current point $\{\mathbf{x}\}^k$ and scale them diagonally as follows:

$$\{\tilde{\mathbf{g}}\} = [\mathbf{D}^s]^{-1}\{\mathbf{g}\}^k = [\mathbf{D}^s]^{-1}\{\mathbf{f}\}_k^T[\nabla\mathbf{f}]^k \tag{8.39}$$

$$[\tilde{\mathbf{B}}]^k = [\mathbf{D}^s]^{-1}[\mathbf{B}]^k[\mathbf{D}^s]^{-1} = [\mathbf{D}^s]^{-1}[\nabla\mathbf{f}]_k^T[\nabla\mathbf{f}]^k[\mathbf{D}^s]^{-1} \tag{8.40}$$

Step 2.1. Calculate the Cauchy step:

$$\{\mathbf{d}_C\} = -\tau_1^k\Delta^k\frac{\{\tilde{\mathbf{g}}\}^k}{\left\|\{\tilde{\mathbf{g}}\}^k\right\|} \tag{8.41}$$

where the coefficient τ_1^k guarantees that the Cauchy step does not exceed the trust region boundary and is expressed as follows:

$$\tau_1^k = \min\left[1, \frac{\left\|\{\tilde{\mathbf{g}}\}^k\right\|^3}{\Delta^k\{\tilde{\mathbf{g}}\}_k^T[\tilde{\mathbf{B}}]^k\{\tilde{\mathbf{g}}\}^k}\right] \tag{8.42}$$

If $\|\{\mathbf{d}_C\}\| = \Delta^k$, set $\{\mathbf{d}\} = [\mathbf{D}^s]^{-1}\{\mathbf{d}_C\}$ and go to *Step 2.4*.

Step 2.2. If $\left\|\{\mathbf{d}_C\}^k\right\| < \Delta^k$, then calculate the Newton step:

$$\{\mathbf{d}_{GN}\} = -[\tilde{\mathbf{B}}]_k^{-1}\{\tilde{\mathbf{g}}\}^k \tag{8.43}$$

If $\|\{\mathbf{d}_{GN}\}\| \le \Delta^k$, set $\{\mathbf{d}\} = [\mathbf{D}^s]^{-1}\{\mathbf{d}_{GN}\}$ and go to *Step 2.4*.

Step 2.3. If $\left\|\{\mathbf{d}_{GN}\}^k\right\| > \Delta^k$, then calculate the double dogleg step:

$$\{\mathbf{d}_D\} = \{\mathbf{d}_C\} + \tau_2^k\left(\tau_3^k\{\mathbf{d}_{GN}\} - \{\mathbf{d}_C\}\right) \tag{8.44}$$

where $\tau_3^k \in \left(\tau_4^k, 1\right)$ and τ_4^k is obtained as follows:

$$\tau_4^k = \frac{\left\|\{\tilde{\mathbf{g}}\}^k\right\|^4}{\left(\{\tilde{\mathbf{g}}\}_k^T[\tilde{\mathbf{B}}]^k\{\tilde{\mathbf{g}}\}^k\right)\left(\{\tilde{\mathbf{g}}\}_k^T[\tilde{\mathbf{B}}]_k^{-1}\{\tilde{\mathbf{g}}\}^k\right)} \tag{8.45}$$

where τ_3^k is usually set to $0.8\tau_4^k + 0.2$. The double dogleg method degenerates into the single dogleg method under the condition where $\tau_3^k = 1$. τ_2^k can be obtained by solving the quadratic equation $\left\|\{\mathbf{d}_C\} + \tau_2^k\left(\tau_3^k\{\mathbf{d}_{GN}\} - \{\mathbf{d}_C\}\right)\right\| = \Delta_k$ as follows:

$$\tau_2^k = \frac{-\{\mathbf{v}\}_k^T\{\mathbf{d}_C\} \pm \sqrt{\left(\{\mathbf{v}\}_k^T\{\mathbf{d}_C\}\right)^2 - \left\|\{\mathbf{v}\}^k\right\|^2\left(\|\mathbf{d}_C\|^2 - \Delta_k^2\right)}}{\left\|\{\mathbf{v}\}^k\right\|^2} \tag{8.46}$$

where $\{\mathbf{v}\}^k = \tau_3^k\{\mathbf{d}_{GN}\} - \{\mathbf{d}_C\}$, and the positive root of Eq. (8.46) is always chosen due to $0 \le \tau_2^k \le 1$.

Finally, set $\{\mathbf{d}\} = [\mathbf{D}^s]^{-1}\{\mathbf{d}_D\}$.

Step 2.4. Exit with $\{\mathbf{x}^*\}^{k+1}=\{\mathbf{x}\}^k + \{\mathbf{d}\}$.

The implementation of the MCC model based on the trust region algorithm is briefly summarized as follows: First, the rate constitutive equations of the MCC model constrained by the loading/unloading equality are discretized into a set of nonlinear algebraic equations by the backward Euler scheme. Then, the nonlinear equations constrained by the loading/unloading equality are transformed into a smooth form by using the *FB* smoothing function instead of the non-smooth *KKT* conditions. Finally, the smooth nonlinear constitutive equations are considered an equivalent unconstrained minimization problem, which is solved further by the double dogleg trust region method (improved by the non-monotonic strategy). The computation procedures of the proposed stress updating algorithm for the MCC model are presented in Fig. 8.3. It should be emphasized that the proposed algorithm can work for all elastoplastic models. For the numerical implementation of soil models based on the critical state concept, there are typically nonlinear and non-smooth characteristics, which can be addressed better by the proposed algorithm. In addition, the smooth function and trust region method can work for the finite strain plasticity (Areias & Rabczuk, 2010; Areias et al., 2015) because they are essentially used to address common computational difficulties in the numerical implementation of the plasticity model, i.e., the non-smoothness and nonlinearity. In this chapter, the MCC model with the small formulation is used as an example.

8.4 Numerical Examples

As mature finite-element simulation software, ABAQUS is an excellent analysis platform for assessing algorithm performance. In this section, the MCC model is implemented in ABAQUS based on the proposed unconstrained stress updating algorithm. Then, the performance of the proposed algorithm is evaluated through numerical analysis at the integration point level and structural level and is compared with the ABAQUS default algorithm. As a necessary statement, the following numerical examples are run on a single processor of the same computer. The computer configuration includes an Intel Core i7-9750 processor @ 2.60 GHz, 16 GB of RAM, and the Windows 10 Ultimate 64-bit operating system. The tolerance of the proposed algorithm is set to $FTOL = 10^{-6}$.

8.4.1 Element Test

First, to evaluate the convergence of the stress update algorithm based on the NMTR method, a simple numerical example with only a single element is performed, where the possibility of calculation interruption caused by newton iteration failure at the global scale is minimized. The parameters of the MCC model used for the numerical example are given as follows: $M = 1$, $\lambda = 0.15$, $\kappa = 0.03$, $\nu = 0.3$, and $e_0 = 0.5$.

Input: $\boldsymbol{\sigma}_n, p_{c,n}$, and $\Delta\boldsymbol{\varepsilon}_{n+1}$
Compute $\boldsymbol{\sigma}_{n+1}^{trial}$ as the initial point of iteration and set $FTOL$
Set $k = 0$, $l = 0$, $p_{n+1}^0 = p_{n+1}^{trial}$, $q_{n+1}^0 = q_{n+1}^{trial}$, $\Delta\phi_{n+1}^0 = 0$, $p_{c,n+1}^0 = p_{c,n}$, and $\beta = FTOL^2/2$
Compute $\psi^0 = \psi\left(\{\mathbf{x}\}_{n+1}^0\right)$, Δ_{max} using Eq. (8.32), and set $\Delta^0 = \Delta_{max}$
do while $\left\|\mathbf{f}\left(\{\mathbf{x}\}^k\right)\right\| \le FTOL$ and $k \le k_{max}$
    Compute $\{\mathbf{x}^*\}_{n+1}^{k+1}$ using the double dogleg method
    Compute ρ_1^k using Eq. (8.35), ρ_2^k using Eq. (8.36) and $\rho^k = \max\left(\rho_1^k, \rho_2^k\right)$
    if $\rho^k < 0.1$, **then**
        Reject the trial point and set $\{\mathbf{x}\}_{n+1}^{k+1} = \{\mathbf{x}\}_{n+1}^k$
    else
        Accept the trial point, i.e., set $\{\mathbf{x}\}_{n+1}^{k+1} = \{\mathbf{x}^*\}_{n+1}^{k+1}$, and set $l = l + 1$
        Storage ψ^l and d^l using Eq. (8.38)
    end if
    Update the region radius Δ^{k+1} using Eq. (8.33)
    Set $k = k + 1$
end do
Update the stress $\boldsymbol{\sigma}_{n+1}$ using Eqs. (8.12) and (8.13)
Compute the consistent tangent operator $\mathbb{D}^{CTO}$
Output: $\boldsymbol{\sigma}_{n+1}, p_{c,n+1}$, and $\mathbb{D}^{CTO}$

Fig. 8.3 Flow chart of the stress updating algorithm

The initial stress state and the pre-consolidation pressure are set to $\sigma_1 = \sigma_2 = \sigma_3 = 200$ kPa and $p_c = 200$ kPa. Note that the $\boldsymbol{\sigma}_n$ is used as the initial point for iteration to create a more demanding iteration condition in this subsection. The loading paths of the element are shown in Fig. 8.4a. During the loading process, a condition requiring equal time step sizes is imposed. Based on the unconstrained stress updating strategy, the stress and strain results for the Newton method and the NMTR method are presented in Fig. 8.4b–f. The minimum number of load steps needed to achieve convergence for the two methods are first explored, and then, for the NMTR method, the prediction results under the same steps as that used by the Newton method are also presented. The N denotes the number of the load increment. Smaller N means greater load increment and higher convergence challenges. As expected, the two different implementations give rise to the same results under the same number of load steps, whereas a larger loading increment size is allowed for the NMTR method on the premise of calculation convergence.

Then, the effect of the non-monotonic technique on the trust region method in terms of improvement is presented by two loading and unloading examples. The initial stress state of the numerical examples is represented by the point A in Fig. 8.5a.

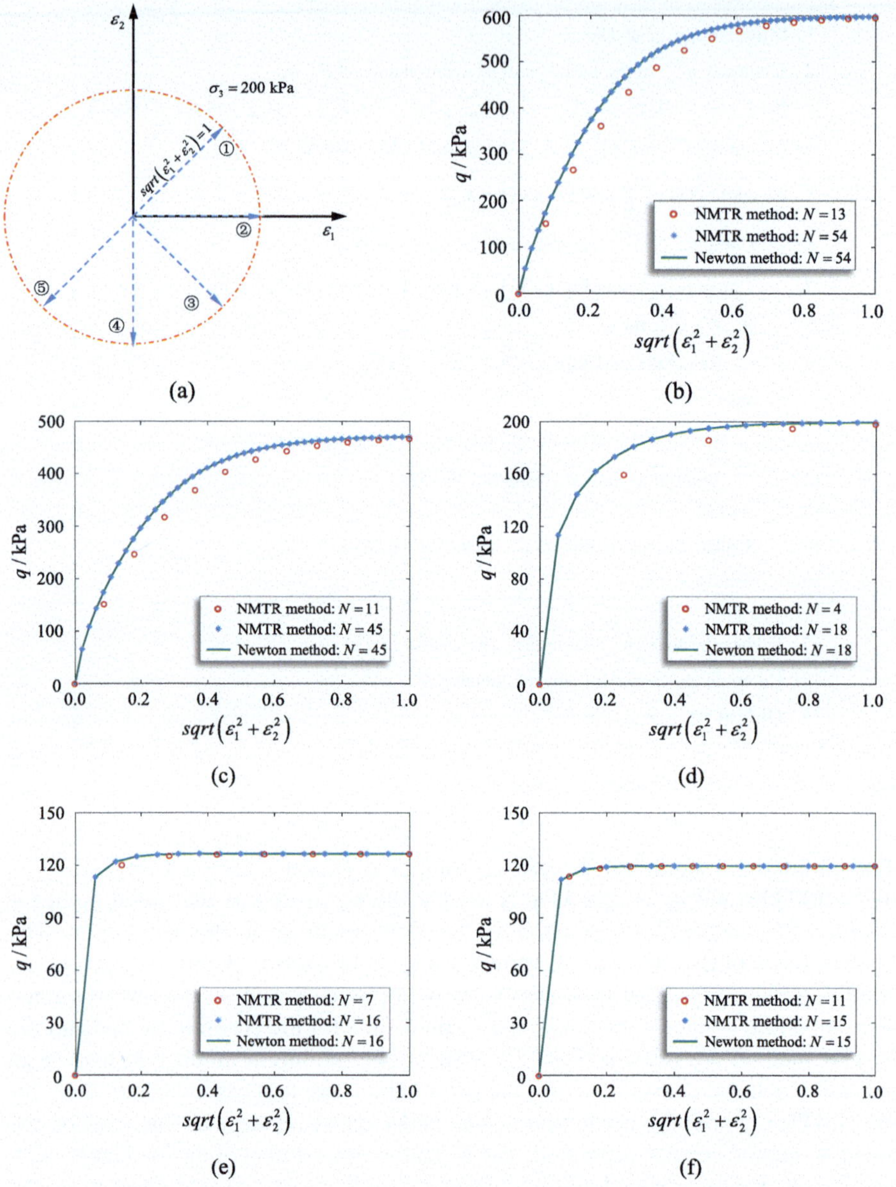

Fig. 8.4 Convergence test with a single element: **a** Considered loading paths; **b** Result of path 1; **c** Result of path 2; **d** Result of path 3; **e** Result of path 4; **f** Result of path 5

The parameters of the MCC model in the last example are also used here, but the preconsolidation pressure is set to $p_c = 218$ kPa to ensure that the current yield curve passes through the initial stress point. The strain increments are set to $\Delta\varepsilon_1 = \Delta\varepsilon_2 = -0.015$ and $\Delta\varepsilon_3 = 0$ for the unloading case and to $\Delta\varepsilon_1 = \Delta\varepsilon_2 = 0.03$ and $\Delta\varepsilon_3 = 0$ for the loading case. Only a single load step is considered. Figure 8.5b and c show the changes in $\left\| \{\mathbf{f}(\mathbf{x})\}_{n+1} \right\|$ as the number of iterations increases for the MTR method and the non-NMTR method under loading and unloading cases, respectively. The two methods converge to results satisfying the tolerance requirement under loading conditions, but a distinct computational challenge arises in the unloading case for the MTR method. Namely, for the unloading situation where the initial stress point is on the yield surface, the smoothing function has a very large curvature due to the fact that $f = 0$ and $\Delta\phi = 0$, which easily causes the search behaviour based on the monotonic strategy to fall into a local optimum. A similar phenomenon in which the MTR method may suffer from computational difficulty near areas of high curvature has also been reported in the reference (Lester & Scherzinger, 2017). The search ability of the NMTR method is enhanced because the merit function is allowed to increase limitedly at the local scope to achieve a better decline at the global scope. Therefore, the convergence of the algorithm can be guaranteed even under the condition of high curvature. Notably, the computational difficulty of the smoothing function near areas of high curvature can also be solved by choosing an appropriate initial iteration point. For instance, the stress update in the unloading cases can be realized directly without the need for second iterations when the elastic trial point is set to the initial iteration point.

8.4.2 Strip Foundation under Inclined Load

In what follows, the convergence of stress update algorithm based on the line search method LSM is investigated by comparing it with the ABAQUS default algorithm. The target example is a strip foundation under an inclined load. Figure 8.6 presents boundary conditions and finite element mesh of example. The elastic model and MCC model are adopted for the strip foundation and soil, respectively. $E = 20$ MPa and $\nu = 0.3$ are chosen for the elastic model. The material parameters of MCC model (Sheng et al., 2000) are given as follows: $M = 0.898$, $\lambda = 0.25$, $\kappa = 0.05$, $\nu = 0.3$, and $e_1 = 1.6$. The total weight of soil is set to $19\,\mathrm{kN}/\mathrm{m}^3$ to generate the initial void ratio and stress field. The top surface of the foundation is subjected to an inclined displacement load ($U_1 = -0.8$ m and $U_1 = -0.2$ m). The total analysis time and initial time increment are set to 1 s and 0.01 s. The size of subsequent time increment is determined by the ABAQUS default step control strategy.

The numerical results from the UMAT and the ABAQUS default algorithm are presented in Fig. 8.7. The UMAT's CPU time is less than half that of the ABAQUS default algorithm on the premise that the simulation results are almost the same. The reason is that the LSM used in the model implementation has a stronger convergence

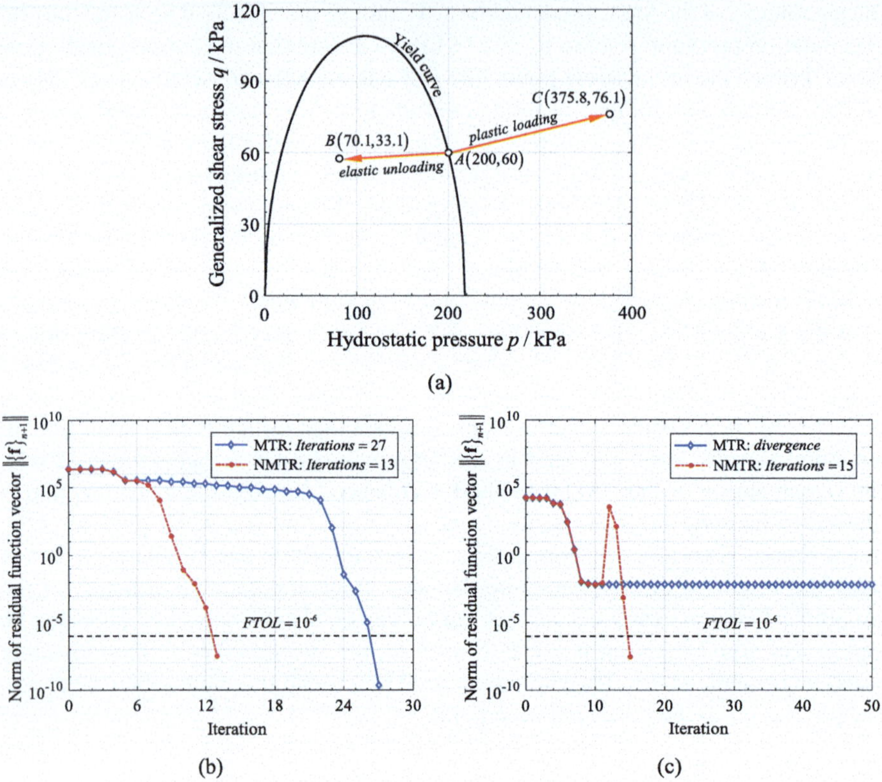

Fig. 8.5 Comparison between the non-monotonic strategy and monotonic strategy: **a** Load paths; **b** Load case; **c** Unload case

than the Newton method. Therefore, a larger load step input is allowed for the stress update algorithm based on the LSM than the ABAQUS default algorithm.

8.4.3 Cylinder under Cyclic Shear Loads

Finally, the performance of the stress update algorithm based on the MTR method is investigated again by comparing it with the ABAQUS default algorithm. The numerical example is that the cylindrical sample is subjected to cyclically shear load. The example is often used to test the numerical implementation of the constitutive model based on the optimization methods (Lester & Scherzinger, 2017; Scherzinger, 2017) since it generates a sufficiently hard stress condition to evaluate the algorithm's robustness. The cylinder in the problem is 0.2 m long with a cross section of diameter 0.1 m, as shown in Fig. 8.8a. Three discretization types are used to evaluate the performance of the proposed algorithm, as shown in Fig. 8.8b. The initial stress

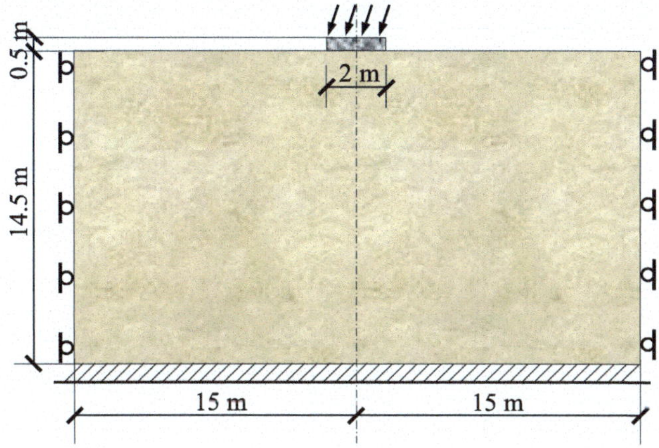

(a)

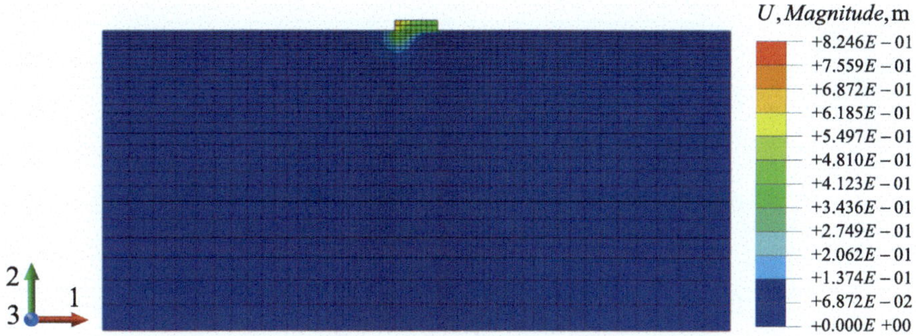

(b)

Fig. 8.6 Strip foundation: **a** Model geometry; **b** Finite element mesh

state, the pre-consolidation pressure, and material parameters of the first example in Sect. 8.4.1 are also used here. The origin of the global coordinate system is at the centre of the bottom face of the cylinder. All degrees of freedom of the bottom face of the cylinder are fixed. The top face of the cylinder is fully coupled to a reference point (0, 0, 0.25).

Figure 8.9a shows the displacement time-history curves. The displacement load in the 1-direction is imposed on the top face of the cylinder by the reference point. During the calculation process, a total of 10 load analysis steps are used, where the new analysis step is activated once the direction of the displacement load changes. During each analysis step, the initial time step is set to 0.1 s. The adjustment of the following time step size is controlled by ABAQUS default *automatic time stepping*, and the maximum time step size is the total time of the current load analysis step.

The reaction displacement curves of the reference point in the vertical direction are demonstrated for the different discretization types in Fig. 8.9b–d. The job information

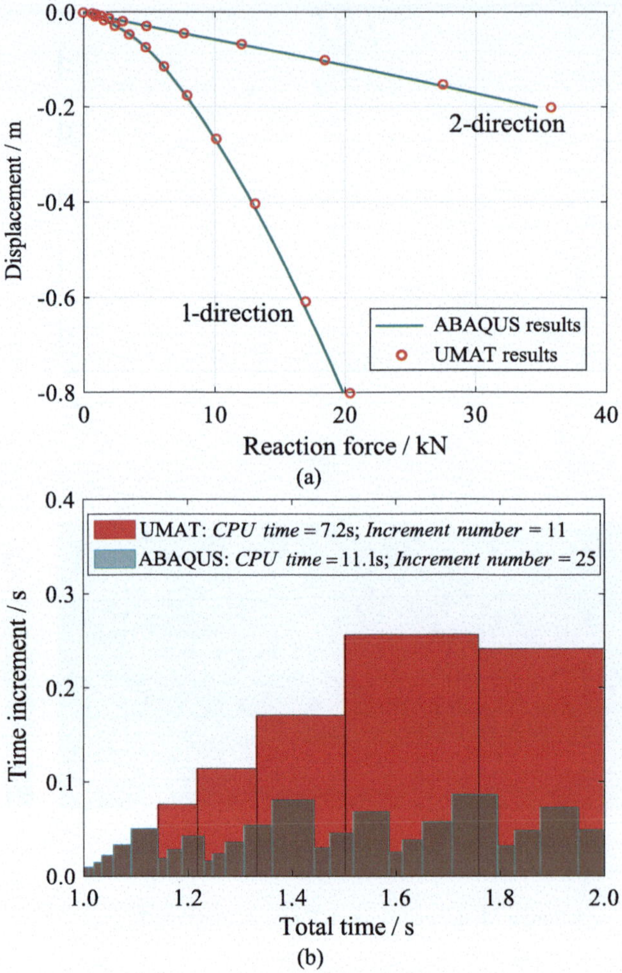

Fig. 8.7 Comparison between ABAQUS default algorithm and the presented numerical implementation with $\mu = 1.0$: **a** Reaction force versus displacement; **b** Change in time increment

with the different discretization is summarized in Table 8.1. It should be noted that, in the case of the discretization *c*, the ABAQUS default algorithm is aborted earlier than the proposed algorithm. Therefore, Table 8.1 additionally supplements the Job information of the proposed algorithm under the same analysis step time as the ABAQUS default algorithm so as to compare the computational efficiency difference between them under the same condition. It can be found that the reaction displacement curves obtained by the proposed algorithm are in good agreement with those obtained by the ABAQUS algorithm, while the former has significant computational savings compared with the latter. In what follows, the result with the discretization *a* is used as an example to present the convergence advantages of the proposed algorithm in more

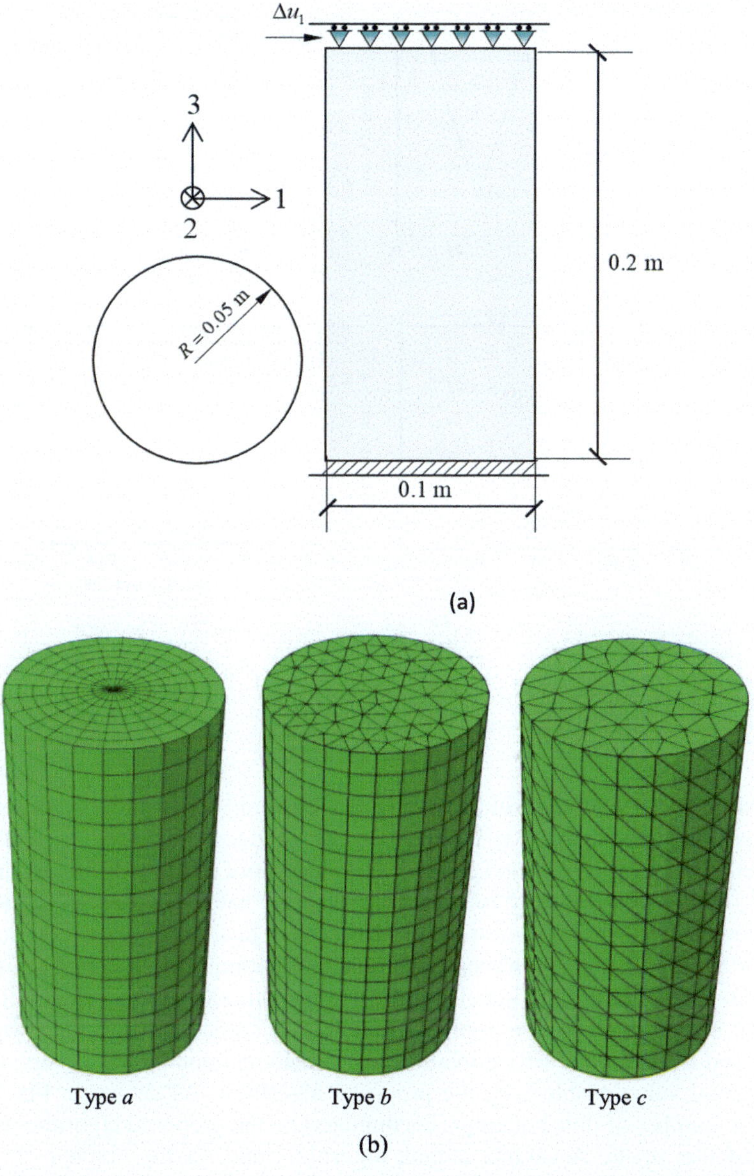

Fig. 8.8 Geometry and discretization of the cylinder sample: **a** Geometry; **b** Discretization

detail. Figure 8.10a shows that the changes in the time step size for the two algorithms for the discretization *a*. During the initial loading period, the time step sizes of the two algorithms are basically the same. Then, compared with those of the ABAQUS default algorithm, larger time steps are allowed for the proposed stress updating algorithm. The CPU time consumed by the proposed algorithm is only 44.9% of that

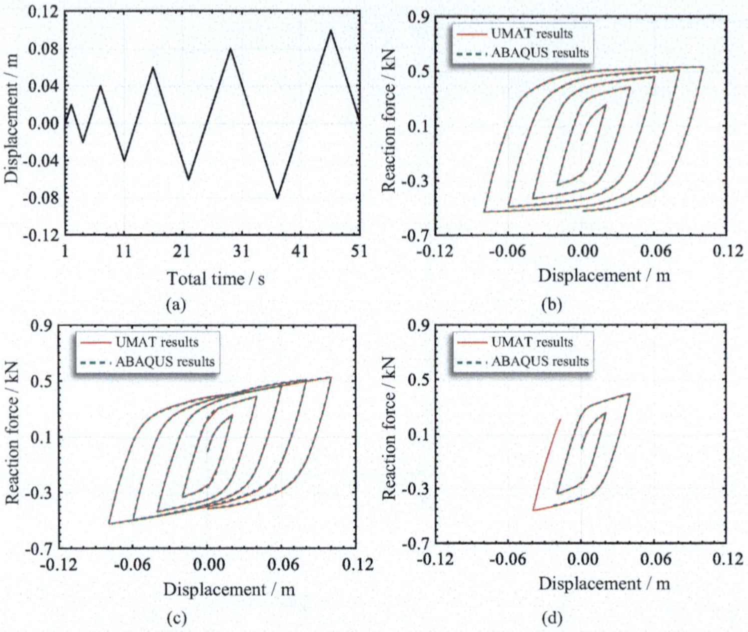

Fig. 8.9 Load and response in case *b*: **a** Displacement load; **b** Reaction force versus displacement for discretization *a*; **c** Reaction force versus displacement for discretization *b*; **d** Reaction force versus displacement for discretization *c*

of the ABAQUS default algorithm. There are a few main reasons for this difference. On the one hand, in contrast to the ABAQUS default algorithm, a local problem may be solved by the NMTR method even under larger, more complex loadings, thereby enabling the global solution to be found. On the other hand, as discussed earlier, the use of the smoothing CTO is capable of effectively guaranteeing the convergence behavior of the global solution even under repeated loading and unloading conditions because the smoothing function eliminates the non-smoothness of the elastoplastic problem. Based on these two improvements, costly cutbacks to the time step size are avoided due to the effective treatment of the non-smoothness and nonlinearity of the initial value problem by the proposed algorithm. As shown in Fig. 8.10b and c, the number of failed attempts encountered by the proposed algorithm is only 11.9% of that of the ABAQUS default algorithm, and for the number of global equilibrium iterations wasted on the failed attempts, the former is only 15.0% of the latter. Therefore, during each load analysis step, the displacement load can be applied in fewer incremental steps by the proposed algorithm, as shown in Fig. 8.10d. During the whole loading process, the number of total load increment steps required by the proposed algorithm is only 25.6% of that required by the ABAQUS default algorithm. Correspondingly, the number of global equilibrium iterations occupying the main computing resources can be reduced by the proposed algorithm, as shown in

Table 8.1 Job information with the different discretization

Mesh	Algorithms	Job status (total time/s)	CPU time/s	Elements number	Total/failed load increments		Total/failed iterations	
a	ABAQUS	Completed (51)	428.2	2520	587	118	793	140
	NMTR		192.4		150	14	376	21
b	ABAQUS	Completed (51)	315.9	2628	1102	240	1370	254
	NMTR		98.3		170	19	467	31
c	ABAQUS	Aborted (10.2964)	376.9	2771	404	97	588	151
	NMTR	Aborted (12.1412/10.2964)	227.8 (133.7)		140 (84)	30 (13)	356 (210)	97 (26)

Fig. 8.10e. The number of total global equilibrium iterations required by the proposed algorithm is only 47.4% of that required by the ABAQUS default algorithm.

8.5 Summary

The smooth nonlinear stress integral equations problem is equivalent to an unconstrained minimization problem, which can be efficiently solved by advanced numerical optimization methods. In this chapter, the inexact LSM and the NMTR method are used to solve the unconstrained minimization problem in the stress updating of the elastoplastic model, developing two stress update algorithms based on numerical optimization methods. They all show better convergence than the standard Newton method, especially under the condition of large strain increment inputs. The performance of the developed stress updating algorithms is evaluated by comparing them with the ABAQUS default algorithm. Here, the MCC model is used as the application object of the algorithm. The main conclusions of this chapter are as follows:

(1) In the proposed stress update algorithm based on the LSM, the Newton direction is used as the direction of the search step. The inexact line search strategy is used to determine the allowed step size, which improves the convergence of the solution. Compared with the exact, it saves the computational cost when obtaining the step size, but does not affect the convergence of the solution too much. The calculation example of strip foundation under inclined load also shows the proposed algorithm has better robustness and computational efficiency than ABAQUS DIM, allowing larger strain increment inputs.

(2) In the proposed stress update algorithm based on the trust region method, the size of the search step is limited by the dynamically updated trust region radius. The double dogleg method is used to determine the search direction which is a combination of Newton direction and steepest descent direction. In addition,

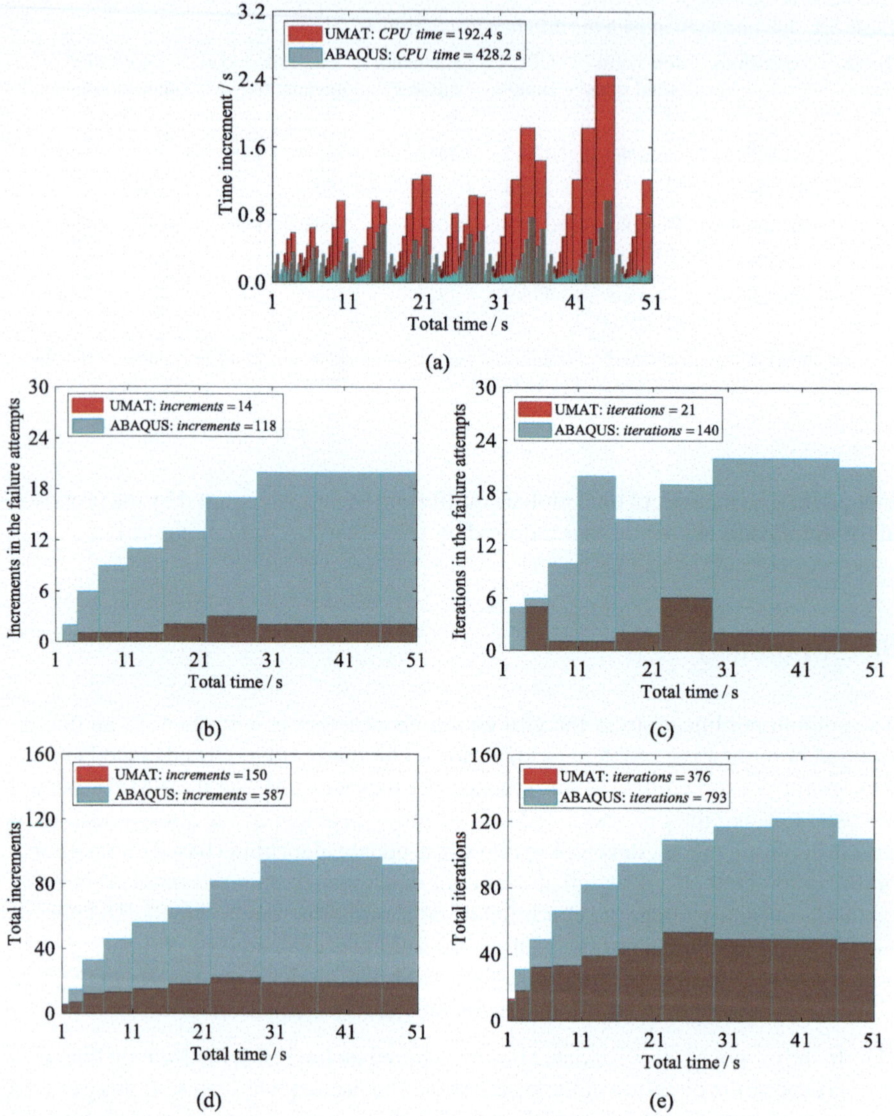

Fig. 8.10 Convergence behaviour at each load analysis step: **a** Change in the time step size; **b** Load increments in the failed attempts; **c** Global equilibrium iterations in the failed attempts; **d** The number of total load increments; **e** Total global equilibrium iterations

the non-monotonic technique is used to further improve the convergence of the trust region method, which allows the merit function to rise in some iteration steps, but the overall trend of change needs to keep declining. Numerical analysis shows that the developed algorithm considerably reduces the possibility of local calculation failure caused by the model nonlinearity and a step size that is too large. The computation cost is reduced since the finite element calculation can be completed in fewer increment steps. This superior performance ensures the efficient numerical analysis of geotechnical engineering problems and brings a prospect worth applying the proposed algorithm in other elastoplastic models.

References

Ahookhosh, M., & Amini, K. (2012). An efficient nonmonotone trust-region method for unconstrained optimization. *Numerical Algorithms, 59*(4), 523–540.

Areias, P., & Rabczuk, T. (2010). Smooth finite strain plasticity with non-local pressure support. *International Journal for Numerical Methods in Engineering, 81*(1), 106–134.

Areias, P., Rabczuk, T., & César De Sá, J. (2015). Semi-implicit finite strain constitutive integration of porous plasticity models. *Finite Elements in Analysis and Design, 104*, 41–55.

Borja, R. I. (1991). Cam-clay plasticity, part ii: Implicit integration of constitutive equation based on a nonlinear elastic stress predictor. *Computer Methods in Applied Mechanics and Engineering, 88*(2), 225–240.

Byrd, R. H., Schnabel, R. B., & Shultz, G. A. (1988). Approximate solution of the trust region problem by minimization over two-dimensional subspaces. *Mathematical Programming, 40*(1), 247–263.

Conn, A. R., Gould, N. I. M., & Toint, P. L. (2000). *Trust-Region Methods.* Society for Industrial and Applied Mathematics, Philadelphia, PA.

Dennis Jr, J. E., & Mei, H. H. (1979). Two new unconstrained optimization algorithms which use function and gradient values. *Journal of Optimization Theory and Applications, 28*(4), 453–482.

Hei, L. (2003). A self-adaptive trust region algorithm. *Journal of Computational Mathematics, 21*(2), 229–236.

Lester, B. T., & Scherzinger, W. M. (2017). Trust region based return mapping algorithm for implicit integration of elastic-plastic constitutive models. *International Journal for Numerical Methods in Engineering, 12*(3), 257–282.

More, J.J. (1977). The Levenberg–Marquardt algorithm: implementation and theory. In: Watson, G.A. (Ed.), *Numerical Analysis*, Lecture Notes in Mathematics (vol. 630, pp. 105–116). Springer, New York.

Moré, J. J. (1983). Recent developments in algorithms and software for trust region methods. In A. Bachen, M. Grötschel, & B. Korte (Eds.), *Mathematical programming: The state of the art* (pp. 258–287). Springer, Berlin.

Nocedal, J., & Wright, S. (2006). *Numerical Optimization.* Springer Science & Business Media, New York.

Pérez-Foguet, A., & Armero, F. (2002). On the formulation of closest-point projection algorithms in elastoplasticity—part II: Globally convergent schemes. *International Journal for Numerical Methods in Engineering, 53*(2), 331–374.

Powell, M.J.D. (1970) A Hybrid Method for Nonlinear Equations. In: Rabinowitz, P., (Ed.), *Numerical Methods for Nonlinear Algebraic Equations* (pp. 87-114). Gordon and Breach, London.

Scherzinger, W. M. (2017). A return mapping algorithm for isotropic and anisotropic plasticity models using a line search method. *Computer Methods in Applied Mechanics and Engineering, 317*, 526–553.

Sheng, D. C., Sloan, S. W., & Yu, H. S. (2000). Aspects of finite element implementation of critical state models. *Computational Mechanics, 26*(2), 185–196.

Shterenlikht, A., & Alexander, N. A. (2012). Levenberg-marquardt vs powell's dogleg method for gurson-tvergaard-needleman plasticity model. *Computer Methods in Applied Mechanics and Engineering, 237–240*, 1–9.

Steihaug, T. (1983). The conjugate gradient method and trust regions in large scale optimization. *Siam Journal on Numerical Analysis, 20*(3), 626–637.

Sun, W. Y. (2004). Nonmonotone trust region method for solving optimization problems. *Applied Mathematics and Computation, 156*(1), 159–174.

Toint, P. L. (1997). Non-monotone trust-region algorithms for nonlinear optimization subject to convex constraints. *Mathematical Programming, 77*(3), 69–94.

Walmag, J. M. B., & Delhez, É. J. M. (2005). A note on trust-region radius update. *Siam Journal on Optimization, 16*(2), 548–562.

Yoon, S. Y., Lee, S. Y., & Barlat, F. (2020). Numerical integration algorithm of updated homogeneous anisotropic hardening model through finite element framework. *Computer Methods in Applied Mechanics and Engineering, 372*, 113449.

Yuan, Y. X. (2015). Recent advances in trust region algorithms. *Mathematical Programming, 151*(1), 249–281.

Zhang, X. S., Zhang, J. L., & Liao, L. Z. (2002). An adaptive trust region method and its convergence. *Science in China Series A: Mathematics, 45*(5), 620–631.

Zhou, X., Lu, D. C., Zhang, Y. N., Du, X. L., & Rabczuk, T. (2022). An open-source unconstrained stress updating algorithm for the modified Cam-clay model. *Computer Methods in Applied Mechanics and Engineering, 390*, 114356.

Chapter 9
Stress Update Algorithm Based on Numerical Differentiation Methods

Implicit calculation of elastoplastic models involves the construction of the Jacobian matrix of nonlinear stress integral equations in local stress update iterations and the consistent tangent operator (CTO) in global equilibrium iterations, both of which require a substantial number of 1st and 2nd derivatives. For simple constitutive models, the derivative terms can be obtained by the manual differentiation method (MDM) that is accurate and efficient. However, the MDM can quickly become difficult, time-consuming, and error-prone as the constitutive models increase in complexity. The numerical differentiation methods (Pérez-Foguet et al., 2000) are increasingly favored by researchers since such methods are easy to implement and not sensitive to the complexity of the functions. Compared with other methods, the FDM (forward difference method) and CDM (central difference method) were used earlier by researchers for the derivative evaluation in stress update problems (Miehe, 1996; Pérez-Foguet et al., 2000) than other ones. These two methods, however, suffer from truncation and subtractive cancellation errors. Even worse, as the perturbation value decreases, the former decreases while the latter increases. The careful selection of the perturbation value is crucial to minimize the overall numerical error for the two methods (Pérez-Foguet et al., 2000). The CSDA only suffers from the truncation error due to the absence of subtraction operation in the 1st differentiation formula (Balzani et al., 2015; Kiran & Khandelwal, 2014). It is a pity that this advantage is lost when computing the 2nd derivative using the complex step derivative approximation (CSDA) (Lai & Crassidis, 2008). As a result, a practical approach to developing implicit stress updating algorithms based on CSDA is to manually derive the 1st derivative of the objective function and employ CSDA to obtain the 2nd derivative (Lu et al., 2023). The numerical differential method which is not affected by the numerical error has been eagerly desired for the implicit calculation of elastoplastic models. The hyper-dual step derivative approximation (HDSDA) developed by Fike and his co-worker (Fike, 2013; Fike & Alonso, 2011) almost achieves error-free differentiation, without truncation or subtractive cancellation errors, so its calculation results are comparable to the MDM (Fujikawa et al., 2022; Tanaka et al.,

D. Lu et al., *Multi-physical and Non-orthogonal Geotechnical Plasticity Modeling and Algorithms*, https://doi.org/10.1007/978-981-96-2369-3_9

2014). Certainly, implementing HDSDA necessitates customization of the operating rules for hyper-dual numbers (HDNs) through Operator Overloading and Function Overloading (Tanaka et al., 2015). Despite its high accuracy and effectiveness, the HDSDA-based stress update algorithms (Fohrmeister et al., 2018), particularly for elastoplastic models, remain largely unexplored, warranting further investigation.

This chapter presents an implicit stress update algorithm that utilizes the HDSDA to address derivative evaluations in elastoplastic problems. Initially, the performance of various numerical differentiation methods is discussed and compared by examining their numerical errors in the representative example. Subsequently, the HDSDA, without truncation and subtractive cancellation errors, is employed to compute the Jacobian matrix and consistent tangent operator, ensuring quadratic convergence in both local and global computations. The size of the Newton search step is optimized by the line search technique, thereby enhancing the convergence in solving nonlinear stress integral equations. Finally, the proposed stress update algorithm is used to implement the non-associated Mohr–Coulomb plasticity model in the ABAQUS software using the UMAT subroutine. The stress update algorithm's performance and its practical application in geotechnical engineering problems are demonstrated using five boundary value problems.

9.1 Numerical Differentiation Methods

This section provides a brief review of basic concepts and properties for some typical numerical differentiation methods. The differentiation formulas for various methods are derived through the Taylor series expansion of the objective function subject to a small perturbation along a specific numerical axis. The performance of different differentiation methods is compared based on a derivative evaluation problem.

9.2 Finite Difference Methods

The FDM and CDM are categorized as finite difference methods. Both methods apply a perturbation value h to the objective function along the real number axis. The Taylor series expansion of the objective function $f(x+h)$ perturbed can be written as follows:

$$f(x+h) = f(x) + \frac{\mathrm{d}f(x)}{\mathrm{d}x}h + \frac{\mathrm{d}^2 f(x)}{\mathrm{d}x^2}\frac{h^2}{2!} + \frac{\mathrm{d}^3 f(x)}{\mathrm{d}x^3}\frac{h^3}{3!} + \ldots \tag{9.1}$$

After some simple algebraic operations, Eq. (9.1) can be re-written as follows:

$$\frac{\mathrm{d}f(x)}{\mathrm{d}x} = \frac{f(x+h) - f(x)}{h} + \mathcal{O}(h) \tag{9.2}$$

where $\mathcal{O}(h)$ denotes higher-order terms. Neglecting $\mathcal{O}(h)$ in Eq. (9.2), one can obtain the differentiation formula of 1st derivative for the FDM.

$$\frac{\mathrm{d}f(x)}{\mathrm{d}x} = \frac{f(x+h) - f(x)}{h} \tag{9.3}$$

Similarly, the $f(x - h)$ is expressed by:

$$f(x-h) = f(x) - \frac{\mathrm{d}f(x)}{\mathrm{d}x}h + \frac{\mathrm{d}^2 f(x)}{\mathrm{d}x^2}\frac{h^2}{2!} - \frac{\mathrm{d}^3 f(x)}{\mathrm{d}x^3}\frac{h^3}{3!} + \cdots \tag{9.4}$$

Subtracting Eq. (9.1) from Eq. (9.4) and ignoring $\mathcal{O}(h^2)$, one can obtain the differentiation formula of 1st derivative for the CDM.

$$\frac{\mathrm{d}f(x)}{\mathrm{d}x} = \frac{f(x+h) - f(x-h)}{2h} \tag{9.5}$$

It is straightforward to calculate the derivatives using the FDM and the CDM as it only involves making multiple calls to the objective function with slightly varied inputs. However, Eqs. (9.3) and (9.5) are approximate that are affected by numerical errors, primarily truncation and subtractive cancellation errors. The truncation error arises from disregarding higher-order error terms, such as $\mathcal{O}(h)$ in the FDM and $\mathcal{O}(h^2)$ in the CDM. It will gradually diminish as the perturbation value decreases, as depicted in Fig. 9.1. The subtractive cancellation error is a specific type of round-off error stemming from the computer's operational rules, where the real numbers are approximated by the floating-point numbers with finite digits. When the perturbation value h is very small, the resulting objective function values can be very similar. Performing a subtraction operation between them will result in a significant reduction of digits. Thus, the subtractive cancellation error increases as the perturbation value decreases, contrary to the behavior of the truncation error. For the FDM and the CDM, the h needs to be meticulously chosen to reduce the overall error.

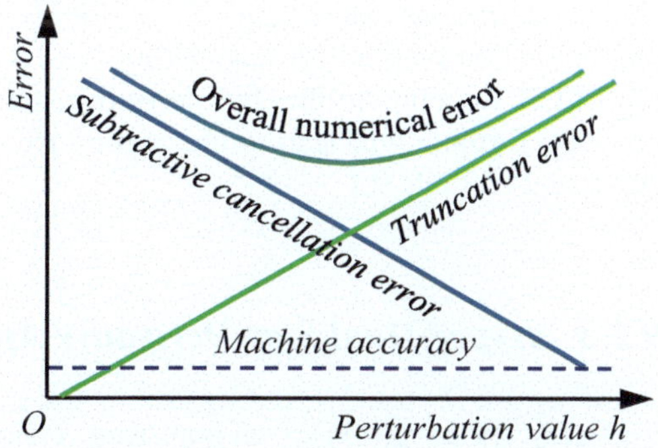

Fig. 9.1 The variation law of numerical errors with the perturbation h

Table 9.1 Differentiation formulas for different methods

Differentiation methods	$\frac{df}{dx} = (\cdot)$	Truncation error	$\frac{d^2f}{dx^2} = (\cdot)$	Truncation error
FDM	$\frac{f(x+h)-f(x)}{h}$	$O(h)$	$\frac{f(x+2h)-2f(x+h)+f(x)}{h^2}$	$O(h)$
CDM	$\frac{f(x+h)-f(x-h)}{2h}$	$O(h^2)$	$\frac{f(x+h)-2f(x)+f(x-h)}{h^2}$	$O(h^2)$
CSDA	$\frac{\Re_{\vec{i}}\left[f\left(x+\vec{i}h\right)\right]}{h}$	$O(h^2)$	$\frac{2}{h^2}\left[f(x) - \Re_{\vec{i}}\left[f\left(x+\vec{i}h\right)\right]\right]$	$O(h^2)$
HDSDA	$\frac{\Re_{\vec{\epsilon}_1}[f(x+h_1\vec{\epsilon}_1+h_2\vec{\epsilon}_2)]}{h_1}$	None	$\frac{R_{\vec{\epsilon}_1\vec{\epsilon}_2}[f(x+h_1\vec{\epsilon}_1+h_2\vec{\epsilon}_2)]}{h_1h_2}$	None

9.2.1 Complex Step Derivative Approximation

The 1st differentiation formula in the CSDA does not involve subtraction operations, thereby effectively eliminating subtractive cancellation error. By perturbing the objective function along the imaginary number axis, $f(x + \vec{i}h)$ is expressed by:

$$f\left(x + h\vec{i}\right) = f(x) + \frac{\mathrm{d}f(x)}{\mathrm{d}x}h\vec{i} - \frac{\mathrm{d}^2 f(x)}{\mathrm{d}x^2}\frac{h^2}{2!} - \frac{\mathrm{d}^3 f(x)}{\mathrm{d}x^3}\frac{h^3}{3!}\vec{i} + \cdots \tag{9.6}$$

where $\vec{i}$ is the unit imaginary number. After neglecting higher-order terms $\mathcal{O}(h^2)$, one can obtain the differentiation formula of 1st derivative for the CSDA by extracting the imaginary part of Eq. (9.6) and dividing it by h.

$$\frac{\mathrm{d}f(x)}{\mathrm{d}x} = \frac{\Re_{\vec{i}}\left[f\left(x + h\vec{i}\right)\right]}{h} \tag{9.7}$$

where $\Re_{\vec{i}}(\cdot)$ denotes the operation of taking the imaginary part. The CSDA does not suffer from the subtractive cancellation error, enabling the truncation error to be minimized below the machine precision through the selection of a sufficiently small perturbation value (e.g., $h = 10^{-20}$). The CSDA offers a highly precise approximation method for calculating the 1st derivative. However, when utilizing the CSDA to compute the 2nd derivative, the subtraction operation in the differentiation formula will lead to the occurrence of subtractive cancellation errors, as illustrated in Table 9.1.

9.2.2 Hyper-Dual Step Derivative Approximation

The HDNs (Fike, 2013) are an improved version of dual numbers proposed by Clifford (1873). The dual numbers ($a = a_0 + a_1\vec{\epsilon}$) can be regarded as a generalized complex number with a special imaginary part, where the unit "imaginary" number

$\vec{\epsilon}$ satisfies the property $\vec{\epsilon}^2 = 0$. This property will make higher-order terms $\mathcal{O}(h^2)$ in the Taylor series expansion of the objective function vanish exactly. The $f(x + h\vec{\epsilon})$ can be expressed by:

$$f(x + h\vec{\epsilon}) = f(x) + \frac{\mathrm{d}f(x)}{\mathrm{d}x} h\vec{\epsilon} \tag{9.8}$$

After some simple algebraic operations, Eq. (9.8) can be re-written by:

$$\frac{\mathrm{d}f(x)}{\mathrm{d}x} = \frac{\Re_{\vec{\epsilon}} f(x + h\vec{\epsilon})}{h} \tag{9.9}$$

where $\Re_{\vec{\epsilon}}(\cdot)$ denotes the operation of taking the "imaginary" part of dual numbers. There are no subtractive cancellation and truncation errors in Eq. (9.9). It should be noted that Eq. (9.8) does not include a 2nd derivative term, making it impossible to use dual numbers for constructing the differentiation formulas of the 2nd derivative. To address this limitation, Fike and Alonso(2011) extended dual numbers to a higher dimensional version known as the HDNs represented as $(a = a_0 + a_1\vec{\epsilon}_1 + a_2\vec{\epsilon}_2 + a_3\vec{\epsilon}_1\vec{\epsilon}_2)$. The HDNs consist of one real part and three "imaginary" parts. The latter is characterized by two "imaginary" number units $\vec{\epsilon}_1$ and $\vec{\epsilon}_2$ with the following properties:

$$\vec{\epsilon}_1^2 = 0,\ \vec{\epsilon}_2^2 = 0,\ \vec{\epsilon}_1\vec{\epsilon}_2 = \vec{\epsilon}_2\vec{\epsilon}_1 \tag{9.10}$$

By applying perturbations h_1 and h_2 to the objective function along the $\vec{\epsilon}_1$ axis and $\vec{\epsilon}_2$ axis, respectively, one can obtain the following Taylor series expansion.

$$\begin{aligned} f(x + h_1\vec{\epsilon}_1 + h_2\vec{\epsilon}_2 + 0\vec{\epsilon}_1\vec{\epsilon}_2) &= f(x) \\ &+ (h_1\vec{\epsilon}_1 + h_2\vec{\epsilon}_2)\frac{\mathrm{d}f(x)}{\mathrm{d}x} + h_1h_2\vec{\epsilon}_1\vec{\epsilon}_2\frac{\mathrm{d}^2f(x)}{\mathrm{d}x^2} \end{aligned} \tag{9.11}$$

where higher order terms $\mathcal{O}(h_1^2 + h_2^2)$ vanish exactly due to the definition of Eq. (9.10). Let us define four operation symbols $\Re(\cdot)$, $\Re_{\vec{\epsilon}_1}(\cdot)$, $\Re_{\vec{\epsilon}_2}(\cdot)$, and $\Re_{\vec{\epsilon}_1\vec{\epsilon}_2}(\cdot)$ for extracting real and "imaginary" parts of HDNs $a = a_0 + a_1\vec{\epsilon}_1 + a_2\vec{\epsilon}_2 + a_3\vec{\epsilon}_1\vec{\epsilon}_2$.

$$\Re(a) = a_0,\ \Re_{\vec{\epsilon}_1}(a) = a_1,\ \Re_{\vec{\epsilon}_2}(a) = a_2,\ \Re_{\vec{\epsilon}_1\vec{\epsilon}_2}(a) = a_3 \tag{9.12}$$

Then, differentiation formulas of 1st and 2nd derivatives for the HDSDA are obtained by:

$$\frac{\mathrm{d}f(x)}{\mathrm{d}x} = \frac{\Re_{\vec{\epsilon}_1}\left[f(x + h_1\vec{\epsilon}_1 + h_2\vec{\epsilon}_2)\right]}{h_1} = \frac{\Re_{\vec{\epsilon}_2}\left[f(x + h_1\vec{\epsilon}_1 + h_2\vec{\epsilon}_2)\right]}{h_2} \tag{9.13}$$

$$\frac{\mathrm{d}^2f(x)}{\mathrm{d}x^2} = \frac{\Re_{\vec{\epsilon}_1\vec{\epsilon}_2}\left[f(x + h_1\vec{\epsilon}_1 + h_2\vec{\epsilon})\right]}{h_1h_2} \tag{9.14}$$

Similar to Eq. (9.9), differentiation formulas for the HDSDA avoid truncation and subtractive cancellation errors. Thus, the perturbation values h_1 and h_2 h_2 can be arbitrarily chosen. By setting $h_1 = h_2 = 1$, one can obtain two simpler differentiation formulas for the HDSDA.

$$\frac{\mathrm{d}f(x)}{\mathrm{d}x} = \Re_{\vec{\epsilon}_1}\left[f(x + \vec{\epsilon}_1 + \vec{\epsilon}_2)\right] = \Re_{\vec{\epsilon}_2}\left[f(x + \vec{\epsilon}_1 + \vec{\epsilon}_2)\right] \tag{9.15}$$

$$\frac{\mathrm{d}^2 f(x)}{\mathrm{d}x^2} = \Re_{\vec{\epsilon}_1\vec{\epsilon}_2}\left[f(x + \vec{\epsilon}_1 + \vec{\epsilon}_2)\right] \tag{9.16}$$

The HDSDA can be readily extended to handle multivariate functions. For a given multivariable function $g(x, y)$ with two independent variables x and y, by applying perturbations h_1 and h_2 to the objective function along the $\vec{\epsilon}_1$ axis of variable x and $\vec{\epsilon}_2$ axis of y, respectively, one can yield:

$$\begin{aligned} g(x + h_1\vec{\epsilon}_1, y + h_2\vec{\epsilon}_2) = g(x, y) + \frac{\partial g}{\partial x}(x, y)h_1\vec{\epsilon}_1 \\ + \frac{\partial g}{\partial y}(x, y)h_2\vec{\epsilon}_2 + h_1 h_2\vec{\epsilon}_1\vec{\epsilon}_2\frac{\partial^2 g}{\partial x \partial y}(x, y) \end{aligned} \tag{9.17}$$

where higher-order terms $\mathcal{O}\left((h_1 + h_2)^2\right)$ also vanish exactly. Then, the 1st and 2nd partial derivatives of the multivariable function with respect to variables x and y can be obtained as follows:

$$\frac{\partial g(x, y)}{\partial x} = \frac{\Re_{\vec{\epsilon}_1}\left[g(x + h_1\vec{\epsilon}_1, y + h_2\vec{\epsilon}_2)\right]}{h_1} \tag{9.18}$$

$$\frac{\partial g(x, y)}{\partial y} = \frac{\Re_{\vec{\epsilon}_2}\left[g(x + h_1\vec{\epsilon}_1, y + h_2\vec{\epsilon}_2)\right]}{h_2} \tag{9.19}$$

$$\frac{\partial g(x, y)}{\partial y} = \frac{\Re_{\vec{\epsilon}_2}\left[g(x + h_1\vec{\epsilon}_1, y + h_2\vec{\epsilon}_2)\right]}{h_2} \tag{9.20}$$

The 2nd derivative of a multivariate function over a single variable can be obtained by:

$$\frac{\partial^2 g(x, y)}{\partial x^2} = \frac{\Re_{\vec{\epsilon}_1\vec{\epsilon}_2}\left[g(x + h_1\vec{\epsilon}_1 + h_2\vec{\epsilon}_2, y)\right]}{h_1 h_2} \tag{9.21}$$

$$\frac{\partial^2 g(x, y)}{\partial y^2} = \frac{\Re_{\vec{\epsilon}_1\vec{\epsilon}_2}\left[g(x, y + h_1\vec{\epsilon}_1 + h_2\vec{\epsilon}_2)\right]}{h_1 h_2} \tag{9.22}$$

9.2.3 Numerical Evaluation

Table 9.1 provides a summary of the differentiation formulas for the 1st and 2nd derivatives in the four numerical differentiation methods. Now, we will demonstrate the error laws of these differential methods by calculating the derivative of $f(x) = xe^x \Big/ \sqrt{\sin^3 x + \cos^3 x}$ at $x = \pi \big/ 12$. The HDSDA is implemented by the Operator Overloading and Function Overloading of Fortran. The source code of this example for implementing these numerical differentiation methods is provided in Appendix B. The relative numerical errors of 1st and 2nd derivatives are defined by:

$$error_1 = \left| \frac{\left.\frac{df(x)}{dx}\right|_{analytical} - \left.\frac{df(x)}{dx}\right|_{numerical}}{\left.\frac{df(x)}{dx}\right|_{analytical}} \right|, \; error_2 = \left| \frac{\left.\frac{d^2f(x)}{dx^2}\right|_{analytical} - \left.\frac{d^2f(x)}{dx^2}\right|_{numerical}}{\left.\frac{d^2f(x)}{dx^2}\right|_{analytical}} \right| \tag{9.23}$$

where $error_1$ and $error_2$ represent the relative errors of the 1st and 2nd numerical derivatives, respectively. The subscripts "analytical" and "numerical" denote the analytical and numerical derivatives, respectively.

Figure 9.2 shows the relative numerical error under different perturbation values for the 1st and 2nd derivatives, respectively. As the h begins to decrease, the numerical errors of FDM, CDM, and CSDA decrease, and their decreasing rates are consistent with the orders of the truncation error presented in Table 9.1. At this stage, the truncation error dominates the overall numerical error. Further decreasing the perturbation value leads to an increase in the numerical errors of CDM and FDM since the subtractive cancellation error begins to dominate the overall numerical error. In the case of the 1st derivative, the numerical error of CSDA approaches the machine accuracy when $h < 10^{-8}$. However, for the 2nd derivative, the numerical error of CSDA increases again as the perturbation value decreases. This is due to the reintroduction of the subtraction operation in the differentiation formula for the 2nd derivative in the CSDA, resulting in the subtractive cancellation error. On the other hand, the HDSDA always maintains the numerical errors of the 1st and 2nd derivatives near the machine accuracy, as the HDSDA does not suffer from truncation error and subtractive cancellation error.

9.3 Implicit Stress Integral Scheme of Mohr–Coulomb Plasticity Model

The HDSDA will be used for the derivative evaluation in the stress update problem. This section presents a brief overview of the non-associated Mohr–Coulomb plasticity model which serves as an application case for the proposed algorithm. The

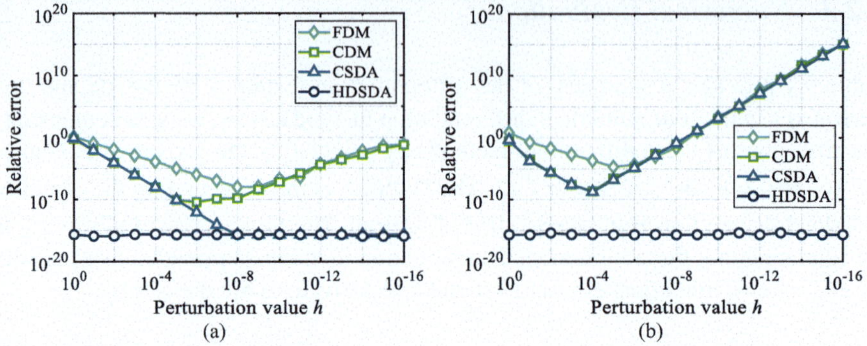

Fig. 9.2 Errors change with the perturbation value h: **a** 1st derivative; **b** 2nd derivative

selection of the Mohr–Coulomb model is motivated by its wide applicability and clear ground of physics.

9.3.1 Non-associated Mohr–Coulomb Plasticity Model

In such a plasticity model, the Mohr–Coulomb strength criterion is developed as the yield function of the material, which is expressed by:

$$f = R_{mc}(\theta)q + Mp - K \tag{9.24}$$

where $R_{mc}(\theta)$ is the shape function which defines the shape of the Mohr–Coulomb criterion in the deviatoric plane, i.e. an irregular hexagon with sides of equal length. The expression of $R_{mc}(\theta)$ is:

$$R_{mc}(\theta) = \frac{(1 - \sin\varphi)}{(3 - \sin\varphi)} \sin\theta + \frac{(3 + \sin\varphi)}{\sqrt{3}(3 - \sin\varphi)} \cos\theta \tag{9.25}$$

where φ is the internal friction angle. θ is the lode angle and can be expressed by:

$$\theta = \frac{1}{3} \arccos\left(\frac{3\sqrt{3}}{2} \frac{J_3}{J_2^{3/2}}\right) \tag{9.26}$$

where $J_2 = \frac{1}{2}\mathbf{s} : \mathbf{s}$ and $J_3 = \det(\mathbf{s})$ are the second and third stress invariants of $\mathbf{s}$, respectively.

The mean stress p and the generalized shear stress q in Eq. (9.24) are defined by:

$$p = \frac{1}{3}\boldsymbol{\sigma} : \mathbf{1} \tag{9.27}$$

$$q = \sqrt{\frac{3}{2}\mathbf{s} : \mathbf{s}} \tag{9.28}$$

where $\mathbf{1}$ is the 2nd order unit tensor. $\boldsymbol{\sigma}$ and $\mathbf{s} = \boldsymbol{\sigma} - \mathbf{1}p/3$ are the 2nd order stress tensor and deviatoric stress tensor, respectively.

The parameters M and K are determined by the internal friction angle φ and cohesion c.

$$M = \frac{6\sin\varphi}{(3-\sin\varphi)\sqrt{3}},\ K = \frac{6c\cos\varphi}{(3-\sin\varphi)\sqrt{3}} \tag{9.29}$$

where the cohesion c is used as the hardening function and is expressed as follows:

$$c = c_0 + H_{\mathrm{p}}\varepsilon_{\mathrm{d}}^{\mathrm{p}} \tag{9.30}$$

where c_0 is the initial cohesion. $\varepsilon_{\mathrm{d}}^{\mathrm{p}} = \sqrt{2/3}\left\|\boldsymbol{\varepsilon}^{\mathrm{p}} - \varepsilon_{\mathrm{v}}^{\mathrm{p}}\mathbf{1}/3\right\|_2$ is the equivalent plastic shear strain and is used as the plastic internal variable. $\varepsilon_{\mathrm{v}}^{\mathrm{p}}$ is the plastic volume strain. H_{p} is the plastic modulus. $H_{\mathrm{p}} > 0$, $H_{\mathrm{p}} = 0$, and $H_{\mathrm{p}} < 0$ denote the strain hardening, the perfect plasticity, and the strain softening, respectively. Among these, strain hardening is a phenomenon where a material becomes stronger and more resistant to further plastic deformation as it undergoes plastic deformation. The strain softening, opposite to strain hardening, represents a mechanical behavior where the material becomes weaker as the plastic deformation accumulates. The perfect plasticity is somewhere between the two. From the point of view of the stress–strain relationship, $H_{\mathrm{p}} > 0, H_{\mathrm{p}} = 0$, and $H_{\mathrm{p}} < 0$ mean that the yield stress will decrease, remain constant, and increase as the plastic strain increases, respectively.

The original Mohr–Coulomb criterion has 6 non-smooth corners on the deviatoric plane, as illustrated in Fig. 9.3b. The singularity of derivatives at corners poses challenges to the numerical implementation of constitutive models. To this end, a smooth function form of R_{mc} proposed by Zhang et al. (2021) is employed as follows:

$$R_{mc} = \frac{\sqrt{3}}{3}\alpha_w \cos\left[\frac{1}{3}\arccos(\beta\cos 3\theta) - \frac{\pi}{6}\gamma\right] \tag{9.31}$$

The $\beta \in [0, 1]$ is a smooth parameter. Equation (9.24) will degenerate into the original Mohr–Coulomb yield criterion for $\beta = 1$ and into the Drucker-Prager yield criterion for $\beta = 0$. $\beta = 0.999$ is employed herein. The parameters α_w and γ are defined by:

$$\alpha_w = \frac{1}{\cos\left[\left(\overline{\gamma}_w + 1\right)\frac{\pi}{6}\right]},\ \gamma = 1 - \overline{\gamma}_w,\ \overline{\gamma}_w = \frac{6}{\pi}\arctan\left(\frac{\sin\varphi}{\sqrt{3}}\right) \tag{9.32}$$

The non-associated flow rule can more reasonably capture the dilatancy behavior of geomaterials. Herein, the hyperbolic Menetrey-Willam function (Menetrey &

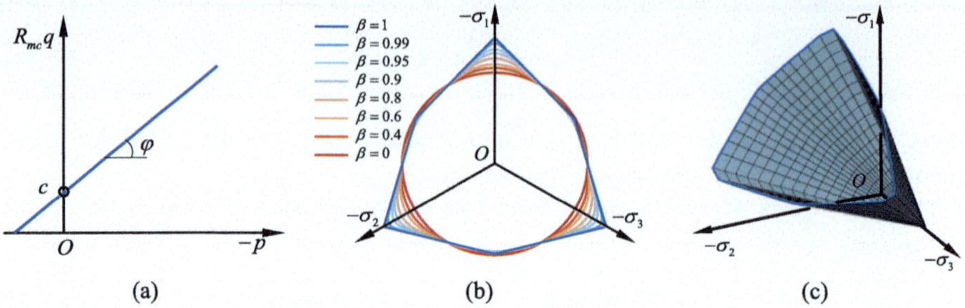

Fig. 9.3 Mohr–Coulomb yield function at: **a** The meridian plane; **b** The deviatoric plane; **c** The principal stress space

Willam, 1995) is employed as the plastic potential function, as shown in Fig. 9.4. This selection aligns with the Mohr–Coulomb model embedded in ABAQUS software.

$$g = \sqrt{(\epsilon c_0)^2 + (R_{mw}q)^2} + p \tan \psi - c_0 \tag{9.33}$$

where ψ is the dilation angle. ϵ is the meridional eccentricity which controls the proximity of the hyperbolic curve to the asymptote line ($q = -p\tan \psi / R_{mw}$), as shown in Fig. 9.4a. The default value $\epsilon = 0.1$ is used in the ABAQUS software. R_{mw} controls the shape of the plastic potential function.

$$R_{mw} = \frac{4(1-e^2)\cos^2\theta + (2e-1)^2}{2(1-e^2)\cos\theta + (2e-1)\sqrt{4(1-e^2)\cos^2\theta + 5e^2 - 4e}} \left(\frac{1}{2\cos\varphi} - \frac{1}{6}\tan\varphi\right) \tag{9.34}$$

where $e \in (0.5, 1]$ is the eccentricity parameter, as shown in Fig. 9.4b expressed by:

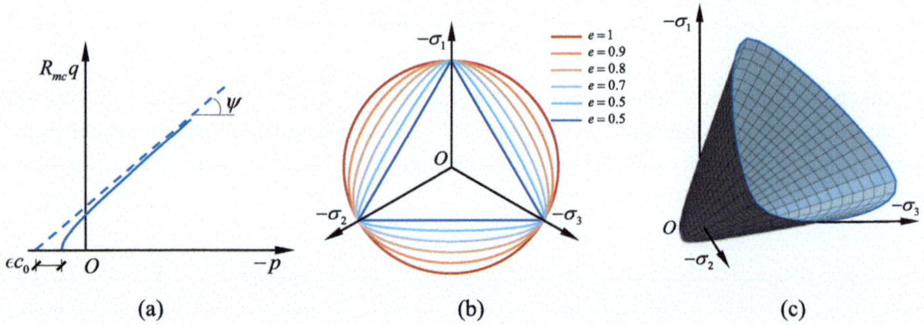

Fig. 9.4 Plastic potential function at: **a** The meridian plane; **b** The deviatoric plane; **c** The principal stress space

$$e = \frac{3 - \sin\varphi}{3 + \sin\varphi} \tag{9.35}$$

9.3.2 Implicit Stress Integration Scheme

Based on the yield function, plastic internal variable, and plastic potential function presented in Sect. 9.4.1, we can give the control differential equations used for the elastoplastic calculation. For the existence of material nonlinearity, the plasticity model generally descript the stress–strain behavior in the form of differential equations.

$$\begin{cases} \mathrm{d}\boldsymbol{\sigma} = \mathbb{D} : (\mathrm{d}\boldsymbol{\epsilon} - \mathrm{d}\boldsymbol{\epsilon}^{\mathrm{p}}) & \textit{Hooke's law} \\ \mathrm{d}\boldsymbol{\epsilon}^{\mathrm{p}} = \mathrm{d}\phi\mathbf{r} & \textit{Non-associated flow rule} \\ \mathrm{d}\kappa = \mathrm{d}\phi h^{\mathrm{p}} & \textit{Hardening law} \\ \mathrm{d}\phi \geq 0,\ f \leq 0,\ \mathrm{d}\phi f = 0 & \textit{Karush-Kuhn-Tucker conditions} \end{cases} \tag{9.36}$$

where $\mathbb{D}$ is the 4th order elastic stiffness tensor. $\boldsymbol{\varepsilon}$, and $\boldsymbol{\varepsilon}^{\mathrm{p}}$ denote 2nd order strain tensor and its plastic part, respectively. The plastic flow direction $\mathbf{r}$ is determined by $\partial g / \partial \boldsymbol{\sigma}$. $\mathrm{d}\phi$ is the plastic multiplier. κ and h^{p} denote the plastic internal variable and its gradient. For the Mohr–Coulomb model presented in Sect. 9.4.2, $\kappa = \varepsilon_{\mathrm{d}}^{\mathrm{p}}$ and $h = \partial g / \partial q$. Equation $(9.36)_4$ is the loading/unloading inequality, also called *Karush–Kuhn–Tucker* (*KKT*) conditions.

In numerical calculations, the ordinary differential equations defined by Eq. (9.36) is integrated over the time domain $[t_n,\ t_{n+1}]$ to be converted to an algebraic system of equations. According to the backward Euler integral scheme, one can yield:

$$\begin{cases} \boldsymbol{\sigma}_{n+1} - \boldsymbol{\sigma}_n - \mathbb{D} : (\Delta\boldsymbol{\epsilon}_{n+1} - \Delta\phi_{n+1}\mathbf{r}_{n+1}) = \mathbf{0} \\ \kappa_{n+1} - \kappa_n - \Delta\phi_{n+1} h_{n+1}^{\mathrm{p}} = 0 \\ \Delta\phi_{n+1} \geq 0,\ f_{n+1} \leq 0,\ \Delta\phi_{n+1} f_{n+1} = 0 \end{cases} \tag{9.37}$$

where the subscript "n" represents that the item is a known quantity and $\Delta\boldsymbol{\varepsilon}_{n+1}$ is also often used as a known input. There are three unknown quantities to be solved in Eq. (9.38), i.e., $\boldsymbol{\sigma}_{n+1}$, κ_{n+1}, and $\Delta\phi_{n+1}$. The other terms r_{n+1}, h_{n+1}^{p}, and f_{n+1} are functions of the three independent variables.

9.4 Return Mapping Algorithm Using Line Search Method and Hyper-Dual Step Derivative Approximation

KKT conditions in Eq. (9.37) restricts the allowable stress and strain states of the material under elastic or elastic–plastic loading cases, i.e., $\Delta\phi_{n+1} = 0$ and $f_{n+1} \leq 0$ for the elastic case, $\Delta\phi_{n+1} > 0$ and $f_{n+1} = 0$ for the plastic case. However, the inequality constraints in *KKT* conditions also result in non-linear stress integral equations not being solved directly. Some stress update schemes have been proposed to deal with the inequality constraints, such as the return-mapping strategy consisting of both the elastic prediction and plastic correction (Simo & Hughes, 2006), the unconstrained stress update strategy based on the smooth function (Zhou et al., 2022), and the projection-contraction strategy rising from finite-dimensional variational inequalities problem (Zheng et al., 2020). In this chapter, return-mapping strategy is employed. Firstly, it assumes that only elastic strain occurs for a given $\Delta\boldsymbol{\varepsilon}_{n+1}$, without plastic strain. Then, trial stress is calculated by Hook's Law, i.e., $\boldsymbol{\sigma}_{n+1}^{trial} = \boldsymbol{\sigma}_n + \mathbb{D} : \Delta\boldsymbol{\varepsilon}_{n+1}$. If $\boldsymbol{\sigma}_{n+1}^{trial}$ is in the elastic domain, i.e., $f\left(\boldsymbol{\sigma}_{n+1}^{trial}, \kappa_n\right) \leq 0$, the trial stress will be considered as the real stress at the $n+1$-th step. This process is also known as the elastic prediction.

$$\boldsymbol{\sigma}_{n+1} = \boldsymbol{\sigma}_{n+1}^{trial}, \; \kappa_{n+1} = \kappa_n, \; \Delta\phi_{n+1} = 0 \tag{9.38}$$

If $\boldsymbol{\sigma}_{n+1}^{trial}$ exceeds the elastic domain, the plastic correction is required with $f_{n+1} = 0$ and $\Delta\phi_{n+1} > 0$. Equation (9.37) will transform into unconstrained nonlinear equations.

$$\{\mathbf{f}(\mathbf{x})\}_{n+1} = \begin{Bmatrix} \boldsymbol{\sigma}_{n+1} - \boldsymbol{\sigma}_n - \mathbb{D} : (\Delta\boldsymbol{\epsilon}_{n+1} - \Delta\phi_{n+1}\mathbf{r}_{n+1}) \\ \kappa_{n+1} - \kappa_n - \Delta\phi_{n+1} h_{n+1}^{\mathrm{p}} \\ f(\boldsymbol{\sigma}_{n+1}, \kappa_{n+1}) \end{Bmatrix} = \begin{Bmatrix} \mathbf{0} \\ 0 \\ 0 \end{Bmatrix} \tag{9.39}$$

where $\{\mathbf{x}\}_{n+1} = \{\boldsymbol{\sigma}_{n+1}, \kappa_{n+1}, \Delta\phi_{n+1}\}$. The plastic correction process means that the stress point is gradually pulled back to the real stress at the n + 1-th step when iteratively solving Eq. (9.37). Generally, the trial stress point $\boldsymbol{\sigma}_{n+1}^{trial}$ can be used as the initial iteration point $\boldsymbol{\sigma}_{n+1}^{0}$, as depicted in Fig. 9.5.

9.4.1 Solution of Nonlinear Equations Based on Line Search Method

The iteration procedure is likely to diverge for the large $\Delta\boldsymbol{\varepsilon}_{n+1}$ if the standard Newton method is directly used to solve Eq. (9.39). The LSM presented in Sect. 8.2 can improve the convergence of the solution by optimizing the search step size. The iterative procedure of LSM is defined as follows:

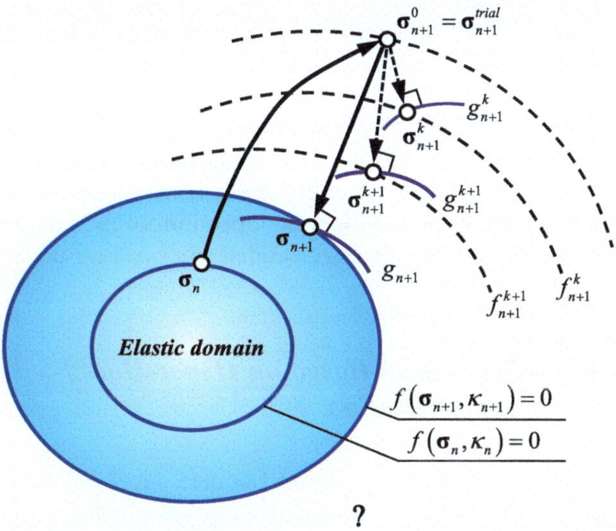

Fig. 9.5 Return-mapping stress update strategy

$$\{\mathbf{x}\}_{n+1}^{k+1} = \{\mathbf{x}\}_{n+1}^{k} + \alpha^k\{\mathbf{d}\}_{n+1}^{k} \tag{9.40}$$

where the subscript "$n+1$" will be omitted hereafter for the convenience of writing. $\{\mathbf{d}\}^k$ and α^k represent the size and direction of the search step at the kth iteration, respectively. The default value for α^k is set to 1 in the Newton method.

In order to ensure the quadratic convergence speed, the Newton search direction is used as follows:

$$\{\mathbf{d}\}^k = -[\mathbf{J}]_k^{-1}\{\mathbf{f}(\mathbf{x})\}^k \tag{9.41}$$

where $[\mathbf{J}]$ is the Jacobian matrix, i.e., the derivatives of the nonlinear equations $\{\mathbf{f}(\mathbf{x})\}$ with respect to independent variables $\{\mathbf{x}\}$. α^k is calculated by the following iterative formula with *Goldstein's* condition.

$$\begin{cases} Accept\ \alpha_j^k\ and\ exit\ IF\ \psi\left(\alpha_j^k\right) < \left(1-2\rho\alpha_j^k\right)\psi(0) \\ \alpha_{j+1}^k = \frac{\psi(0)}{\psi(0)+\psi\left(\alpha_j^k\right)} \qquad ELSE \end{cases} \tag{9.42}$$

where α_0^k is set to 1. ρ is the parameter in *Goldstein's* condition and its recommended value is 10^{-4} (Lu et al., 2023; Scherzinger, 2017). $\psi\left(\alpha_j^k\right)$ is the merit function and is defined by:

$$\psi\left(\alpha_j^k\right) = \frac{1}{2}\left\{\mathbf{f}\left(\mathbf{x}^k + \alpha_j^k\mathbf{d}^k\right)\right\}^T\left\{\mathbf{f}\left(\mathbf{x}^k + \alpha_j^k\mathbf{d}^k\right)\right\} \tag{9.43}$$

To avoid slow convergence due to too small the step size, α_{j+1}^k updated by Eq. (9.43) is required to not be less than a specific value.

$$\alpha_{j+1}^k = \max\left\{\varsigma\alpha_j^k, \alpha_{j+1}^k\right\} \tag{9.44}$$

where $\varsigma = 0.1$ is recommended (Lu et al., 2023; Scherzinger, 2017). It is worth noting that since Eq. (9.42) only involves the operation of the scalar function, the computational cost of optimizing step size is relatively small for the LSM.

9.4.2 Derivative Evaluation Based on Hyper-Dual Step Derivative Approximation

The HDSDA is used to calculate the 1st and 2nd derivatives involved in the above stress update process. These derivative terms mainly exist in the Jacobian matrix of Eq. (9.39). The Jacobian matrix [**J**] can be expressed by:

$$[\mathbf{J}] = \left[\frac{\partial \mathbf{f}(\mathbf{x})}{\partial \mathbf{x}}\right] = \begin{bmatrix} \mathbb{I} + \Delta\phi\mathbb{D} : \frac{\partial \mathbf{r}}{\partial \boldsymbol{\sigma}} & \Delta\phi\mathbb{D} : \frac{\partial \mathbf{r}}{\partial \kappa} & \mathbb{D} : \mathbf{r} \\ -\Delta\phi\frac{\partial h^{\mathrm{p}}}{\partial \boldsymbol{\sigma}} & 1 - \Delta\phi\frac{\partial h^{\mathrm{p}}}{\partial \kappa} & -h^{\mathrm{p}} \\ \frac{\partial f}{\partial \boldsymbol{\sigma}} & \frac{\partial f}{\partial \kappa} & 0 \end{bmatrix} \tag{9.45}$$

where $\mathbb{I}$ denotes the 4th order unit tensor. In Eq. (9.45), 1st derivative terms include $\mathbf{r} = \partial g/\partial\boldsymbol{\sigma}, h^{\mathrm{p}} = \partial g\big/\partial q, \partial f/\partial\boldsymbol{\sigma}$, and $\partial f\big/\partial\kappa$. 2nd derivative terms include $\partial\mathbf{r}/\partial\boldsymbol{\sigma} = \partial^2 g/\partial\boldsymbol{\sigma}^2$, $\partial\mathbf{r}/\partial\kappa = \partial^2 g/(\partial\boldsymbol{\sigma}\partial\kappa)$, $\partial h^{\mathrm{p}}/\partial\boldsymbol{\sigma}$, and $\partial h^{\mathrm{p}}\big/\partial\kappa$. In the matrix operation, the 2nd order tensors are transformed into a vector with dimension 6×1 or 1×6, and 4th order tensors are transformed into a matrix with dimension 6×6. It is worth emphasizing that a small amount of simple manual differentiation operations are required for the proposed algorithm, such as the derivation of general expressions for the Jacobian matrix. Then, the yield function and plastic potential function will be used as the objective functions of HDSDA to determine the specific forms of 1st and 2nd derivative terms in [**J**]. For convenience, the perturbation values are set to 1. Firstly, by perturbing the f along the $\vec{\epsilon}_1$ axis of σ_i and $\vec{\epsilon}_2$ axis of κ, $\{\partial f/\partial\boldsymbol{\sigma}\}$ and $\partial f\big/\partial\kappa$ can be obtained as follows:

$$\frac{\partial f}{\partial \sigma_i} = \Re_{\vec{\epsilon}_1}\left[f(\boldsymbol{\sigma} + \mathbf{e}_i\vec{\epsilon}_1, \kappa + \vec{\epsilon}_2)\right] \tag{9.46}$$

$$\frac{\partial f}{\partial \kappa} = \Re_{\vec{\epsilon}_2}\left[f(\boldsymbol{\sigma} + \mathbf{e}_i\vec{\epsilon}_1, \kappa + \vec{\epsilon}_2)\right] \tag{9.47}$$

where $\sigma_i (i = 1, 2, \ldots, 6)$ is the ith component of $\{\boldsymbol{\sigma}\}$ in vector form. $\{\mathbf{e}_i\}$ is a unit vector with dimension 6×1 where the ith component of $\{\mathbf{e}_i\}$ is 1 and the remaining components are 0, for example $\{\mathbf{e}_4\} = \{0, 0, 0, 1, 0, 0\}^T$ and $\mathbf{e}_4\vec{\epsilon}_1 = \{0, 0, 0, \vec{\epsilon}_1, 0, 0\}^T$.

Similarly, by perturbing the g along the $\vec{\epsilon}_1$ axis of σ_i and $\vec{\epsilon}_2$ axis of κ, $\{\mathbf{r}\}$ and $\{\partial \mathbf{r}/\partial \kappa\}$ can be obtained as follows:

$$r_i = \frac{\partial g}{\partial \sigma_i} = \Re_{\vec{\epsilon}_1}\left[g(\boldsymbol{\sigma} + \mathbf{e}_i\vec{\epsilon}_1, \kappa + \vec{\epsilon}_2)\right] \tag{9.48}$$

$$\frac{\partial r_i}{\partial \kappa} = \frac{\partial^2 g}{\partial \sigma_i \partial \kappa} = \Re_{\vec{\epsilon}_1\vec{\epsilon}_2}\left[g(\boldsymbol{\sigma} + \mathbf{e}_i\vec{\epsilon}_1, \kappa + \vec{\epsilon}_2)\right] \tag{9.49}$$

where r_i and $\partial r_i / \partial \kappa$ are ith components of $\{\mathbf{r}\}$ and $\{\partial \mathbf{r}/\partial \kappa\}$, respectively.

By perturbing the g along the $\vec{\epsilon}_1$ axis of σ_i and $\vec{\epsilon}_2$ axis of σ_j, $[\partial \mathbf{r}/\partial \boldsymbol{\sigma}]$ can be obtained as follows:

$$\frac{\partial r_i}{\partial \sigma_j} = \frac{\partial^2 g}{\partial \sigma_i \partial \sigma_j} = \Re_{\vec{\epsilon}_1\vec{\epsilon}_2}\left[g\left(\boldsymbol{\sigma} + \mathbf{e}_i\vec{\epsilon}_1 + \mathbf{e}_j\vec{\epsilon}_2, \kappa\right)\right] \tag{9.50}$$

where $\partial r_i / \partial \sigma_j$ is the component of $[\partial \mathbf{r}/\partial \boldsymbol{\sigma}]$ in the ith row and jth column

The stress invariants (p, q, θ) and κ should be defined directly as the HDNs when the HDSDA is used to calculate h^{p} and $\partial h^{\mathrm{p}} / \partial \kappa$. By perturbing the g along the $\vec{\epsilon}_1$ axis of q and $\vec{\epsilon}_2$ axis of κ, h^{p} and $\partial h^{\mathrm{p}} / \partial \kappa$ can be obtained as follows:

$$h^{\mathrm{p}} = \frac{\partial g}{\partial q} = \Re_{\vec{\epsilon}_1}\left[g(p, q + \vec{\epsilon}_1, \theta, \kappa + \vec{\epsilon}_2)\right] \tag{9.51}$$

$$\frac{\partial h^{\mathrm{p}}}{\partial \kappa} = \frac{\partial^2 g}{\partial q \partial \kappa} = \Re_{\vec{\epsilon}_1\vec{\epsilon}_2}\left[g(p, q + \vec{\epsilon}_1, \theta, \kappa + \vec{\epsilon}_2)\right] \tag{9.52}$$

It is worth noting that $\{\partial \mathbf{r}/\partial \kappa\} = \{\mathbf{0}\}$ and $\partial h^{\mathrm{p}} / \partial \kappa = 0$ for the present Mohr–Coulomb model since the g defined by Eq. (9.33) is independent of κ. In comparison to other 2nd derivative terms, $\partial h^{\mathrm{p}}/\partial \boldsymbol{\sigma} = \partial^2 g/(\partial q \partial \boldsymbol{\sigma})$ is somewhat unique due to the fact that q and $\boldsymbol{\sigma}$ are not independent of each other. According to the properties of continuous 2nd mixed partial derivative, where the order of differentiation does not affect the result, one can obtain:

$$\frac{\partial h^{\mathrm{p}}}{\partial \boldsymbol{\sigma}} = \frac{\partial^2 g}{\partial q \partial \boldsymbol{\sigma}} = \frac{\partial^2 g}{\partial \boldsymbol{\sigma} \partial q} \tag{9.53}$$

Firstly, the perturbation is applied to the first "imaginary" part of q^*. The perturbed stress $\boldsymbol{\sigma}^*$ is used to calculate the stress invariants (p^*, q^*, θ^*).

$$\boldsymbol{\sigma}^* = \boldsymbol{\sigma} + \mathbf{e}_i\vec{\epsilon}_1 \Rightarrow p^*, q^*, \text{ and } \theta^* \tag{9.54}$$

Then, the perturbation is applied to the second "imaginary" part of q^*. $\partial h^{\mathrm{p}}/\partial \boldsymbol{\sigma}$ can be calculated by extracting the third "imaginary" part of the plastic potential function.

$$\frac{\partial h^p}{\partial \sigma_i} = \frac{\partial^2 g}{\partial \sigma_i \partial q} = \Re_{\vec{\epsilon}_1\vec{\epsilon}_2}\left[g\left(p^*,\ q^* + \vec{\epsilon}_2,\ \theta^*,\ \kappa\right)\right] \tag{9.55}$$

where $\partial h^p/\partial \sigma_i$ is the jth component of $\{\partial h^{\mathrm{p}}/\partial \boldsymbol{\sigma}\}$.

So far, all the derivatives involved in the Jacobian matrix have been calculated by the HDSDA. To some extent, at the integral point level, the Jacobian matrix defined by Eq. (9.45) ensures the quadratic convergence speed in stress update iterations. At the structure level, the quadratic convergence speed in global equilibrium iterations is guaranteed by the CTO. Taking the total differentiation of Eq. (9.39), one can yield:

$$\begin{Bmatrix} \mathrm{d}\boldsymbol{\sigma} \\ \mathrm{d}\kappa \\ \mathrm{d}\Delta\phi \end{Bmatrix}_{n+1} = [\mathbf{J}]_{n+1}^{-1} \begin{bmatrix} \mathbb{D} & \mathbf{0} & \mathbf{0} \\ \mathbf{0} & 0 & 0 \\ \mathbf{0} & 0 & 0 \end{bmatrix} \begin{Bmatrix} \mathrm{d}\boldsymbol{\varepsilon} \\ 0 \\ 0 \end{Bmatrix}_{n+1} = \begin{bmatrix} \mathbb{A}_{11} & \mathbf{A}_{12} & \mathbf{A}_{13} \\ \mathbf{A}_{21} & A_{22} & A_{23} \\ \mathbf{A}_{31} & A_{32} & A_{33} \end{bmatrix} \begin{Bmatrix} \mathrm{d}\boldsymbol{\varepsilon} \\ 0 \\ 0 \end{Bmatrix}_{n+1} \tag{9.56}$$

where $\boldsymbol{\sigma}_{n+1}$, κ_{n+1}, $\Delta\phi_{n+1}$, and $\boldsymbol{\varepsilon}_{n+1}$ are all considered variables. $[\mathbb{A}_{11}]$ is a matrix with dimension 6×6, and it is also called the CTO.

$$\mathbb{D}^{CTO} = \frac{\mathrm{d}\boldsymbol{\sigma}_{n+1}}{\mathrm{d}\boldsymbol{\epsilon}_{n+1}} = \mathbb{A}_{11} \tag{9.57}$$

It has come to our notice that the $\mathbb{D}^{CTO}$ can be determined through some matrix operations of the Jacobian matrix, in which there are no new derivative terms that need to be calculated. The global equilibrium iteration can achieve quadratic convergence speed when the global stiffness matrix is integrated using $\mathbb{D}^{CTO}$. Figure 9.6 shows the flowchart of the proposed algorithm. Note that the algorithm flow includes three judgment conditions: the first is the loading/unloading judgment for the elasticity and the plasticity, the second is to judge whether the size of the search step meets the relative optimal condition to sufficiently reduce the merit function, and the third is to judge whether the solution for the nonlinear equations is convergent. The *FTOL* is a small tolerance and is set to 10^{-7} in the proposed algorithm.

9.5 Numerical Evaluations

This section verifies first the convergence speed of the HDSDA-based return mapping algorithm in global equilibrium iterations. The verification is done by a tension problem of the 3D notched plate, in which the convergence behavior of stress update

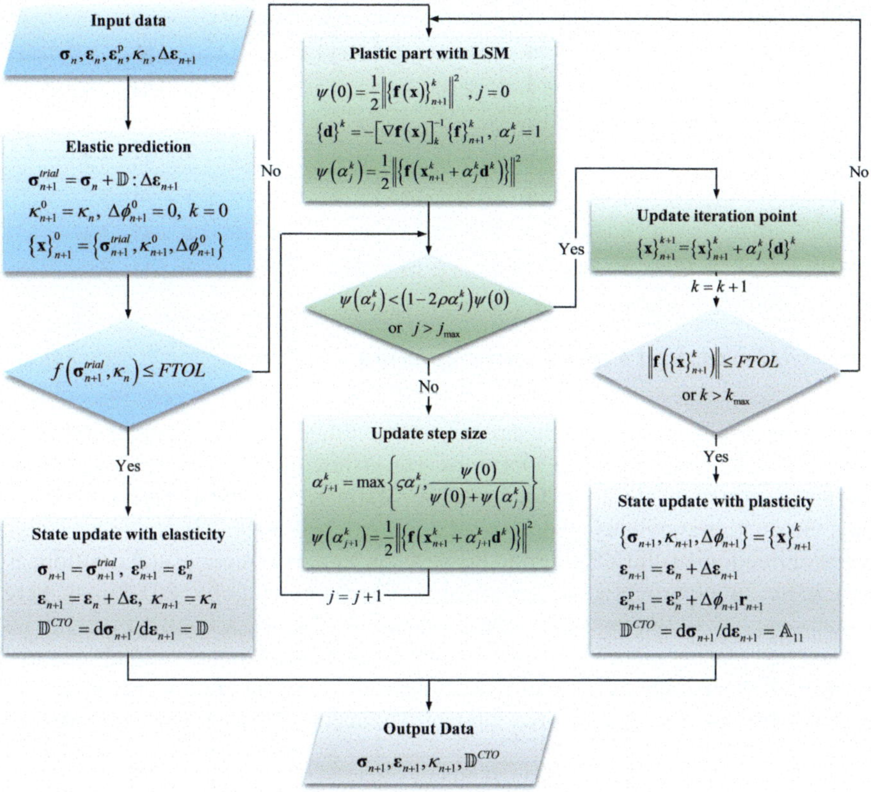

Fig. 9.6 Algorithm flow chart

algorithms is also investigated through the logarithm of residual force. Then, the algorithm's robustness is evaluated by contrast with the one based on the Newton method. Finally, the proposed algorithm is used to analyze common geotechnical engineering problems, such as the bearing capacity test of the strip footing, the slope stability analysis, and the tunnel excavation. The calculation results are further compared with the outcomes obtained from the Mohr–Coulomb model embedded in the ABAQUS software, in which the classic return mapping algorithm is used for the model implementation. The stress update algorithm differs from the proposed algorithm in that the nonlinear stress integral equations are solved by the Newton method instead of the LSM and the derivative terms used for the model implementation are obtained analytically rather than numerically. The parameters of the Mohr–Coulomb model used for the following numerical examples are provided in Table 9.2.

Table 9.2 Number of global equilibrium iterations required for different CTOs

Increment No	1	2	3	4	5	6	7	8	9	10	11	12	13	14	15	Σ	CPU time/s
MDM	1	3	4	3	2	2	2	2	2	3	2	2	2	2	2	34	2.5
CDM ($h = 10^1$)	1	4	6	6	4	4	4	3	3	4	4	4	3	3	3	56	5.2
CDM ($h = 10^{-2}$)	1	3	4	3	2	2	2	2	2	3	2	2	2	2	2	34	3.4
CDM ($h = 10^{-4}$)	1	3	4	3	2	2	2	2	2	3	2	2	2	2	2	34	3.5
CDM ($h = 10^{-6}$)	1	3	5	5	4	4	3	3	3	3	3	2	2	2	2	45	4.2
HDSDA	1	3	4	3	2	2	2	2	2	3	2	2	2	2	2	34	3.9

9.5.1 3D Notched Plate Under Tension

The tension problem of the notched plate is often utilized to investigate the influence of continuum tangent operator (CON) and CTO on the convergence behavior of global equilibrium equations (Simo & Hughes, 2006; Starman et al., 2014). Figure 9.7a illustrates the quarter-symmetry geometry of a 3D notched plate, where the horizontal degree of freedom on the left side and the vertical degree of freedom at the bottom are fixed. In the simulation process, a vertical displacement load of $u = 0.25$m is incrementally applied to the top of the notched plate in 15 equal-sized time increments. The notched plate is modeled as a hardening-type plastic Mohr–Coulomb material. Figure 9.7b depicts the equivalent plastic shear strain nephogram of the plate. It is observed that the plastic zone predominantly concentrates around the bottom-right notch.

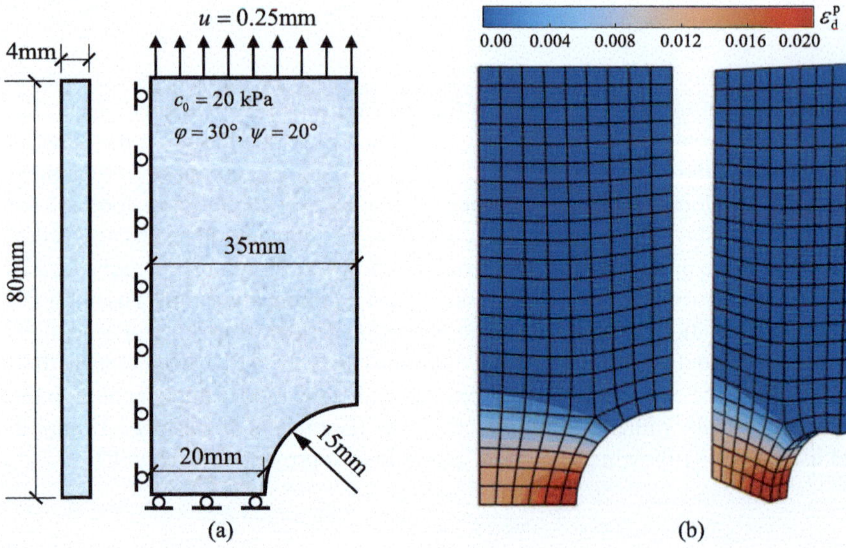

Fig. 9.7 3D notched plate: **a** Geometry and boundary conditions; **b** FE model

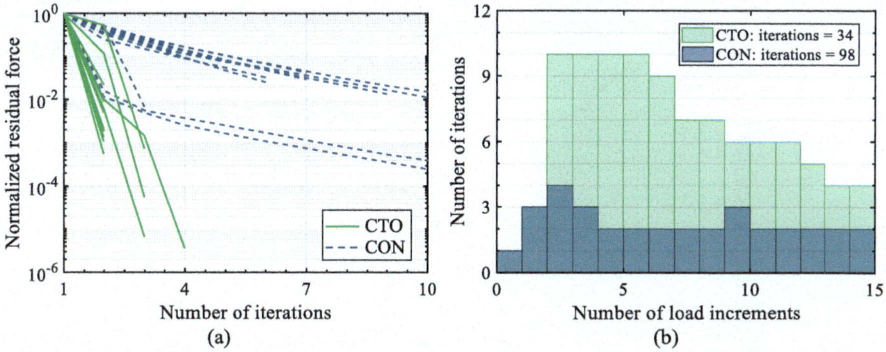

Fig. 9.8 Comparison of convergence behavior between CTO and CON: **a** Normalized residual force at critical nodes; **b** Number of iterations

Figure 9.8 illustrates the convergence behavior of the CON and CTO. The convergence law of the residual force, as shown in Fig. 9.8a, reveals that the convergence speed of the CON is notably slower compared to that of the CTO, leading to a higher number of iterations, as shown in (b). The HDSDA-based CTO ensures the quadratic convergence speed at the global level, validating the effectiveness of the proposed algorithm. Importantly, the ABAQUS software provides access to the residual force information at the critical nodes through the "Job Diagnostic" tab in the toolbar of the visualization module.

The convergence behavior of CTOs obtained by different differentiation methods is further examined, as shown in Table 9.1. In numerical examples based on the CDM, the Jacobian matrix obtained by the HDSDA is employed to prevent numerical errors that could impede the convergence of local stress update iterations. It can be found that the HDSDA-based CTO and the analytical CTO exhibit an equal number of iterations, reaffirming the asymptotic quadratic convergence property of the HDSDA-based CTO. The convergence behavior of the CDM-based CTO relies heavily on the chosen perturbation value. Inadequate perturbation values, either too large or too small, will result in the loss of quadratic convergence, as depicted in Fig. 9.9. Extreme cases (e.g., $h \leq 10^{-7}$ or $h \geq 10^2$) may even cause failure in the global equilibrium iterations. It is noteworthy that the four curves corresponding to the legend 'MDM', 'CDM ($h = 10^{-2}$)', 'CDM ($h = 10^{-4}$)', and 'HDSDA' in Fig. 9.9 overlap with each other because the CTOs in these cases can all provide the quadratic convergence rate. Tanaka et al. (2015) have reported that the computational time for the stress update algorithm utilizing numerical differentiation methods is generally higher compared to the algorithm based on analytical derivatives, but the ratio of the two will decrease with the increase of the number of elements because the amount of time for solving the nonlinear equilibrium equations becomes dominant in large-scale examples compared to the local stress update calculation (Table 9.3).

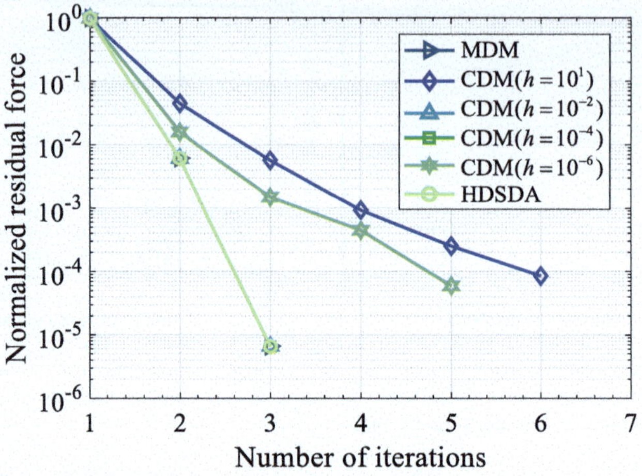

Fig. 9.9 Convergence behavior of different CTOs at increment No. 4

Table 9.3 Model parameters used in the boundary problems

Boundary value problems	E/kPa	ν	c_0/kPa	$\varphi/^\circ$	$\psi/^\circ$	H_p/E
3D notched plate	30,000	0.30	20	30	20	$1E-2$
Cylinder under shear load	40,000	0.20	40	20	20	$1E-8$
Strip footing on the slope	20,000	0.30	40	20	10	$1E-8$
Stability analysis of the side slope	20,000	0.49	50	20	6.667	$1E-8$, $-0.5E-3$, $-1E-3$, $-1.5E-3$
Tunnel excavation	30,000	0.32	6	30	0.1	$1E-8$

9.5.2 *Cylinder Under Shear Loads*

This section considers a case study of a cylinder sample subjected to the shear load to assess the proposed algorithm's robustness for large load increments. Similar cases have also been considered in the literature (Lester & Scherzinger, 2017; Shterenlikht & Alexander, 2012) because the load and boundary conditions presented in Fig. 9.10a can generate sufficiently complex and diverse stress paths to test the algorithm's performance. The cross-sectional radius of the cylinder is 0.05 m, and its height is 0.2 m. The bottom face of the sample is fully fixed, while the top face is allowed to move in the horizontal direction (i.e., the 1-direction). The sides of the cylinder are free without any constraint pressure. A horizontal displacement load of 0.03 m is applied to the top face through a reference point. The cylinder is

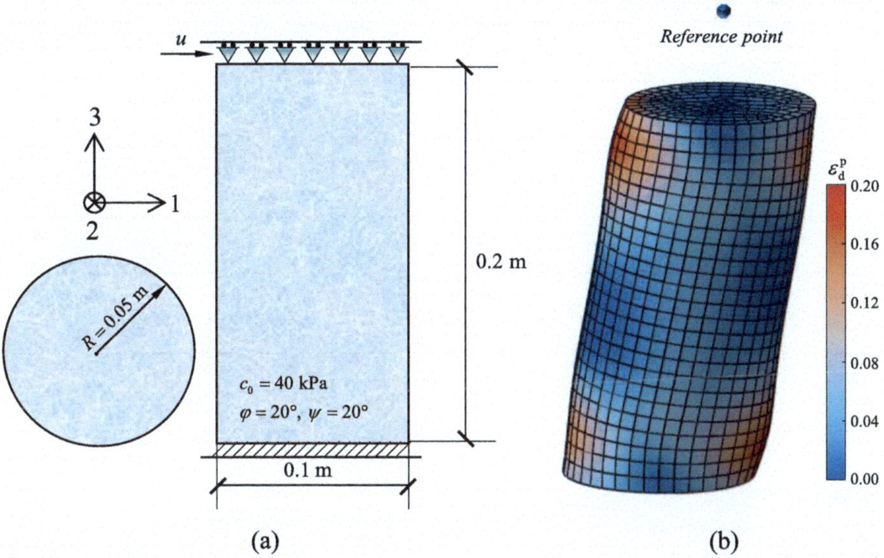

Fig. 9.10 3D cylinder: **a** Geometry and boundary conditions; **b** Numerical model

modeled as a perfectly plastic material by setting a very small plastic modulus, i.e., $H_p/E = 10^{-8}$. Figure 9.10b depicts the final configuration and the equivalent plastic strain nephogram of the cylinder at the end of loading.

In the numerical simulation, an initial time increment of 0.01 s and a total analysis time of 1 s are set. Figure 9.11 shows the variation of the time increment size for two algorithms and the required number of global equilibrium iterations for each time increment. It is worth noting that the HDSDA-based CTO is used for both stress update algorithms to ensure the quadratic convergence speed at the global level. Both stress update algorithms converge during the initial loading stage. As the size of the time increment further increases, however, the Newton method-based stress update algorithm encounters iteration failures in increments No. 8 and No. 12 due to excessively large strain increment input that prevent local stress update iterations from converging, as shown in Fig. 9.11a. According to "*automatic increment control*" in the ABAQUS/Standard, a smaller time increment, 25% of its previous value, is requested for increments No. 8 and No. 12. A total of 17-time increments are required for the Newton method-based stress update algorithm throughout the analysis process. The LSM-based stress update algorithm does not encounter iteration failures throughout the loading process, and only 11-time increments are needed in total, as shown in Fig. 9.11b. This suggests that the stress update algorithm based on the LSM is more robust compared to the algorithm based on the Newton method. It has been reported that the LSM requires more computational effort in local calculations (Scherzinger, 2017). From an overall computational cost perspective, however, the stress update algorithm based on the LSM offers the advantage of accommodating large load increments, leading to a reduced number of time increments and global

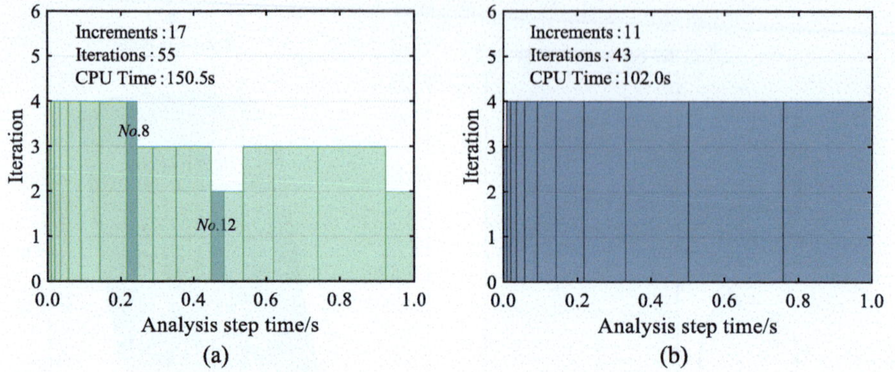

Fig. 9.11 Comparison of convergence behavior between LSM and Newton method: **a** Newton method-based stress update algorithm; **b** LSM-based stress update algorithm

Fig. 9.12 Displacement-reaction force curves of reference point

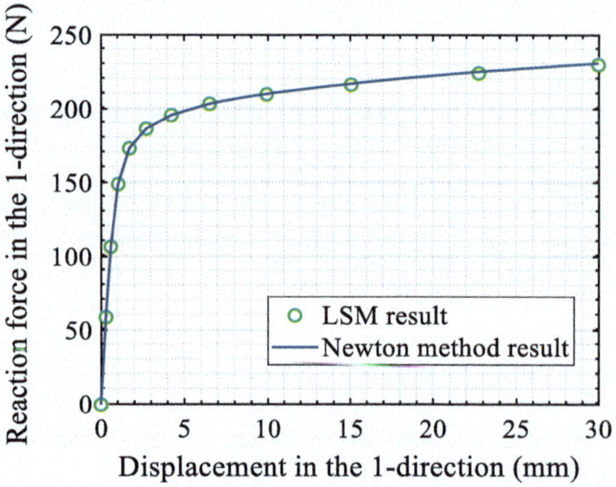

equilibrium iterations. Therefore, the CPU time consumed by the LSM is significantly lower than that of the Newton method. For this study case, the ratio of CPU times between the two algorithms is 67.8%. On the other hand, Fig. 9.11 shows that the computational results of both algorithms are in good agreement, indicating that the LSM-based stress update algorithm does not introduce distortions in the results due to large load increment input Fig. 9.12.

9.5.3 Strip Footing on Slope

In what follows, the Mohr–Coulomb model implemented by the proposed algorithm is used to analyze the behavior of a rigid strip footing resting on the side slope. In this example, the singularity at the footing edges and the asymmetric shear bands make the simulation of boundary value problem more challenging. Figure 9.13a shows the geometry and boundary conditions of the study case. The strip footing is considered a linear elastic material with Young's modulus $E = 20$ MPa and Poisson's ratio $v = 0.2$. The numerical simulation process includes the geostatic analysis step and the static general one. The geostatic step is used to generate the initial geostress field. The unit weight of soil mass is 20 kN/m^3. The vertical displacement load of $u = 0.8$ m is applied at the center of the footing top in the static general step. Figure 9.13b depicts the equivalent plastic strain nephograms of the side slope in the case of when the vertical displacement load reaches 0.08 and 0.8 m. During the initial loading stage, the plastic zone develops downward from both sides of the footing, gradually forming symmetrical shear bands. With the increase of vertical displacement load, the shear band towards the slope side gradually becomes dominant.

Figure 9.14 displays the displacement-reaction force curves under different plastic moduli. The minimum value of cohesion c is set to 0.001 kPa for the cases of $H_\mathrm{p} < 0$ to avoid negative cohesion without physical significance. This treatment may cause slight fluctuations in the curves of $H_\mathrm{p} < 0$ after the peak value. It is observed that as the plastic modulus decreases, the peak reaction force gradually decreases while the softening rate of the curves gradually increases. The simulation results verify the effectiveness of the proposed algorithm for the numerical implementation of softening models. The Mohr–Coulomb model embedded in ABAQUS and the one in Sect. 9.4.1 adopt different plastic internal variables, and thus their calculation results are only comparable in the case of ideal plasticity. Excellent consistency is observed for the proposed algorithm and ABAQUS. Finally, it should be emphasized that the numerical applications of softening models are often combined with a certain regularization method, such as the integral nonlocal method (Su et al., 2023), the gradient method (Xue et al., 2023), the viscoplastic method (Niazi et al., 2013), etc., to avoid potential mesh dependence problem caused by the ellipticity loss of partial differential equations. This aspect is not included in the research interest scope of the paper but can be found in the literature (Su et al., 2023).

9.5.4 Slope Stability Analysis

This numerical example demonstrates the application of the proposed algorithm in slope stability analysis. Figure 9.15a shows the geometry and boundary information of the slope. In the simulation process, the instability of the slope is caused by the increasing gravity load, in which the weight of the soil is 20 kN/m^3. Figure 9.15a depicts the nephogram of the equivalent plastic shear strain. It is observed that the sliding surface represented by the plastic zone is roughly arc-shaped and passes through the slope foot. Figure 9.16 illustrates the vertical displacement-gravity load factor curves of point A and the safety factor obtained from the limit analysis method. It can be found that the calculation results from the Mohr–Coulomb model embedded in ABAQUS software exhibit excellent agreement with the results from

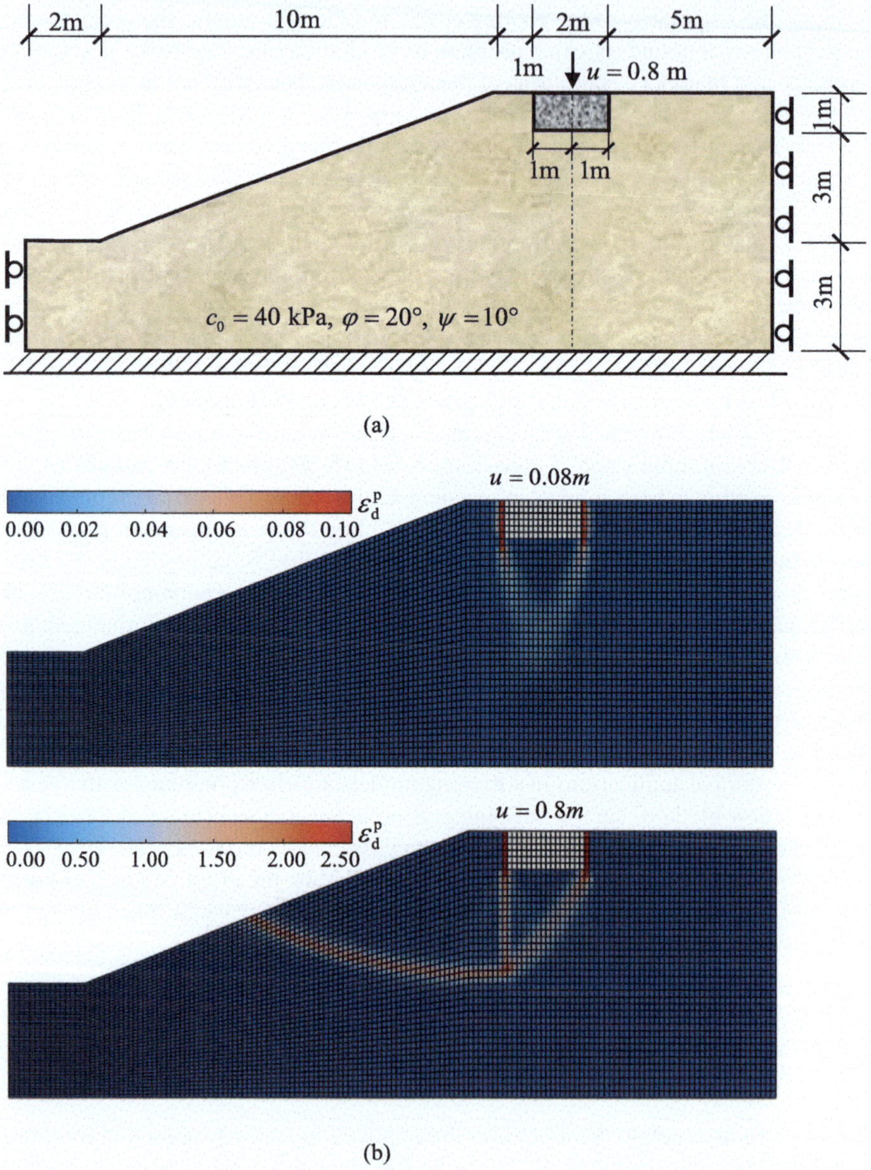

Fig. 9.13 Strip footing on the slope: **a** Geometry and boundary conditions; **b** Distribution of equivalent plastic shear strain in the case of $H_p = -30$ kPa

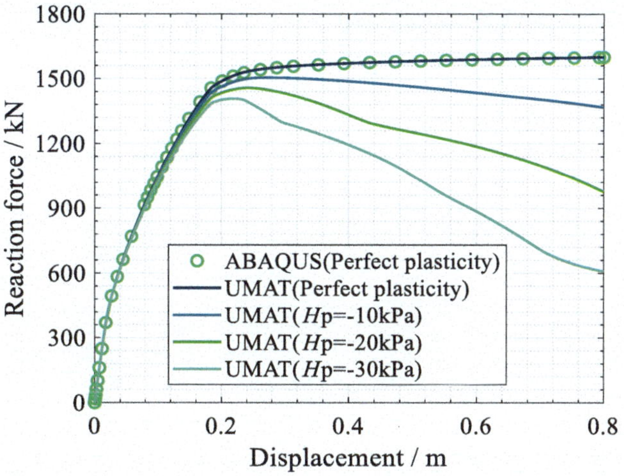

Fig. 9.14 Displacement-reaction force curves

the proposed algorithm. The safety factor determined by the finite element analysis is about 4.01, which is 0.87% lower than the limit analysis solution of 4.045.

9.5.5 Tunnel Excavation Simulation

The final case study examines a tunnel excavation problem involving lining construction. The progress softening method is employed to account for stress release following tunnel excavation and prior to lining support. Figure 9.17a illustrates the geometry and boundary conditions of the case study. The simulation process comprises one geostress analysis step and two static general analysis steps. The initial stress field is generated during the geostress analysis step, with a soil unit weight of 20 kN/m^3. In the first static general step, the elastic modulus of the soil in the excavation area is reduced from 30 to 15 MPa to simulate stress release of the surrounding soil after tunnel excavation. The lining material is modeled as linear elastic, with Young's modulus (E) of 2.13 GPa and Poisson's ratio (ν) of 0.2. The lining structure is activated in the second static general analysis step. Figure 9.17b presents the vertical displacement field of the ground, where the settlement and rebound phenomena are observed at the top and bottom of the tunnel, respectively. Figure 9.17 shows the settlement curve of the ground surface. It is observed that the maximum settlement is concentrated near the centerline of the tunnel, gradually decreasing as the distance from the centerline increases. The simulation results obtained from the proposed algorithm and ABAQUS once again exhibit good consistency. The numerical examples presented in Sects. 9.5.3 to 9.5.5 illustrate that the constitutive model implemented numerically by the proposed algorithm can be effectively used for the numerical analysis of practical engineering problems Fig. 9.18.

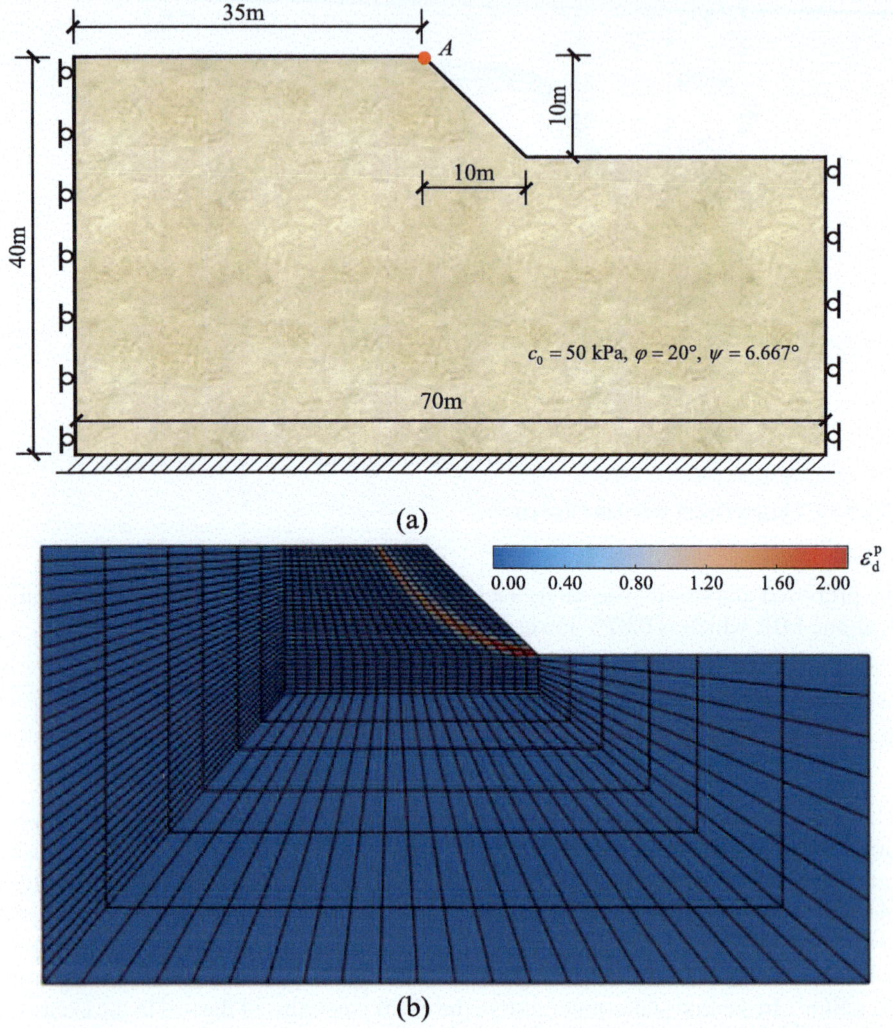

Fig. 9.15 Side slope: **a** Geometry and boundary conditions; **b** Distribution of equivalent plastic shear strain

The numerical analysis of the above boundary value problems has well verified the effectiveness of the proposed algorithm for Mohr–Coulomb plasticity model. The proposed algorithm can also handle more complex constitutive models because its main advantage lies in its ability to evaluate the derivative of the objective function with high precision using HDSDA, without being restricted by the function's complexity. Additionally, the algorithm exhibits well robust convergence when solving nonlinear equations, thanks to the use of the LSM. These advantages become more evident when the proposed algorithm is applied to complex constitutive models.

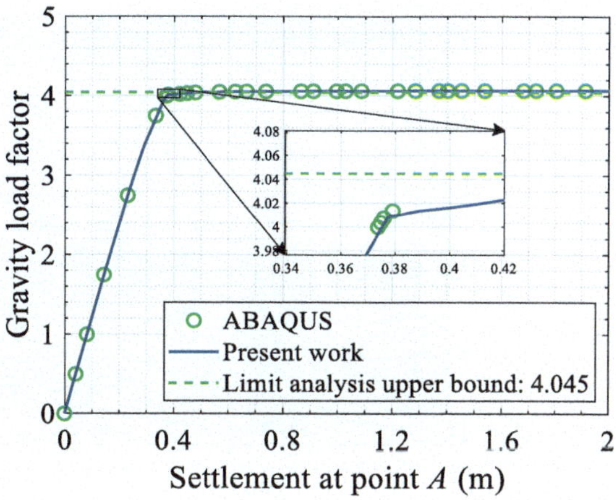

Fig. 9.16 Displacement-reaction force curves

Certainly, some modifications are necessary based on the characteristics of the selected constitutive models. For instance, for damage models or plastic damage models with softening constitutive laws, the numerical implementation of the model is often combined with a certain regularization method to obtain calculation results with mesh objectivity. If the selected constitutive model is already a nonlocal model, there is no need to introduce regularization techniques. For the hypoplastic model, its governing equations do not include the *KKT* conditions, and thus the loading and unloading judgments in the proposed algorithm need to be omitted. Additionally, for some nonlinear elastic models, where the stress update calculation may not entail solving implicit nonlinear equations, the LSM in the proposed algorithm can be removed, retaining solely the derivative evaluation ability facilitated by the HDSDA.

9.6 Summary

The precise evaluation of derivatives and convergence of solutions for nonlinear equations have always been significant challenges in the implicit numerical implementation of constitutive models. In particular, the increasing complexity of models poses growing difficulties for manual differentiation-based derivative evaluation. In this chapter, a novel return mapping algorithm is proposed to overcome the above two issues existing in the implicit model implementation by combining HDSDA and LSM. The main conclusions are given as follows:

(1) The research results demonstrate the robustness and accuracy of the derivatives obtained using HDSDA. They remain unaffected by subtraction cancellation and truncation errors, while also displaying insensitivity to perturbation values. Consequently, reliable and robust stress update calculations are ensured. The convergence analysis of residual force for the boundary value problem reveals that the HDSDA-based CTO can provide the quadratic convergence speed in global equilibrium iterations, comparable to the manual differentiation-based CTO. It is worth noting that the utilization of numerical differentiation, as

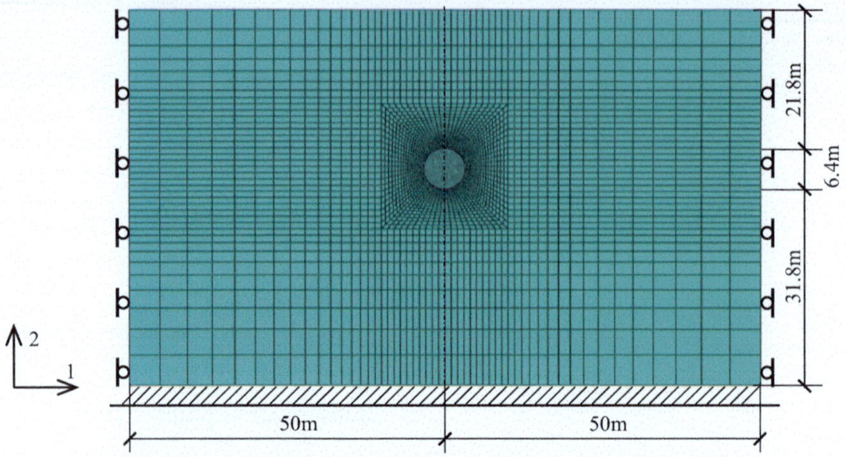

(a)

(b)

Fig. 9.17 Tunnel excavation: **a** Numerical model; **b** Vertical displacement field

opposed to manual differentiation, will require more computational efforts in the local stress update, further leading to an increase in overall analysis time. But it cannot conceal the convenience it brings for the numerical implementation of complex models.

(2) The HDNs are currently not integrated into the math libraries of mainstream programming languages. Consequently, HDSDA implementation commonly involves Operator Overloading and Function Overloading techniques. For the convenience of readers, an example source code is provided in Appendix B. These resources facilitate the convenient adoption of the tool for research purposes. Furthermore, the LSM represents a substantial enhancement over the classical return mapping algorithm, ensuring convergence of solutions

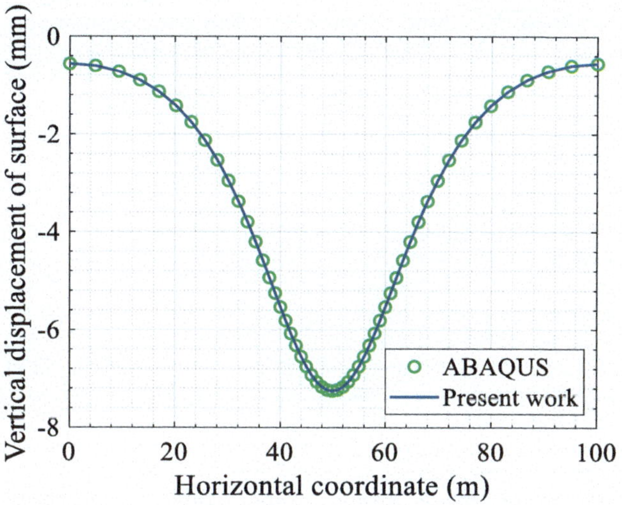

Fig. 9.18 Settlement curves of the ground surface

for nonlinear equations, particularly in challenging scenarios involving high-curvature constitutive functions or large load increment steps. The numerical analysis of typical geotechnical engineering problems has demonstrated the effectiveness of the proposed algorithm for the Mohr–Coulomb plasticity model. These promising results inspire confidence in the algorithm's potential for handling more challenging models such as anisotropy and coupled physics cases. The relevant explorations shall be pursued in future efforts.

References

Balzani, D., Gandhi, A., Tanaka, M., & Schröder, J. (2015). Numerical calculation of thermo-mechanical problems at large strains based on complex step derivative approximation of tangent stiffness matrices. *Computational Mechanics, 55*(5), 861–871.

Clifford, C. (1873). Preliminary sketch of biquaternions. *Proceedings of the London Mathematical Society, 4*(64), 381–395.

Fike, J. & Alonso, J. (2011). The development of hyper-dual numbers for exact second-derivative calculations. In: *49th AIAA Aerospace Sciences Meeting Including the New Horizons Forum and Aerospace Exposition,* Orlando Florida (pp. 886).

Fike, J. A. (2013) Multi-objective optimization using hyper-dual numbers. PhD thesis, Stanford University.

Fohrmeister, V., Bartels, A., & Mosler, J. (2018). Variational updates for thermomechanically coupled gradient-enhanced elastoplasticity—implementation based on hyper-dual numbers. *Computer Methods in Applied Mechanics and Engineering, 339*, 239–261.

Fujikawa, M., Tanaka, M., Mitsume, N., & Imoto, Y. (2022). Hyper-dual number-based numerical differentiation of eigensystems. *Computer Methods in Applied Mechanics and Engineering, 390*, 114452.

Kiran, R., & Khandelwal, K. (2014). Complex step derivative approximation for numerical evaluation of tangent moduli. *Computers & Structures, 140*, 1–13.

Lai, K., & Crassidis, J. L. (2008). Extensions of the first and second complex-step derivative approximations. *Journal of Computational and Applied Mathematics, 219*(1), 276–293.

Lester, B. T., & Scherzinger, W. M. (2017). Trust-region based return mapping algorithm for implicit integration of elastic-plastic constitutive models. *International Journal for Numerical Methods in Engineering, 12*(3), 257–282.

Lu, D. C., Zhang, Y. N., Zhou, X., Su, C. C., Gao, Z. W., & Du, X. L. (2023). A robust stress update algorithm for elastoplastic models without analytical derivation of the consistent tangent operator and loading/unloading estimation. *International Journal for Numerical and Analytical Methods in Geomechanics, 47*(6), 1022–1050.

Menetrey, P., & Willam, K. J. (1995). Triaxial failure criterion for concrete and its generalization. *Structural Journal, 92*(3), 311–318.

Miehe, C. (1996). Numerical computation of algorithmic (consistent) tangent moduli in large-strain computational inelasticity. *Computer Methods in Applied Mechanics and Engineering, 134*(3–4), 223–240.

Niazi, M. S., Wisselink, H. H., & Meinders, T. (2013). Viscoplastic regularization of local damage models: Revisited. *Computational Mechanics, 51*, 203–216.

Pérez-Foguet, A., Rodríguez-Ferran, A., & Huerta, A. (2000). Numerical differentiation for local and global tangent operators in computational plasticity. *Computer Methods in Applied Mechanics and Engineering, 189*(1), 277–296.

Scherzinger, W. M. (2017). A return mapping algorithm for isotropic and anisotropic plasticity models using a line search method. *Computer Methods in Applied Mechanics and Engineering, 317*, 526–553.

Shterenlikht, A., & Alexander, N. A. (2012). Levenberg-marquardt vs powell's dogleg method for gurson-tvergaard-needleman plasticity model. *Computer Methods in Applied Mechanics and Engineering, 237–240*, 1–9.

Simo, J. C., & Hughes, T. J. R. (2006). Computational inelasticity. Springer Science & Business Media, New York.

Starman, B., Halilovič, M., Vrh, M., & Atok, B. (2014). Consistent tangent operator for cutting-plane algorithm of elasto-plasticity. *Computer Methods in Applied Mechanics and Engineering, 272*, 214–232.

Su, C. C., Lu, D. C., Zhou, X., Wang, G. S., Zhuang, X. Y., & Du, X. L. (2023). An implicit stress update algorithm for the plastic nonlocal damage model of concrete. *Computer Methods in Applied Mechanics and Engineering, 414*, 116189.

Tanaka, M., Sasagawa, T., Omote, R., Fujikawa, M., Balzani, D., & Schröder, J. (2014). Calculation of stresses and consistent tangent moduli from automatic differentiation of hyerelastic strain energy functions through the use of hyper dual numbers. *Pamm, 14*(1), 429–430.

Tanaka, M., Sasagawa, T., Omote, R., Fujikawa, M., Balzani, D., & Schröder, J. (2015). A highly accurate 1st-and 2nd-order differentiation scheme for hyperelastic material models based on hyper-dual numbers. *Computer Methods in Applied Mechanics and Engineering, 283*, 22–45.

Xue, D. W., Lü, X. L., Lim, K., & Buscarnera, G. (2023). Nonlocal implicit gradient enhancements for strain localization informed by controllability criteria for plastic solids. *Computer Methods in Applied Mechanics and Engineering, 415*, 116275.

Zhang, N., Li, X., & Wang, D. (2021). Smoothed classic yield function for C2 continuities in tensile cutoff, compressive cap, and deviatoric sections. *International Journal of Geomechanics, 21*(3), 04021005.

Zheng, H., Zhang, T., & Wang, Q. S. (2020). The mixed complementarity problem arising from non-associative plasticity with non-smooth yield surfaces. *Computer Methods in Applied Mechanics and Engineering, 361*, 112756.

Zhou, X., Lu, D. C., Zhang, Y. N., Du, X. L., & Rabczuk, T. (2022). An open-source unconstrained stress updating algorithm for the modified Cam-clay model. *Computer Methods in Applied Mechanics and Engineering, 390*, 114356.

Chapter 10
Stress Update Algorithm Based on Integral Nonlocal Method

The above-mentioned chapters of this book detail the computational challenges associated with elastoplastic models at the local level (also referred to as the material point or integral point level) and propose corresponding numerical algorithms. At the structural level, however, when constitutive models with softening laws are directly used for structural failure analysis, the simulation results will suffer from serious mesh dependency (Gao et al., 2022; Mánica et al., 2018; Sarkar et al., 2022). This issue arises because the governing partial differential equations lose ellipticity in certain regions following strain localization (Mazzucco et al., 2019). Generally, regularization methods are necessary for the numerical implementation of the strain-softening model, introducing the material's intrinsic characteristic length to maintain the mesh's objectivity in the calculation results. Among various regularization methods, the integral nonlocal method is more cost-effective, as it does not necessitate modifying PDEs or adding extra degrees of freedom.

In this chapter, we first reviewed the basic idea of nonlocal integration methods. Then, a nonlocal implicit stress update algorithm combining the line search and integral nonlocal methods is proposed, which is then applied to a non-orthogonal Mohr-Coulomb damage model. In plastic calculations, nonlinear stress integral equations are addressed using the LSM. In damage calculations, the variable governing the material's strain softening is converted into a nonlocal variable, regularizing the finite element solution at the structural level. Additionally, complex step derivative approximation is employed to derive a consistent tangent operator, ensuring quadratic convergence speed in the global equilibrium iterations. The performance of the proposed algorithm is assessed using five boundary value problems.

D. Lu et al., *Multi-physical and Non-orthogonal Geotechnical Plasticity Modeling and Algorithms*, https://doi.org/10.1007/978-981-96-2369-3_10

10.1 Integral Type Nonlocal Method

When addressing finite element solutions for structural analysis, several effective regularization measures are usually utilized to mitigate the influence of the mesh sensitivity problem. This is often achieved by introducing characteristic scales associated with the material. For instance, the method based on the Cosserat theories (Rubin & Cardon, 2002) characterizes the motion of a solid body by incorporating additional fields that are independent of the displacement field and providing supplementary information on the small-scale kinematics while significantly increasing computational effort. The strain-gradient model (Lü et al., 2020; Mazzucco et al., 2019) solves gradient-enhanced terms with a separate partial differential equation that contains the material scale parameter. Due to the necessity of solving additional partial differential equations or considering extra degrees of freedom, these methods that effectively regularize finite element solutions also significantly increase the numerical computational costs of the solution, which is more pronounced for structural-level analysis. Comparatively, integral nonlocal methods only deal with the weighted averaging of state variables in the constitutive relations, and the cost-effectiveness of solving the problem is remarkable.

Generally, in integral nonlocal methods, the state variable κ that requires weighted averaging is often associated with the variable in the constitutive model used to describe material softening behavior. In a continuum, $\mathbf{x}$ represents the coordinates of a material point, and the state variable that needs to be weighted averaging at $\mathbf{x}$ is denoted by $\kappa(\mathbf{x})$. In calculations of the local constitutive model, the update of the state variable $\kappa(\mathbf{x})$ often occurs locally at material points within the continuum domain. When nonlocal methods are employed, the update of the state variable at each point requires a weighted averaging process to consider the state variables within the current point's neighborhood.

The processed nonlocal variable is denoted as $\hat{\kappa}(\mathbf{x})$ and replaces the locally computed variable $\kappa(\mathbf{x})$ (Bažant & Jirásek, 2002). The corresponding nonlocal variable $\hat{\kappa}(\mathbf{x})$ is defined by:

$$\hat{\kappa}(\mathbf{x}) = \frac{\int_V w_\infty(\mathbf{x}_0, \xi)\kappa(\xi)\mathrm{d}\xi}{\int_V w_\infty(\mathbf{x}, \xi)\mathrm{d}\xi} \tag{10.1}$$

where w_∞ is function without limit, the weighted averaging and ξ is the coordinates of the neighboring points in the domain. It is usually required that the nonlocal operator should not alter a uniform field, which means that the weight function must satisfy the normalizing condition.

$$\int_V w_\infty(\mathbf{x}, \xi)\mathrm{d}\xi = 1 \quad \forall \mathbf{x} \in V \tag{10.2}$$

To ensure the efficiency of the numerical computation, a cut-off radius R must be introduced for integral nonlocal methods (Gao et al., 2021). $R = 6\,l$ is usually recommended, where l is the characteristic length of the material. The computational cost of the integral nonlocal method largely depends on the efficiency of search neighborhood integration points. The authors proposed an efficient nonlocal computation method based on the octree algorithm to reduce the computational cost of search neighborhood integration points. In the octree-nonlocal computation method, the computational domain is covered by a series of hierarchically related node spaces. A large number of integration points far from the interaction domain is directly excluded by comparing the distance between the node spaces and the current integration point, thus significantly improving the search efficiency. Then, the weight function w_∞ can be expressed as:

$$w_\infty = \begin{cases} w & \text{if } \|\xi - \mathbf{x}\| \le R \\ 0 & \text{if } \|\xi - \mathbf{x}\| \ge R \end{cases} \tag{10.3}$$

where the effective weight function w takes the form proposed by Galavi and Schweiger Helmut (2010) is as follows:

$$w = \frac{\|\xi - \mathbf{x}\|}{l^2} \exp\left(-\frac{\|\xi - \mathbf{x}\|^2}{l^2}\right) \tag{10.4}$$

where $\|\xi - \mathbf{x}\|$ is the distance between the current integration point and the neighborhood integration point used for calculating the averaged value. And the literature (Bažant & Jirásek, 2002; Huang et al., 2018; Jirasek & Rolshoven, 2003) has also suggested several of different weight functions, such as bell-shaped and Gaussian-shaped. The weighting function in Eq. (10.4) and the range of the nonlocal variable's integration into the finite element mesh are depicted in Fig. 10.1. Two distinct methodologies exist for calibrating the characteristic length l. The first methodology treats the characteristic length as a material parameter calibrated through matching experimentally observed structural responses, such as shear band width or load-displacement curve. The second methodology utilizes the characteristic length as a numerical parameter specifically regulating shear band width. Based on the route of the second methodology, the authors proposed the scaled nonlocal method insensitive to the characteristic length. It analytically reveals the governing mechanism of characteristic length on the nonlocal response of the structure: variations in characteristic length proportionally rescale load-displacement curves. In this method, the influence of characteristic length on the load-displacement response is eliminated via a dimensionless scaling factor, ultimately producing response curves insensitive to both the element size and the characteristic length size. Irrespective of calibration methodology, the characteristic length is required to exceed element size to ensure the regularization efficacy of the nonlocal method.

Generally, to ensure the efficiency of the nonlocal regularization, an over-nonlocal formulation needs to be used (Strömberg & Ristinmaa, 1996), in which the nonlocal

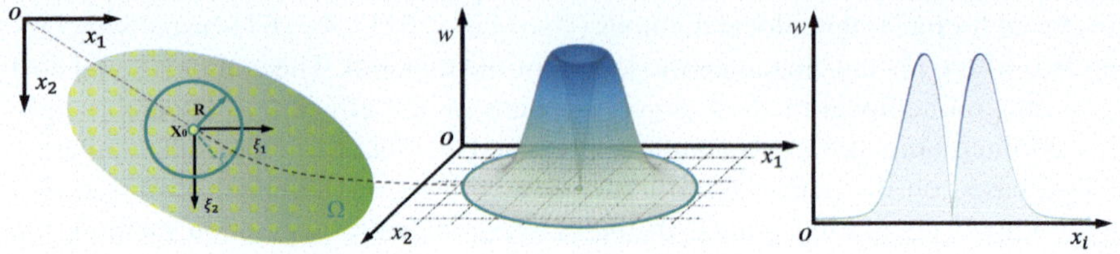

Fig. 10.1 The scope of the integration of the nonlocal variables

variable was defined as a linear combination of the local plastic strain and its nonlocal average. This formulation was validated to be able to act as a localization limiter and prevent the localized plastic zone from being zero:

$$\tilde{\kappa}(\mathbf{x}) = (1 - m)\kappa(\mathbf{x}) + m\hat{\kappa}(\mathbf{x}) \tag{10.5}$$

and the value of $m \in [0, 2]$ is related to the form of the selected effective weight function. Under the specific values of m, such as 0 and 1, Eq. (10.5) will degenerate into the local and the purely nonlocal formulations.

For plastic damage modeling, thanks to one of the classic methods for constructing plastic damage models, the effective stress method, which decouples the damage behavior from the plastic behavior, based on the effective stress method, the ideal variables that need weighted averaging are damage variables because they directly affect the softening behavior and do not participate in the iterative calculation of nonlinear stress integral equations in the plastic part. The stress updating algorithm of elastoplastic model can be directly used in the numerical calculation of plastic part without any adjustment.

10.2 Nonlocal Implicit Stress Update Algorithm

As mentioned above, based on the effective stress method, the stress update algorithm is separated into two parts: the plastic part in the effective stress space and the damage part in the nominal stress space. The relationship between the nominal stress and the undamaged effective stress is expressed as:

$$\sigma = (1 - \omega)\overline{\sigma} \tag{10.6}$$

for effective stress $\overline{\sigma}$, the implicit stress integral equations can be solved by using the optimization algorithm, which has superior convergence to the classic Newton method. The damage variable ω can be taken as the nonlocal variable. In addition, the evaluation of derivatives, including the consistent tangent operator, can be assessed by the high-precision numerical differentiation technique.

10.2.1 Plastic Part

In the effective stress space, the plastic stress–strain relation can be defined by a set of ODEs.

$$\begin{cases} \mathrm{d}\overline{\sigma} = \mathbb{D}_0 : (\mathrm{d}\boldsymbol{\varepsilon} - \mathrm{d}\boldsymbol{\varepsilon}^{\mathrm{p}}) \\ \mathrm{d}\boldsymbol{\varepsilon}^{\mathrm{p}} = \mathrm{d}\phi \mathbf{r} \\ \mathrm{d}\kappa = \mathrm{d}\phi h^{\mathrm{p}} \\ \mathrm{d}\phi \geq 0,\ f(\overline{\sigma}, \kappa^{\mathrm{p}}) \leq 0,\ \text{ and } \mathrm{d}\phi f(\overline{\sigma}, \kappa) = 0 \end{cases} \tag{10.7}$$

where κ is the plastic internal variable. According to the backward Euler scheme, Eq. (10.7) can be converted into the following integral form.

$$\begin{cases} \overline{\sigma}_{n+1} = \overline{\sigma}_n + \mathbb{D}_0 : (\Delta\boldsymbol{\varepsilon}_{n+1} - \Delta\phi_{n+1}\mathbf{r}_{n+1}) \\ \kappa_{n+1} = \kappa_n + \Delta\phi_{n+1} h^{\mathrm{p}}_{n+1} \\ \Delta\phi_{n+1} \geq 0,\ f(\overline{\sigma}_{n+1}, \kappa_{n+1}) \leq 0,\ \text{ and } \Delta\phi_{n+1} f(\overline{\sigma}_{n+1}, \kappa_{n+1}) = 0 \end{cases} \tag{10.8}$$

where only three unknowns need to be solved, i.e., $\overline{\sigma}_{n+1}$, κ_{n+1}, and $\Delta\phi_{n+1}$. The strain increment $\Delta\boldsymbol{\varepsilon}_{n+1}$ and the variable values $(\cdot)_n$ at the beginning of the increment step are considered known.

To make the calculation result of the stress update meet the inequality constraints in Eq. $(10.7)_3$, the return-mapping stress update strategy (Simo & Hughes, 2006) is used. The elastic trial point $\left(\overline{\sigma}^{trial}_{n+1}, \kappa_n\right)$ is calculated first by the Hooke law, i.e., $\overline{\sigma}^{trial}_{n+1} = \overline{\sigma}_n + \mathbb{D}_0 : \Delta\boldsymbol{\varepsilon}_{n+1}$. If the trial point $\left(\overline{\sigma}^{trial}_{n+1}, \kappa_n\right)$ is within the yield surface, it can be directly accepted, Eq. (10.7) can be rewritten as follows:

$$\begin{cases} \overline{\sigma}_{n+1} = \overline{\sigma}_n + \mathbb{D}_0 : \Delta\boldsymbol{\varepsilon}_{n+1} \\ \kappa_{n+1} = \kappa_n \\ \Delta\phi_{n+1} = 0 \end{cases} \tag{10.9}$$

If the trial point $\left(\overline{\sigma}^{trial}_{n+1}, \kappa_n\right)$ is outside the yield surface, the plastic correction is activated by solving:

$$\begin{cases} F_{\overline{\sigma}} = \overline{\sigma}_{n+1} - \overline{\sigma}_n - \mathbb{D}_0 : (\Delta\boldsymbol{\varepsilon}_{n+1} - \Delta\phi_{n+1}\mathbf{r}_{n+1}) = 0 \\ F_{\kappa^{\mathrm{p}}} = \kappa_{n+1} - \kappa_n - \Delta\phi_{n+1} h^{\mathrm{p}}_{n+1} = 0 \\ F_{\Delta\phi} = f(\overline{\sigma}_{n+1}, \kappa_{n+1}) = 0 \end{cases} \tag{10.10}$$

can also be expressed as:

$$\mathbf{F}(\mathbf{x}_{n+1}) = \left\{F_{\overline{\sigma}}, F_{\kappa^{p}}, F_{\Delta\phi}\right\}^{T} = \mathbf{0} \tag{10.11}$$

where $\mathbf{x}_{n+1} = \{\overline{\sigma}_{n+1}, \kappa_{n+1}, \Delta\phi_{n+1}\}^{T}$ denotes the independent variable of nonlinear equations.

Generally, the Newton method with quadratic convergence speed is used for solving the nonlinear equations. The LSM proposed in Chap. 8 can effectively improve the efficiency of the stress update algorithm, which can be regarded as a slight modification of Newton method, and is adopted to solve the stress integral equations with strong nonlinear characteristics. The process of solution can be expressed as follows:

$$\mathbf{x}_{n+1}^{k+1} = \mathbf{x}_{n+1}^{k} + \alpha^{k}\mathbf{d}^{k} \tag{10.12}$$

where the counter "k" represents the iterations. α^{k} represents the step size and $\mathbf{d}^{k}$ denotes the search direction at kth iteration. The kernel idea of the inexact LSM is to find a relative optimal α^{k} for a given $\mathbf{d}^{k}$. $\mathbf{d}^{k}$ is generally determined by the Newton search direction:

$$\mathbf{d}^{k} = -\left[\nabla\mathbf{F}\left(\mathbf{x}_{n+1}^{k}\right)\right]^{-1}\mathbf{F}\left(\mathbf{x}_{n+1}^{k}\right) \tag{10.13}$$

where $\nabla\mathbf{F}\left(\mathbf{x}_{n+1}^{k}\right)$ is the gradient of $\mathbf{F}(\mathbf{x})$, also known as the Jacobian matrix, which can be calculated by the CSDA presented in Chap. 9.

10.2.2 Damage Part

The effective stress needs to be converted into the nominal stress space by the damage variable. The local damage variable ω_{n+1} can be calculated after completing the stress update of the plastic part. According to the definition of nonlocal methods in Sect. 10.1, the nonlocal damage variable can be expressed by:

$$\hat{\omega}_{n+1}(\mathbf{x}) = \frac{\sum_{i=1, i\neq j}^{N} v_i w_i \omega_{i,n}(\mathbf{x}) + v_j w_j \omega_{j,n+1}(\mathbf{x})}{\sum_{i=1}^{N} v_i w_i} \tag{10.14}$$

where N denotes the integration point number within the averaging area. $\mathbf{x}$ represents the coordinate of the integration point. v and w are the volume and weight values of the integration point, respectively. The "j" and "I" represent the current integration point and neighboring points within the averaging area, respectively. Note that only the damage variable of the current integration point is the value at the end of the increment step $(\cdot)_{n+1}$, and the damage variables of other integration points are frozen at the beginning of the increment step $(\cdot)_{n}$. The weight function adopted in Eq. (10.4) is:

$$w_i = \frac{r_i}{l^2}\exp\left(-\frac{r_i^2}{l^2}\right) \tag{10.15}$$

where $r_i = \|\xi_i - \mathbf{x}\|$ denotes the distance between the current integration point and the integration point i (Galavi & Schweiger Helmut, 2010; Gao et al., 2022). The over-nonlocal formulation, which is the linear interpolation of local and nonlocal formulations (Strömberg & Ristinmaa, 1996), are employed in the proposed algorithm.

$$\tilde{\omega}_{n+1} = (1-m)\omega_{n+1} + m\hat{\omega}_{n+1} \tag{10.16}$$

and the effective stress tensor $\overline{\boldsymbol{\sigma}}_{n+1}$ can be transformed into the nominal stress tensor $\boldsymbol{\sigma}_{n+1}$ by:

$$\boldsymbol{\sigma}_{n+1} = (1-\tilde{\omega}_{n+1})\overline{\boldsymbol{\sigma}}_{n+1} \tag{10.17}$$

In addition to updating the stress states of the material, the tangent operator for material stiffness is also required in the local calculation, which is used to form the overall stiffness of the structure.

10.2.3 Consistent Tangent Operator

The implicit calculation of nonlinear finite elements is performed at the global and local levels (Potts et al., 2021; Ren et al., 2021; Yu et al., 2016). The quadratic convergence speed of the solution can be expected if the Newton-type methods are used to solve nonlinear equations at these two levels. To some extent, the quadratic convergence of solution at the local calculation is based on the Jacobian matrix $\nabla\mathbf{F}(\mathbf{x})$, and the quadratic convergence of solution at the global calculation is preserved by the CTO, i.e., $\mathrm{d}\boldsymbol{\sigma}_{n+1}/\mathrm{d}\boldsymbol{\varepsilon}_{n+1}$. Taking the total differential of Eq. (10.11) to the four variables $\overline{\boldsymbol{\sigma}}_{n+1}$, $\kappa^{\mathrm{p}}_{n+1}$, $\Delta\phi_{n+1}$, and $\boldsymbol{\varepsilon}_{n+1}$, one can yield:

$$\begin{bmatrix} \frac{\partial F_{\overline{\sigma}}}{\partial\overline{\boldsymbol{\sigma}}} & \frac{\partial F_{\overline{\sigma}}}{\partial\kappa^{\mathrm{p}}} & \frac{\partial F_{\overline{\sigma}}}{\partial\Delta\phi} & \frac{\partial F_{\overline{\sigma}}}{\partial\boldsymbol{\varepsilon}} \\ \frac{\partial F_{\kappa^{\mathrm{p}}}}{\partial\overline{\boldsymbol{\sigma}}} & \frac{\partial F_{\kappa^{\mathrm{p}}}}{\partial\kappa^{\mathrm{p}}} & \frac{\partial F_{\kappa^{\mathrm{p}}}}{\partial\Delta\phi} & \frac{\partial F_{\kappa^{\mathrm{p}}}}{\partial\boldsymbol{\varepsilon}} \\ \frac{\partial F_{\Delta\phi}}{\partial\overline{\boldsymbol{\sigma}}} & \frac{\partial F_{\Delta\phi}}{\partial\kappa^{\mathrm{p}}} & \frac{\partial F_{\Delta\phi}}{\partial\Delta\phi} & \frac{\partial F_{\Delta\phi}}{\partial\boldsymbol{\varepsilon}} \end{bmatrix}_{n+1} \begin{Bmatrix} \mathrm{d}\overline{\boldsymbol{\sigma}} \\ \mathrm{d}\kappa^{\mathrm{p}} \\ \mathrm{d}\Delta\phi \\ \mathrm{d}\boldsymbol{\varepsilon} \end{Bmatrix}_{n+1} = 0 \tag{10.18}$$

where $\partial F_{\overline{\sigma}}/\partial\boldsymbol{\varepsilon} = -\mathbb{D}$, $\partial F_{\kappa^{\mathrm{p}}}/\partial\boldsymbol{\varepsilon}_{n+1} = 0$ and $\partial F_{\Delta\phi}/\partial\boldsymbol{\varepsilon}_{n+1} = \mathbf{0}$ can be obtained easily by Eq. (10.10). The rest of the left matrix of Eq. (10.18) is the same as the Jacobi matrix $\nabla\mathbf{F}(\mathbf{x}_{n+1})$ of Eq. (10.13). The following expression can be obtained after some necessary matrix operations:

$$\begin{Bmatrix} \mathrm{d}\overline{\sigma} \\ \mathrm{d}\varepsilon_{\mathrm{d}}^{\mathrm{p}} \\ \mathrm{d}\Delta\phi \end{Bmatrix}_{n+1} = \begin{bmatrix} \frac{\partial F_{\overline{\sigma}}}{\partial \overline{\sigma}} & \frac{\partial F_{\overline{\sigma}}}{\partial \varepsilon_{\mathrm{d}}^{\mathrm{p}}} & \frac{\partial F_{\overline{\sigma}}}{\partial \Delta\phi} \\ \frac{\partial F_{\varepsilon_{\mathrm{d}}^{\mathrm{p}}}}{\partial \overline{\sigma}} & \frac{\partial F_{\varepsilon_{\mathrm{d}}^{\mathrm{p}}}}{\partial \varepsilon_{\mathrm{d}}^{\mathrm{p}}} & \frac{\partial F_{\varepsilon_{\mathrm{d}}^{\mathrm{p}}}}{\partial \Delta\phi} \\ \frac{\partial F_{\Delta\phi}}{\partial \overline{\sigma}} & \frac{\partial F_{\Delta\phi}}{\partial \varepsilon_{\mathrm{d}}^{\mathrm{p}}} & \frac{\partial F_{\Delta\phi}}{\partial \Delta\phi} \end{bmatrix}_{n+1}^{-1} \begin{bmatrix} \mathbb{D}_0 & 0 & 0 \\ 0 & 0 & 0 \\ 0 & 0 & 0 \end{bmatrix}_{n+1} \begin{Bmatrix} \mathrm{d}\boldsymbol{\varepsilon} \\ 0 \\ 0 \end{Bmatrix}_{n+1}$$
$$= \begin{bmatrix} \mathbb{A}_{11} & \mathbf{A}_{12} & \mathbf{A}_{13} \\ \mathbf{A}_{21} & A_{22} & A_{23} \\ \mathbf{A}_{31} & A_{32} & A_{33} \end{bmatrix} \begin{Bmatrix} \mathrm{d}\boldsymbol{\varepsilon} \\ 0 \\ 0 \end{Bmatrix}_{n+1} \tag{10.19}$$

where $\mathbb{A}$, $\mathbf{A}$, and A represent the matrix, the vector, and the scalar. The derivatives of $\overline{\boldsymbol{\sigma}}_{n+1}$, $\varepsilon_{\mathrm{d},n+1}^{\mathrm{p}}$, and $\Delta\phi_{n+1}$ with respect to $\boldsymbol{\varepsilon}_{n+1}$ can be extracted from Eq. (10.19) as follows:

$$\begin{cases} \dfrac{\mathrm{d}\overline{\boldsymbol{\sigma}}_{n+1}}{\mathrm{d}\boldsymbol{\varepsilon}_{n+1}} = \mathbb{A}_{11} \\ \dfrac{\mathrm{d}\varepsilon_{\mathrm{d},n+1}^{\mathrm{p}}}{\mathrm{d}\boldsymbol{\varepsilon}_{n+1}} = \mathbf{A}_{21} \\ \dfrac{\mathrm{d}\Delta\phi_{n+1}}{\mathrm{d}\boldsymbol{\varepsilon}_{n+1}} = \mathbf{A}_{31} \end{cases} \tag{10.20}$$

Furthermore, taking the derivative of Eq. (10.17) to $\boldsymbol{\varepsilon}_{n+1}$ and considering Eq. (10.20), the CTO for the plastic loading case is given as follows:

$$\mathbb{D}^{CTO} = \frac{\mathrm{d}\boldsymbol{\sigma}_{n+1}}{\mathrm{d}\boldsymbol{\varepsilon}_{n+1}} = (1-\tilde{\omega}_{n+1})\mathbb{A}_{11} - \overline{\boldsymbol{\sigma}}_{n+1} \otimes \left(\frac{\partial \tilde{\omega}_{n+1}}{\partial \overline{\boldsymbol{\sigma}}_{n+1}} : \mathbb{A}_{11} + \frac{\partial \tilde{\omega}_{n+1}}{\partial \varepsilon_{\mathrm{d},n+1}^{\mathrm{p}}} \mathbf{A}_{21} \right) \tag{10.21}$$

and the CTO for the elastic case is expressed by:

$$\mathbb{D}^{CTO} = \frac{\mathrm{d}\boldsymbol{\sigma}_{n+1}}{\mathrm{d}\boldsymbol{\varepsilon}_{n+1}} = (1-\tilde{\omega}_{n+1})\mathbb{D}_0 \tag{10.22}$$

The analytical derivative operation for obtaining the analytical CTO $\mathrm{d}\boldsymbol{\sigma}_{n+1}/\mathrm{d}\boldsymbol{\varepsilon}_{n+1}$ and Jacobi matrix $\nabla\mathbf{F}(\mathbf{x}_{n+1})$ is extremely cumbersome, which will hinder the proposed algorithm from solving other advanced but complex plastic damage models. Therefore, the numerical differentiation methods in Chap. 9 are recommended to approximate $\mathrm{d}\boldsymbol{\sigma}_{n+1}/\mathrm{d}\boldsymbol{\varepsilon}_{n+1}$ and $\nabla\mathbf{F}(\mathbf{x}_{n+1})$.

10.3 Application to Mohr–Coulomb Plasticity Damage Model

In order to evaluate the performance of the proposed algorithm, the non-orthogonal Mohr-Coulomb plastic damage model reported in literature (Su et al., 2023) is used as the application object of the algorithm. In this section, we outline the details of the model and its nonlocal calculation process.

10.3.1 Basic Constitutive Elements

In the non-orthogonal Mohr-Coulomb plastic damage model, a decomposition of the effective stress tensor into positive (tensile stress is positive) and negative components is carried out as described in (Faria et al., 1998; Ju, 1989) in order to account for the different nonlinear performances of concrete under tension and compression. The nominal stress $\boldsymbol{\sigma}$ is determined by:

$$\boldsymbol{\sigma} = \left(1 - \omega^{+}\right)\overline{\boldsymbol{\sigma}}^{+} + \left(1 - \omega^{-}\right)\overline{\boldsymbol{\sigma}}^{-} \tag{10.23}$$

where ω^{+} and ω^{-} are the variables in tension "+" and compression "–" conditions, respectively. $\overline{\boldsymbol{\sigma}}^{+}$ and $\overline{\boldsymbol{\sigma}}^{-}$ denote the tension and the compression parts of $\overline{\boldsymbol{\sigma}}$, respectively, which can be calculated by fourth-order projection tensors $\mathbb{P}^{\pm} : \overline{\boldsymbol{\sigma}}$. The model uses the damage variables with the same function form for describing the stiffness degradation laws for both tension and compression:

$$\omega^{\pm} = 1 - \exp\left[-\chi^{\pm}\kappa^{\psi}\right] \tag{10.24}$$

where the parameter ψ affects the curve form of the damage variable.

In the effective stress space, the evolution of plastic strain is described by the MC yield function, the cohesion-friction hardening combined law, and the on-orthogonal plastic flow rule. The MC criterion is expressed by:

$$f = \overline{M} \cdot \overline{p} - \overline{K} + \overline{R}_{\mathrm{MC}}\left(\overline{\theta}\right) \cdot \overline{q} = 0 \tag{10.25}$$

where $\overline{p}$, $\overline{q}$, and $\overline{\theta}$ represent the effective mean stress, the effective generalized shear stress, and the effective Lode angle (differing from the definition in Eq. (9.26) by -π/6), respectively. $\overline{M}$ and $\overline{K}$ define the slope and intercept of the MC criterion in the meridian plane, respectively. $\overline{R}_{\mathrm{MC}}$ controls the shape of MC locus on the deviatoric plane. The expressions for $(\overline{M}, \overline{K}, \overline{R}_{\mathrm{MC}})$ are:

$$\overline{M} = \frac{6\sin\overline{\varphi}}{\sqrt{3}(3 - \sin\overline{\varphi})} \tag{10.26}$$

$$\overline{K} = \frac{6\overline{c}\cos\overline{\varphi}}{\sqrt{3}(3 - \sin\overline{\varphi})} \tag{10.27}$$

$$\overline{R}_{MC}\left(\overline{\theta}\right) = \frac{2\cos\overline{\theta}}{(3 - \sin\overline{\varphi})} - \frac{2\sin\overline{\varphi}\sin\overline{\theta}}{\sqrt{3}(3 - \sin\overline{\varphi})} \tag{10.28}$$

where $\overline{c}$ and $\overline{\varphi}$ are the effective cohesion and the effective internal friction angle, respectively, which control the size and position of the MC strength criterion. It is the key to constructing the MC yield function to develop $\overline{c}$ and $\overline{\varphi}$ as functions of the plastic internal variables:

$$\begin{cases} \overline{c} = \overline{c}(\kappa) \\ \overline{\varphi} = \overline{\varphi}(\kappa) \end{cases} \tag{10.29}$$

where κ is the normalized internal variable used to record the plastic and damage process.

$$\kappa = \frac{\varepsilon_{\mathrm{d}}^{\mathrm{p}}}{\varepsilon_{\mathrm{ds}}^{\mathrm{p}}} \tag{10.30}$$

where $\varepsilon_{\mathrm{d}}^{\mathrm{p}}$ denotes the equivalent plastic shear strain and $\varepsilon_{\mathrm{d}}^{\mathrm{p}} = \sqrt{2/3}\|\mathbf{e}^{\mathrm{p}}\|$, where $\mathbf{e}^{\mathrm{p}}$ is the plastic shear strain tensor. $\varepsilon_{\mathrm{ds}}^{\mathrm{p}}$ is the equivalent plastic shear strain at the peak stress state. And the following formula is fitted by a large amount of experimental data to describe the dependence of $\varepsilon_{\mathrm{ds}}^{\mathrm{p}}$ to the stress state.

$$\frac{\varepsilon_{\mathrm{ds}}^{\mathrm{p}}}{\varepsilon_{\mathrm{ds1}}^{\mathrm{p}}} = \frac{14.70}{1 + \exp\left[\ln(13.70) + 2.30 - 2.30 \cdot \left(\frac{q_{\max}}{f_{\mathrm{c}}}\right)\right]} \tag{10.31}$$

where $\varepsilon_{\mathrm{ds1}}^{\mathrm{p}}$ is the value of $\varepsilon_{\mathrm{d}}^{\mathrm{p}}$ corresponding to uniaxial compression strength. $q_{\max} = \max\limits_{0 \le t \le \tau}(\overline{q})$ represents the maximum generalized shear stress during loading history. Figure 10.2 shows the black curve based on Eq. (10.31) and the experimental data (Candappa et al., 2001; Imran & Pantazopoulou, 1996; Kupfer et al., 1969; Lu & Hsu, 2007) used to fit this formula.

The detailed derivation process of the hardening functions can be found in the literature (Su et al., 2023). Here, we directly provide expressions for $\overline{c}$ and $\overline{\varphi}$

$$\begin{cases} \overline{c} = \dfrac{c\cos\varphi(1 + \sin\overline{\varphi})}{\cos\overline{\varphi}(1 + \sin\varphi)(1 - \omega^{+})} \\ \overline{\varphi} = \arcsin\left\{\left[1 - \dfrac{(1 - \sin\varphi)(1 - \omega^{-})}{(1 + \sin\varphi)(1 - \omega^{+})}\right] \Big/ \left[1 + \dfrac{(1 - \sin\varphi)(1 - \omega^{-})}{(1 + \sin\varphi)(1 - \omega^{+})}\right]\right\} \end{cases} \tag{10.32}$$

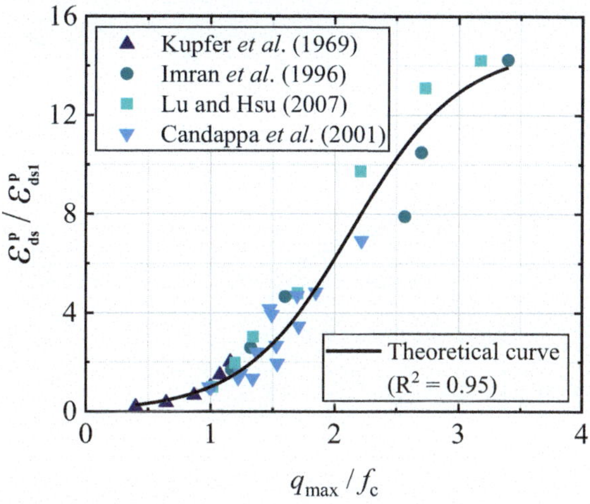

Fig. 10.2 Stress dependence of parameter ε^p_{ds}

where the nominal hardening functions of both $c(\kappa)$ and $\varphi(\kappa)$ from the literature are defined as follows:

$$\begin{cases} c(\kappa) = c_p \cdot H_c(\kappa) \\ \varphi(\kappa) = \varphi_p \cdot H_\varphi(\kappa) \end{cases} \tag{10.33}$$

$$\begin{cases} H_c = \dfrac{A_c(\kappa/\kappa_c)}{(A_c - 1) + (\kappa/\kappa_c)^{A_c}} \\ H_\varphi = \dfrac{A_\varphi(\kappa/\kappa_\varphi)}{(A_\varphi - 1) + (\kappa/\kappa_\varphi)^{A_\varphi}} \end{cases} \tag{10.34}$$

where A_c and A_φ are shape parameters of hardening/softening curves. κ_c and κ_φ are the values of κ when $H_c = 1$ and $H_\varphi = 1$ respectively.

The derivative of the original MC yield function at the corners is not unique, which will reduce the stability in numerical implementation. To eliminate the non-smoothness caused by corners, the original MC yield function is approximated by:

$$f = \overline{M} \cdot \overline{p} - \overline{K} + \sqrt{\overline{R}(\overline{\theta})^2 \cdot \overline{q}^2 + \overline{a}^2 \cdot \overline{M}^2} = 0 \tag{10.35}$$

where the parameter $\overline{a} = e \cdot \overline{K}/\overline{M}$ and $e \in [0, 1]$ control the approximation degree of the smooth yield curve to the original MC yield curve on the meridian plane. As shown in Fig. 10.3a, $e = 1$ denotes that the material has no cohesive strength and $e = 0$ means $\overline{a} = 0$ denotes that the smooth yield function degenerates into the original MC yield function.

On the deviatoric plane, the smooth shape function $\overline{R}(\overline{\theta})$ (Lagioia & Panteghini, 2016) used to replace $\overline{R}_{MC}(\overline{\theta})$ is as follows:

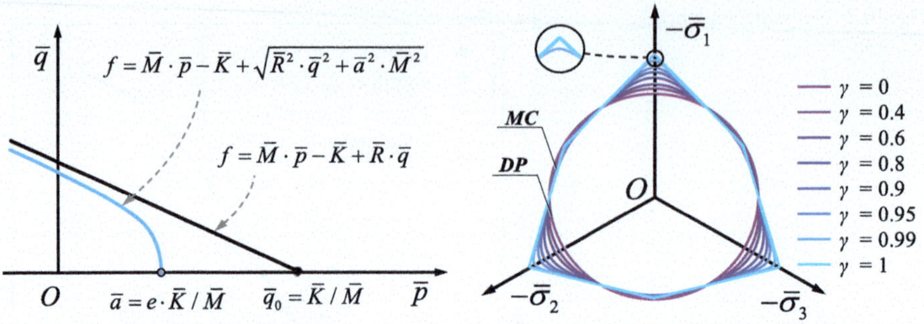

Fig. 10.3 Smooth MC yield curves

$$\bar{R} = \frac{\sqrt{3}}{3}\bar{\alpha}\cos\left[\frac{1}{3}\arccos(-\gamma\sin 3\bar{\theta}) - \frac{\pi}{6}\bar{\beta}\right] \tag{10.36}$$

where the parameter $\gamma \in [0, 1]$ controls the approximation degree of the smooth yield curve to the original MC yield curve, as can be seen in Fig. 10.3b. Under the condition of $e = 0$, $\gamma = 0$ and $\gamma = 1$ make Eq. (10.35) degenerate into Drucker-Prager curve and MC curve, respectively. The parameters $(\bar{\alpha}, \bar{\beta})$ are the functions of the friction angle $\bar{\varphi}$ and are expressed as:

$$\bar{\alpha} = \sec\left[(\gamma_w + 1)\frac{\pi}{6}\right], \bar{\beta} = 1 - \gamma_w \tag{10.37}$$

$$\gamma_w = \frac{6}{\pi}\arctan\left(\frac{\sin\varphi}{\sqrt{3}}\right) \tag{10.38}$$

The new MC yield surface is continuous and smooth anywhere, as shown in Fig. 10.4b, which is helpful for stable numerical calculation of the constitutive model.

The non-orthogonal flow rule is used in the current model to determine the plastic flow direction.

$$\mathbf{r} = \frac{\partial^\mu f}{\partial \bar{p}^\mu}\frac{\partial \bar{p}}{\partial \bar{\boldsymbol{\sigma}}} + \frac{\partial^\mu f}{\partial \bar{q}^\mu}\frac{\partial \bar{q}}{\partial \bar{\boldsymbol{\sigma}}} \tag{10.39}$$

where $\partial^\mu f / \partial(\cdot)^\mu$ denotes the Caputo type fractional derivative operator in this model. $\partial\bar{p}/\partial\bar{\boldsymbol{\sigma}} = \mathbf{1}/3$ and $\mathbf{1}$ represents the 2nd-order unit tensor. $\partial\bar{q}/\partial\bar{\boldsymbol{\sigma}} = 1.5\bar{\mathbf{s}}/\bar{q}$ and $\bar{\mathbf{s}}$ represents the effective deviatoric stress tensor. And $\partial^\mu f/\partial p^\mu$ and $\partial^\mu f/\partial q^\mu$ are as follows:

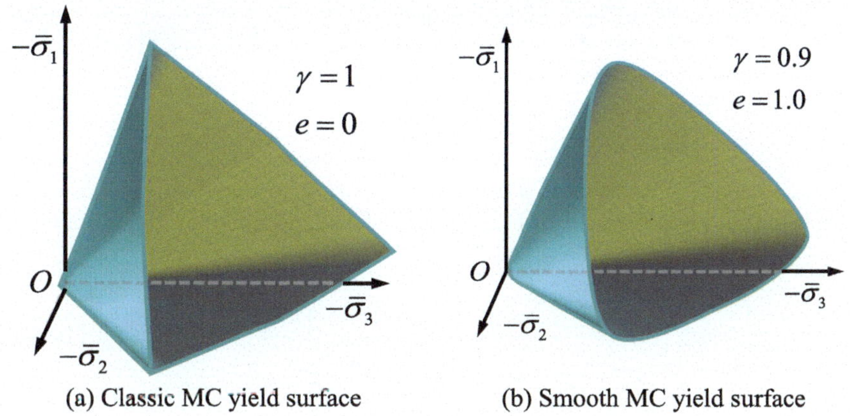

Fig. 10.4 MC yield surface in the principal stress space

$$\begin{cases} \dfrac{\partial^\mu f}{\partial \bar{p}^\mu} = \dfrac{\overline{M} \cdot |\bar{p}|^{1-\mu}}{\Gamma(2-\mu)} \\ \dfrac{\partial^\mu f}{\partial \bar{q}^\mu} = \dfrac{\overline{R} \cdot \bar{q}^{1-\mu}}{\Gamma(2-\mu)} \end{cases} \tag{10.40}$$

substituting Eq. (10.40) into Eq. (10.39), one can obtain:

$$\mathbf{r} = \frac{\overline{M} \cdot |\bar{p}|^{1-\mu}}{\Gamma(2-\mu)} \frac{1}{3}\mathbf{1} + \frac{\overline{R} \cdot \bar{q}^{1-\mu}}{\Gamma(2-\mu)} \frac{3\bar{\mathbf{s}}}{2\bar{q}} \tag{10.41}$$

With the combination of the nonlocal stress update algorithm proposed in Sect. 10.2, a numerical algorithm for the MC plastic damage model can be established for application in structural analysis.

10.3.2 Nonlocal Numerical Implementation

Finally, based on Eq. (10.23), the effective stress tensor $\bar{\boldsymbol{\sigma}}_{n+1}$ can be transformed into the nominal stress tensor $\boldsymbol{\sigma}_{n+1}$ by:

$$\boldsymbol{\sigma}_{n+1} = \left(1 - \tilde{\omega}^+_{n+1}\right)\mathbb{P}^+_{n+1} : \bar{\boldsymbol{\sigma}}_{n+1} + \left(1 - \tilde{\omega}^-_{n+1}\right)\mathbb{P}^-_{n+1} : \bar{\boldsymbol{\sigma}}_{n+1} \tag{10.42}$$

as suggested in the stress update algorithm in Sect. 10.2, damage variables are considered nonlocal variables as key variables affecting the softening behavior of the material. The stress updating in each Gauss point is performed independently by assuming that the state variables for all other points are frozen within the current global iteration. Three unknowns $\bar{\boldsymbol{\sigma}}_{n+1}$, $\varepsilon^{\mathrm{p}}_{\mathrm{d},n+1}$, and $\Delta\phi_{n+1}$ can be solved by the nonlocal stress update algorithm in Sect. 10.2.1. Once the updated effective stress $\overline{\sigma}_{n+1}$ and internal variable $\varepsilon^{\mathrm{p}}_{\mathrm{d},n+1}$ are obtained, the damage variable $\omega^{\pm}_{n+1}$ should be calculated as:

$$\omega^{\pm}_{n+1} = 1 - \exp\left[-\chi^{\pm}\kappa^{\psi}_{n+1}\right] \tag{10.43}$$

where κ_{n+1} can be obtained by substituting $\left(\overline{\sigma}_{n+1}, \varepsilon^{\mathrm{p}}_{\mathrm{d},n+1}\right)$ and Eq. (10.29) into Eq. (10.30). According to Eq. (10.14), the nonlocal damage variables can be obtained as:

$$\hat{\omega}^{\pm}_{n+1}(\mathbf{x}) = \frac{\sum_{i=1,i\neq j}^{N} v_i w_i \omega^{\pm}_{i,n}(\mathbf{x}) + v_j w_j \omega^{\pm}_{j,n+1}(\mathbf{x})}{\sum_{i=1}^{N} v_i w_i} \tag{10.44}$$

and the over nonlocal damage variables can be expressed as:

$$\begin{cases} \tilde{\omega}^{+}_{n+1} = (1-m)\omega^{+}_{n+1} + m\hat{\omega}^{+}_{n+1} \\ \tilde{\omega}^{-}_{n+1} = (1-m)\omega^{-}_{n+1} + m\hat{\omega}^{-}_{n+1} \end{cases} \tag{10.45}$$

Furthermore, taking the derivative of Eq. (10.17) to $\boldsymbol{\varepsilon}_{n+1}$, the CTO for the plastic loading case is given as follows:

$$\begin{aligned} \mathbb{D}^{CTO} = \frac{\mathrm{d}\boldsymbol{\sigma}_{n+1}}{\mathrm{d}\boldsymbol{\varepsilon}_{n+1}} &= \left(1-\tilde{\omega}^{+}_{n+1}\right)\mathbb{Q}^{+}_{n+1} : \mathbb{A}_{11} + \left(1-\tilde{\omega}^{-}_{n+1}\right)\mathbb{Q}^{-}_{n+1} : \mathbb{A}_{11} \\ &- \overline{\sigma}^{+}_{n+1} \otimes \left(\frac{\partial \tilde{\omega}^{+}_{n+1}}{\partial \overline{\sigma}_{n+1}} : \mathbb{A}_{11} + \frac{\partial \tilde{\omega}^{+}_{n+1}}{\partial \varepsilon^{\mathrm{p}}_{\mathrm{d},n+1}} \mathbf{A}_{21}\right) \\ &- \overline{\sigma}^{-}_{n+1} \otimes \left(\frac{\partial \tilde{\omega}^{-}_{n+1}}{\partial \overline{\sigma}_{n+1}} : \mathbb{A}_{11} + \frac{\partial \tilde{\omega}^{-}_{n+1}}{\partial \varepsilon^{\mathrm{p}}_{\mathrm{d},n+1}} \mathbf{A}_{21}\right) \end{aligned} \tag{10.46}$$

and $\mathbb{Q}^{\pm}$ are the corresponding projection tensors of the incremental form of the effective stress tensor $\mathrm{d}\overline{\sigma}$, refer to the definitions of $\mathbb{A}$ and $\mathbf{A}$ in Eq. (10.20). The CTO for the elastic case is expressed by:

$$\mathbb{D}^{CTO} = \frac{\mathrm{d}\boldsymbol{\sigma}_{n+1}}{\mathrm{d}\boldsymbol{\varepsilon}_{n+1}} = \left(1-\tilde{\omega}^{+}_{n+1}\right)\mathbb{Q}^{+}_{n+1} \otimes \mathbb{D}_0 + \left(1-\tilde{\omega}^{-}_{n+1}\right)\mathbb{Q}^{+}_{n+1} \otimes \mathbb{D}_0 \tag{10.47}$$

Finally, the algorithm flow chart is illustrated in Fig. 10.5. For plastic models with softening constitutive laws, and the nonlocal variables need to be adjusted to control the plastic internal variables governing softening behavior. According to the above content, at the integration point level, the local constitutive variables controlling the softening behavior are replaced by nonlocal variables which are the weighted average of the constitutive variables in the current integration point neighborhood. It should be noted that the nonlocal interaction induces the coupling of integration points. To promote the point-by-point stress update and the effectiveness of the nonlocal method, the nonlocal indirect coupling between the integration points is realized by adopting the method of "freezing" the field variables in the neighborhood. This simplified computation is a compromise made to avoid the dependence of consistent tangent operator on the quantities in the neighboring integration points. Similar treatments can also be found in some research on integral type approaches (Gao et al., 2022; Mánica et al., 2018).

Combined with the secondary development of the user material subroutine of ABAQUS, it can be used for the finite element analysis of the structure, and the analysis results can effectively avoid mesh sensitivity. Note that the UMAT subroutine of ABAQUS software can provide the coordinate information of the integration point but fail to provide the element volume of the integration point. The USDFLD subroutine needs to be utilized, in which the utility routine GETVRM can be called using the keyword "IVOL" to obtain the integration point volume. The state information of each integration point is stored in two common block arrays VOLINT and ENCD, among which the VOLINT (NEL, NIP, 1) is used to store the integration point volume, and the ENCD (NEL, NIP, 1–3) and ENCD (NEL, NIP, 4–5) are used to store the coordinate and local tension/compression damage variables of integration points. The NIP is the element integration point number, and the NEL represents the total element number of the part of the finite element model. The UMAT of different integration points can transmit information through the common block array. The neighboring integration points of the current integration point are found by estimating the distance between resting integration points and current integration point. In this process, all integration points are systematically enumerated and looped to check if the condition is met. This is a relatively time-consuming method, but it has been adopted by many nonlocal models due to its ease of implementation since it is simple to implement (Chow et al., 2011; Singh et al., 2021). To extend the proposed algorithm's applicability to larger-scale structural cases, it is helpful to select a more suitable domain search method. Grid-based or tree-based algorithms can be employed to enhance this algorithm, like spatial partitioning or octree-based methods could further enhance the algorithm's applicability (Howard et al., 2016; Kamaludin & Thamburaja, 2023).

10.4 Numerical Examples

The proposed plastic damage model has been implemented based on the proposed algorithm presented in Sect. 10.3 with the aid of the UMAT subroutine provided by the ABAQUS software. In the subsequent sections, we assess both the model's ability to reasonably describe the mechanical responses of structures and the performance of the stress update algorithm through five boundary value problems. The integral nonlocal method used is evaluated through the mesh sensitivity analysis. The quadratic convergence of the CTO evaluated by the CSDA is investigated by the convergence behavior of the logarithm of the residual force. The robustness benefit from the LSM is evaluated by comparing it with the standard Newton method. The concrete parameter settings of all examples are presented in Table 10.1. Besides, the

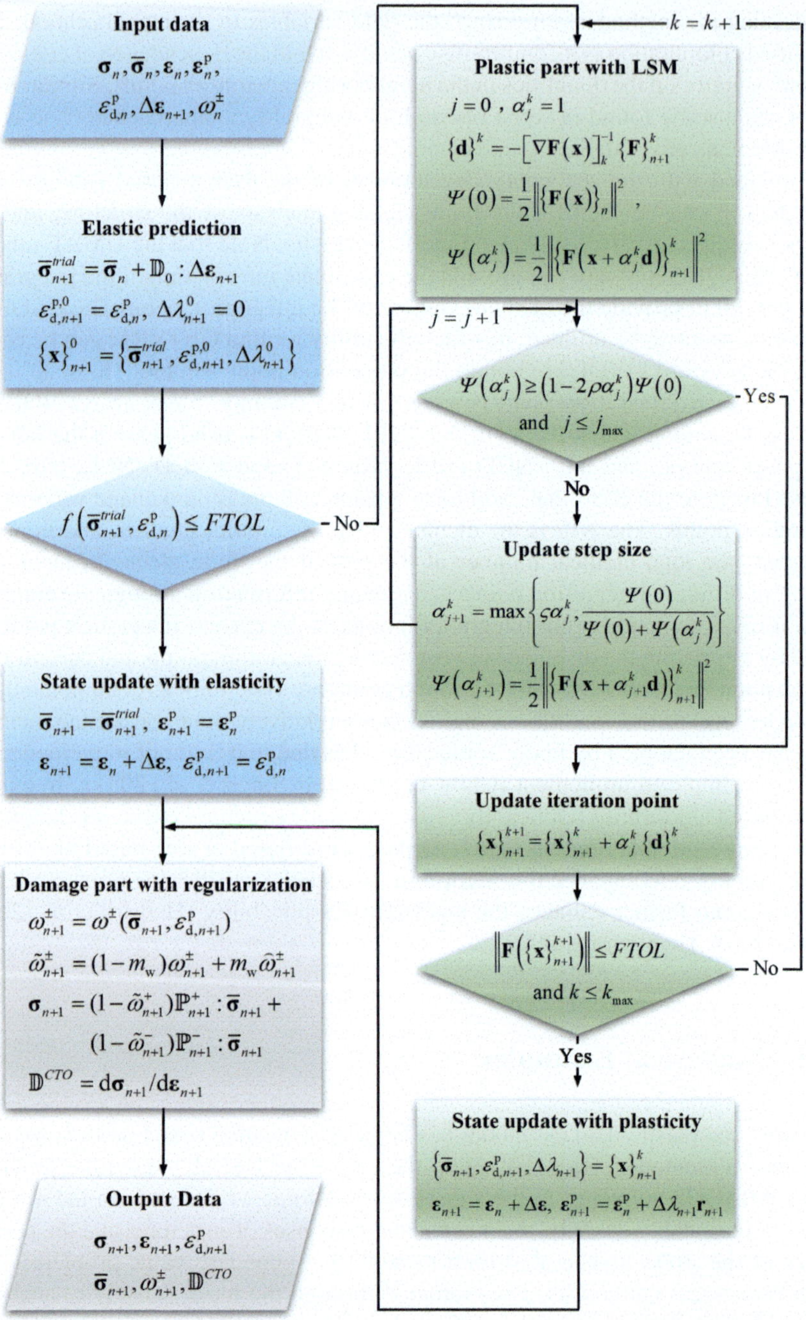

Fig. 10.5 Algorithm flow chart

density of concrete is set as 2400 kg · m^{-3}, and the density of steel is 7850 kg · m^{-3}. The gravity acceleration of 9.81 m · s^{-2} is considered in two steel–concrete structures in Sects. 10.4.3 and 10.4.4. For the parameters of nonlocal formulation, $m = 0.8$ is used for all numerical examples and the material length l can be set to 5 times the maximum mesh size.

10.4.1 2D Plate Plane Strain Compression Test

Firstly, to verify the effectiveness of nonlocal formulation in the proposed algorithm, the compression test of a 2D plate is employed to simulate the structural softening response caused by strain localization, which is also one of the representative boundary problems of investigating nonlocal methods (Gao et al., 2021; Lü et al., 2020). The top of the plane strain specimen is loaded by vertical displacement. In the remaining constraint settings, the bottom of the specimen is constrained and free on both sides. The specimen size is 100 mm × 200 mm, as shown in Fig. 10.6a. Besides, Fig. 10.6b shows four mesh densities used for the simulation, including 561 elements, 1250 elements, 2211 elements and 5000 elements. To trigger strain localization, a 12 × 12mm defect in the form of weak elements is inserted on the left side of the specimen.

As shown in Fig. 10.7a, the F vs. δ curves obtained from the local model show more softened behaviors in the post-peak stage with the refinement of the mesh, while good consistency in Fig. 10.7b is observed in the results obtained by the nonlocal model. The comparison shows that the integral nonlocal method eliminates the mesh dependence of the simulation results to a large extent.

Another manifestation of mesh sensitivity is that the narrower the shear bandwidth of the finer mesh, as shown in Fig. 10.8a. For the results utilizing the integration nonlocal method, the formation of shear bands caused by strain concentration is affected by the nonlocal softening variable, resulting in mesh-independent solutions. The width of the shear band predicted by the integration nonlocal method is similar for different mesh sizes, as illustrated in Fig. 10.8b.

10.4.2 3D Notched Plate Compression Test

The second case study is to verify the quadratic convergence speed of the CTO evaluated by the CSDA. As shown in Fig. 10.9a, a 4 mm-thick notched concrete plate is subjected to the vertical displacement load. The proposed stress update algorithms with the CTO and one with the CON are used for the comparison. Figure 10.9b shows the simulation results in the form of plastic internal variables, in which the element type is set to C3D8R.

Table 10.2 and Fig. 10.10 show the load increment number, the global iteration number, and the CPU time for the two tangent operators in this numerical example.

Table 10.1 Material parameters of concrete for UMAT

Example	f_c (MPa)	$\varepsilon^{\mathrm{p}}_{\mathrm{ds1}}$	E_{c} (GPa)	ν_{c}	$\varphi_{\mathrm{p}}(°)$	κ_φ	c_{p} (MPa)	A_c	κ_c	χ^+	χ^-	ψ	μ
Section 10.4.1	40	0.0011	40	0.2	38	1.8	17	1.40	0.80	0.55	0.36	0.90	0.9
Section 10.4.2	30	0.0030	30	0.2	37	2.0	13	1.21	0.65	0.45	0.20	0.20	0.5
Section 10.4.3	41	0.0050	36	0.2	44	2.0	15	1.36	0.38	0.60	0.20	0.30	0.1
Section 10.4.4	30	0.0025	30	0.2	33	3.5	18	1.19	0.46	0.68	0.14	0.30	0.8
Section 10.4.5	30	0.0025	30	0.2	30	3.0	20	1.60	0.50	0.71	0.50	0.40	0.8

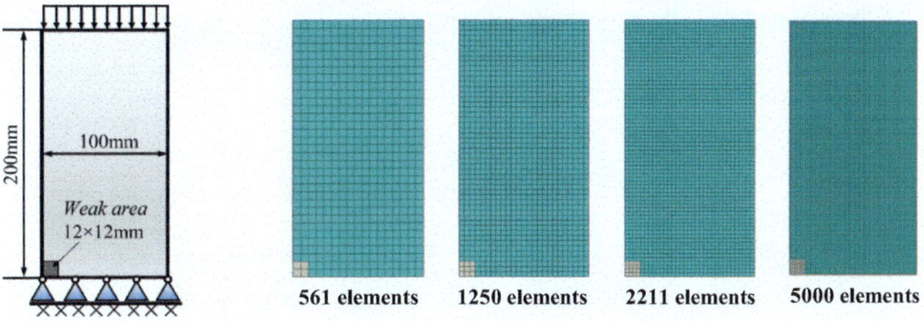

(a) Geometry and loading setup (b) Finite element mesh

Fig. 10.6 Numerical example information of the plane strain compression test

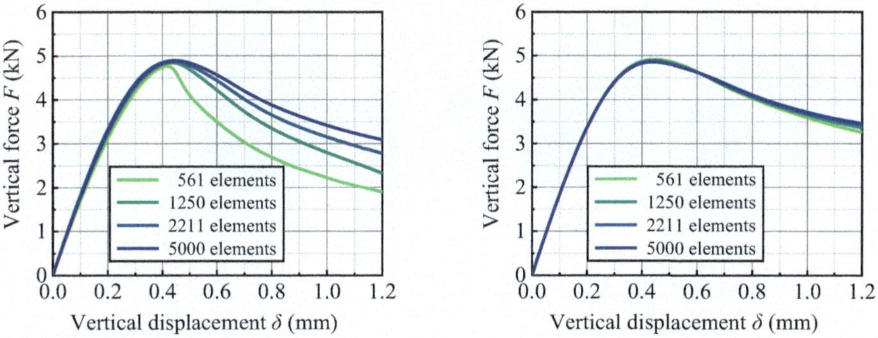

(a) Local model (b) Nonlocal model

Fig. 10.7 F vs. δ curves of the plane strain compression results

As expected, the stress update algorithm using the CTO is more efficient than that using the CON. The load increment number required by CON is about 2 times that of CTO since the former encountered failure in the global balance iteration NO.4, resulting in the subsequent load increment step size decrease. On the other hand, the number of iterations required by CON is about 5 times that of CTO due to the loss of quadratic convergence. The significant difference in globally balanced iteration directly leads to the CPU time consumed by CON being much higher than that of CTO. Furthermore, the convergence behavior of the two tangent operators is compared in the form of the normalized largest residual force (NLRF) logarithm, as illustrated in Fig. 10.11. The results show that poor convergence behavior is observed for CON, but the global quadratic convergence speed is achieved by the CTO, which verifies the effectiveness of using CSDA to evaluate CTO.

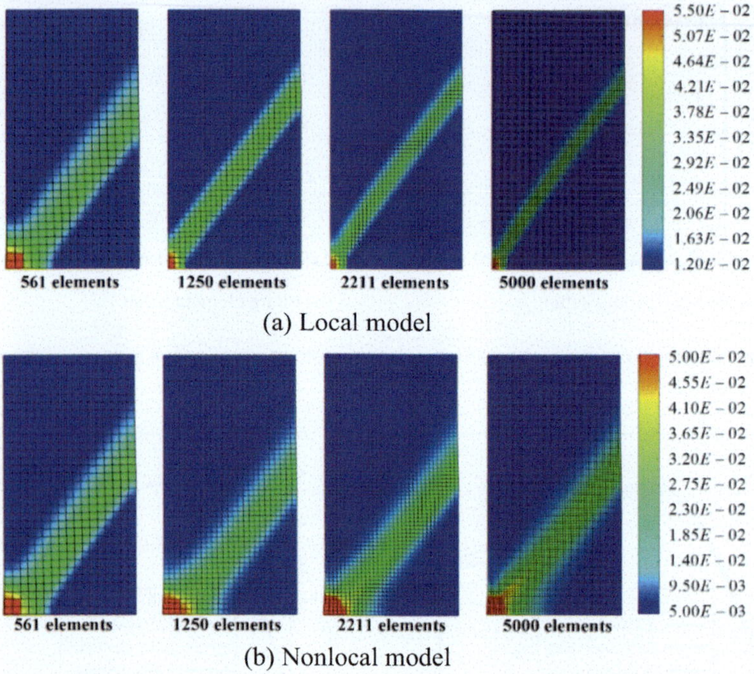

Fig. 10.8 Distribution of the equivalent plastic shear strain

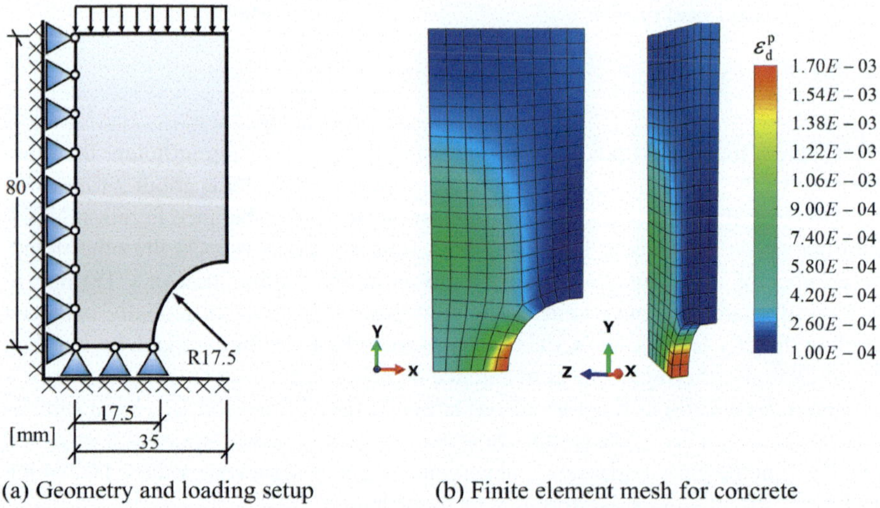

Fig. 10.9 Numerical example information of the 3D plate notched specimen

Table 10.2 Iterations for achieving global equilibrium of each increment

Increment no	1	2	3	4	5	6	7	8	9
Iterations for CON	4	4	7	10	10	12	9	6	1
Iterations for CTO	3	3	3	3	–	–	–	–	–

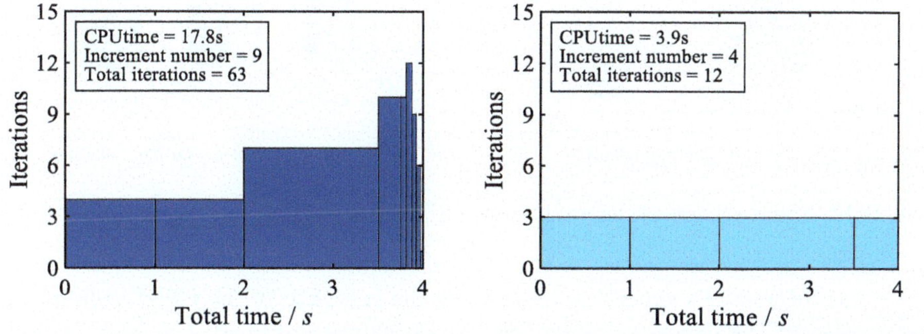

(a) Stress update algorithm with the CON (b) Stress update algorithm with the CTO

Fig. 10.10 Convergence behavior at each increment

Fig. 10.11 NLRF vs. iterations at the critical node

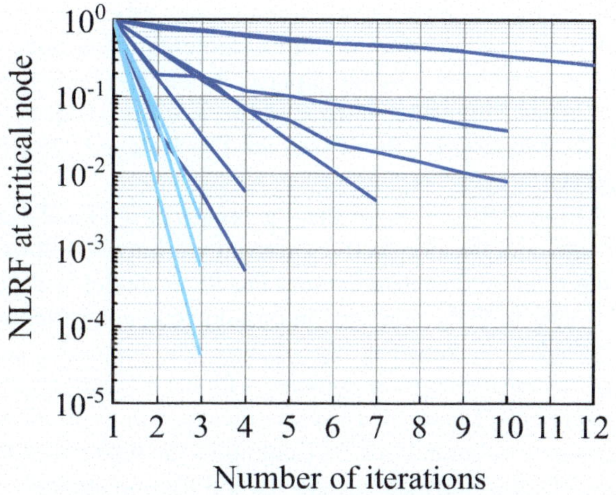

10.4.3 Concrete-Filled Steel-Tube Column under Central Compression

Compression is usually the most favorable stress condition for concrete because it is at a disadvantage of tensile performance. In this subsection, the concrete-filled steel tube (CFST) column experiment conducted by Sakino et al. (2004) is simulated

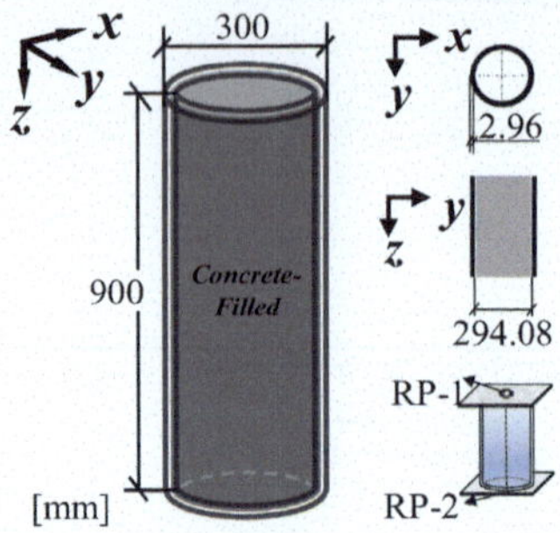

(a) Geometry and loading setup

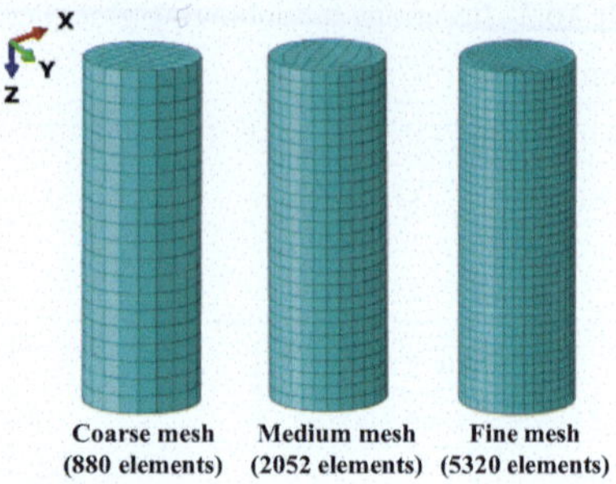

(b) Finite element mesh for concrete

Fig. 10.12 Numerical example information of CFST column

to assess the model's capability to describe the response of the CFST column in compression. The geometry of the specimen is shown in Fig. 10.12a. The top and bottom of the CFST column are fixed through two rigid plates, and the vertical displacement load is applied to the surface of the top rigid plate by a reference point. The contact between the rigid plate and the CFST column is set to surface-to-surface. Similar to the previous examples, the three different mesh densities are used to discrete the concrete filled in the steel tube, as shown in Fig. 10.12b. The C3D8R element is adopted for the finite element model of the concrete and steel tube. Steel uses the bilinear hardening plasticity model (Jukić et al., 2013; Shi et al., 2012) for its material model, and the modulus is set as $0.01E$ in the hardening stage. The yield and ultimate strengths of steel are 170 MPa and 279MPa, respectively (Sakino et al., 2004).

Figure 10.13a shows that the results of the local model are mesh dependent even in the compression load-dominated working condition. It is observed that the mesh dependency disappears when the local model is replaced by the nonlocal model because both the tension/compression damage variables are developed into nonlocal variables, as shown in Fig. 10.13b. The force–displacement curves from different mesh densities almost coincide with each other and are in good agreement with the test results. Figure 10.14 shows the equivalent plastic shear strain distribution of the concrete column. For CFST columns, the restraint of the steel tube gives full play to the plastic deformation capacity and friction strength of the concrete.

The CFST column example with 880 elements is also used to compare the convergence difference between the LSM employed in the proposed stress update algorithm and the standard Newton method. Correspondingly, the simulation results of the two algorithms are shown in Fig. 10.15 in the form of F versus δ curves. Note that the slight unsmooth phenomenon in the F vs. δ curves is caused by the relatively large load increment steps. To highlight the convergence advantages of the LSM-based algorithm, the example corresponding to Fig. 10.15 does not limit the size of the input load step. Based on the automatic step control scheme of ABAQUS, Fig. 10.16 shows the change law of load incremental step size from the two algorithms. The

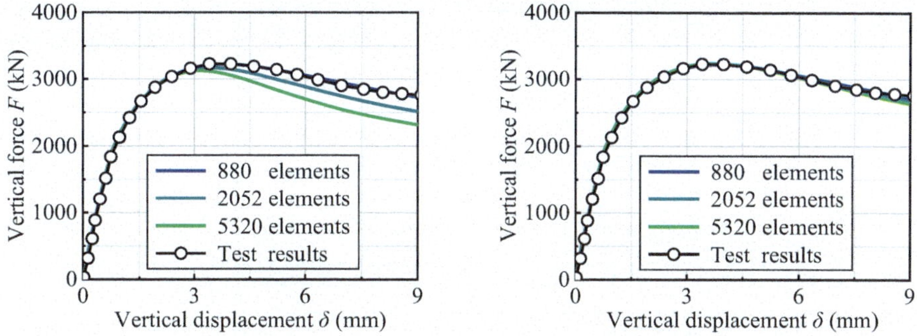

(a) Local model (b) Nonlocal model

Fig. 10.13 F vs. δ curves of RC columns under eccentric compression loading

Fig. 10.14 Equivalent plastic shear strain distribution of concrete

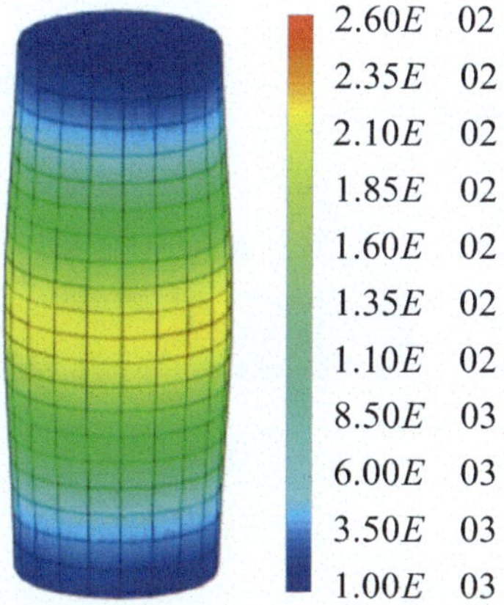

"Total time" in the ABAQUS software refers to a specific time metric representing the cumulative simulation time for a given analysis. It is an essential parameter that defines the time span of the simulation, from the initial state to the final state. Compared with the algorithm based on the Newton method, the algorithm based on LSM reduces the total CPU time by about 24% under the condition of basically consistent results.

For CFST column under central compression loading, the dilatancy behavior will change the constraint stress conditions of concrete, thereby affecting its compressive performance (Lai et al., 2020; Ouyang et al., 2017). In the proposed model, the dilatancy behavior predicted can be adjusted by the parameter μ. Figure 10.17 shows the effect of μ on the plastic flow direction on the meridional plane, which essentially reflects the proportional relationship between $d\varepsilon_v^p$ and $d\varepsilon_d^p$. As μ increases within the range of 0 to 1, the plastic flow direction on the meridian plane gradually turns to the "tension direction" of hydrostatic axis, which indicates that the dilatancy behavior

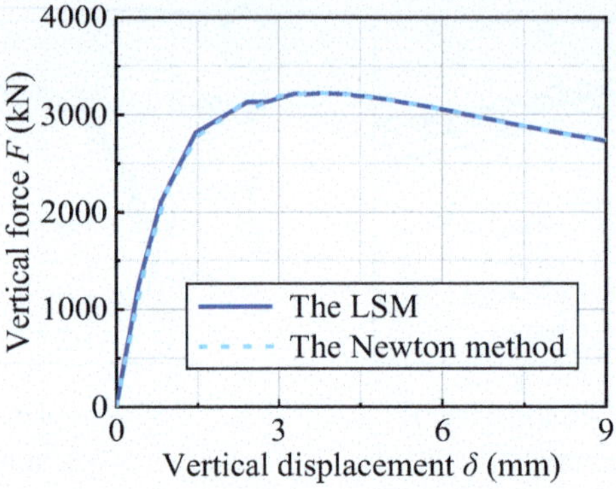

Fig. 10.15 F vs. δ curves based on the LSM and the Newton method

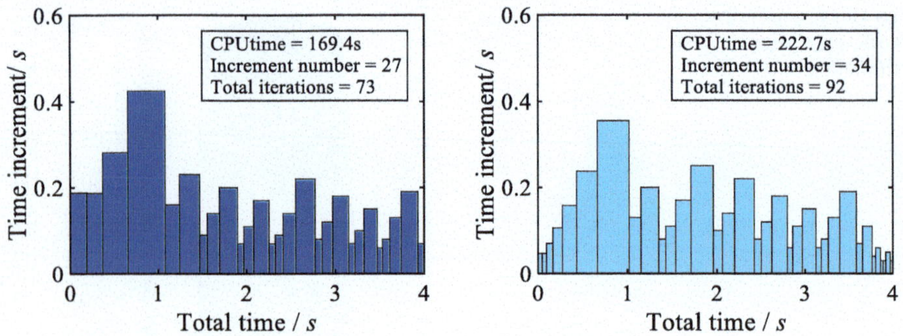

(a) The LSM (b) The Newton method

Fig. 10.16 Change of incremental step size for stress update algorithms

predicted by the non-orthogonal model is more significant. When μ equals 1, the non-orthogonal model degenerates into the orthogonal model, resulting in an overestimation of the dilatancy behavior. Under passive confinement conditions, the unrealistic volume expansion leads to stronger confinement stresses acting on the concrete, resulting in higher but unreasonable simulated structural load-carrying capacity, as shown in Fig. 10.18. The proposed model can reasonably capture concrete's dilatancy behavior by selecting an appropriate parameter μ, further resulting in a reasonable prediction of concrete structures' bearing capacity under passive constraints.

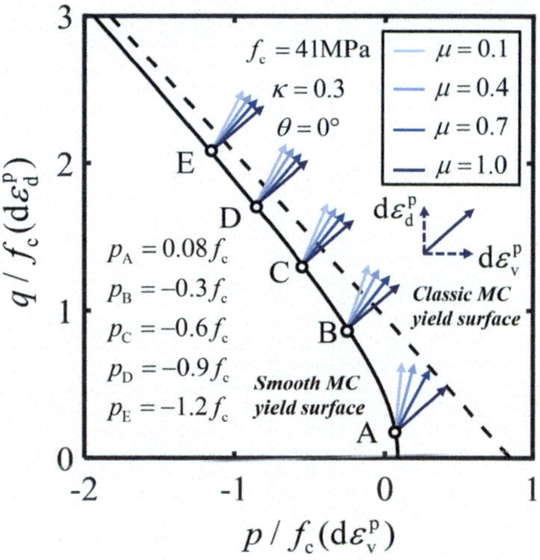

Fig. 10.17 Effect of μ on the plastic flow direction

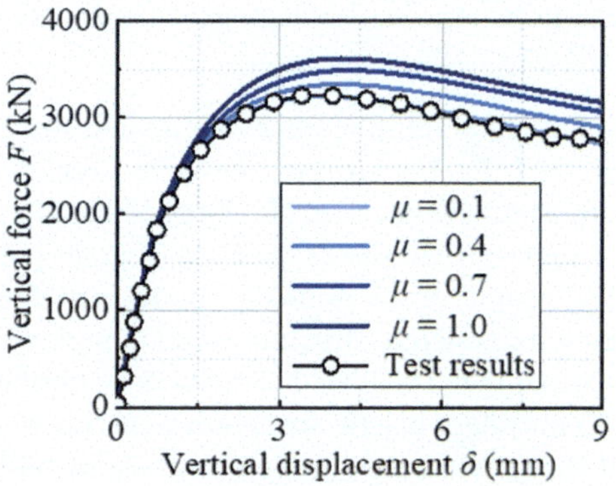

Fig. 10.18 Effect of μ on the simulation results of CFST

10.4.4 Reinforced Concrete Column under Eccentric Compression

The last example with the published test data (Němeček et al., 2005) is to apply the current model to the failure analysis of the reinforced concrete (RC) column under eccentric compression loading. Figure 10.19a illustrates boundary conditions and geometric information of the RC column, in which the spacing of stirrups is 100 mm. In the numerical simulation, the C3D8R element and T3D2 (2-node linear 3-D truss) element are employed for the concrete column and the reinforcing bars, respectively. The interaction between the concrete column and reinforcing bars is an embedded region. It is configured that RP-1 couples with the FE model's upper surface and RP-2 couples with the model's lower surface. The vertical eccentric displacement load is applied at the RP-1. Three different finite element mesh settings are presented in Fig. 10.19b. The bilinear hardening plasticity model is also used for steel here. The

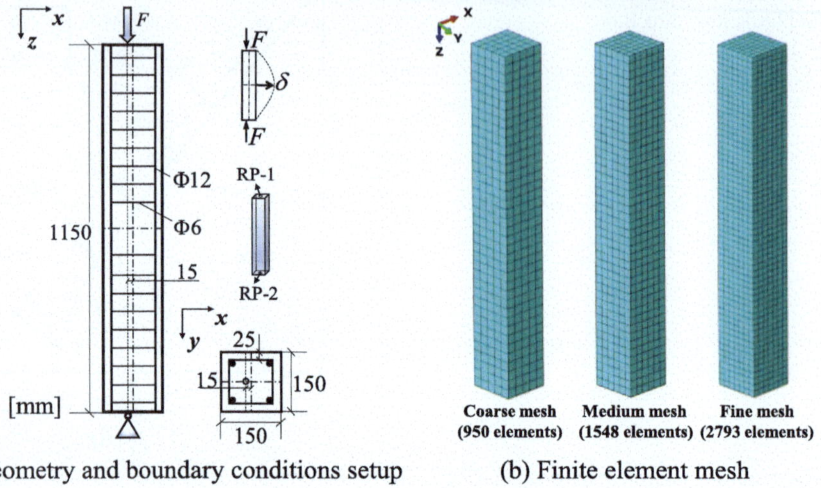

Fig. 10.19 Numerical example information of RC column

longitudinal reinforcement has a yield strength of 400 MPa and an ultimate strength of 550 MPa. For stirrups, the yield strength is 235 MPa, and the ultimate strength is 310 MPa (Němeček et al., 2005; Shi et al., 2012).

Figure 10.20 shows the F vs. δ curves predicted by the local and nonlocal models. The results of the nonlocal model still demonstrate excellent mesh independence even for concrete structures with complex stress states, but the mesh dependence is obvious in the results of local models. Furthermore, Fig. 10.21 shows the distribution of equivalent plastic shear strain of concrete and Mises stress of steel. The wedge-shaped strain concentration zone is observed on the side near the eccentric load in the middle of the column. The failure mode of the eccentrically compressed RC column (Němeček et al., 2005) is well reproduced by the current model.

The RC column example with fine mesh compares the convergence difference between the LSM used in the proposed stress update algorithm and the standard Newton method. The simulation results of different algorithms are also shown in Fig. 10.22. Due to the fact that the difference between the two algorithms is reflected in the computational efficiency rather than the computational results, it can be seen that the results of different algorithms are basically consistent under this loading condition. This numerical example with a complex stress state has more stringent requirements for the convergence of local iterative calculation, so it can highlight the convergence advantage of the LSM compared with the Newton method, as depicted in Fig. 10.23a and b. At the initial stage of loading, the allowable load steps of the two algorithms are similar. From 8 s to 8.4 s (analysis step time) shown in Fig. 10.23c, both algorithms encountered convergence difficulties, but the allowable load increment size of the LSM-based stress update algorithm is significantly larger than that of the algorithm based on the Newton method. For the two algorithms, the total iteration of

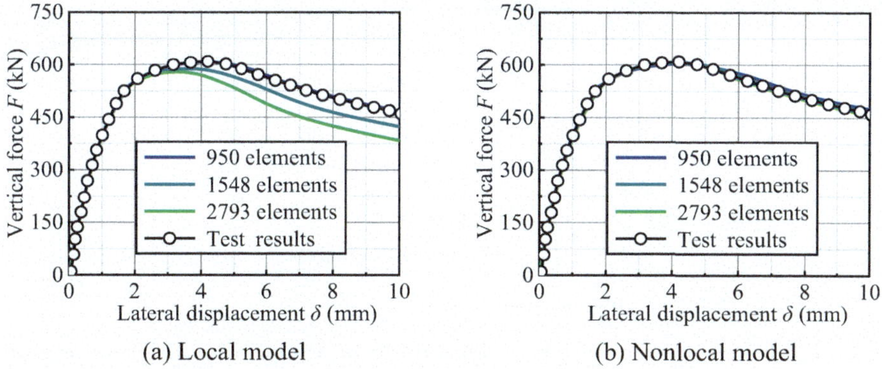

Fig. 10.20 F vs. δ curves of RC columns under eccentric compression loading

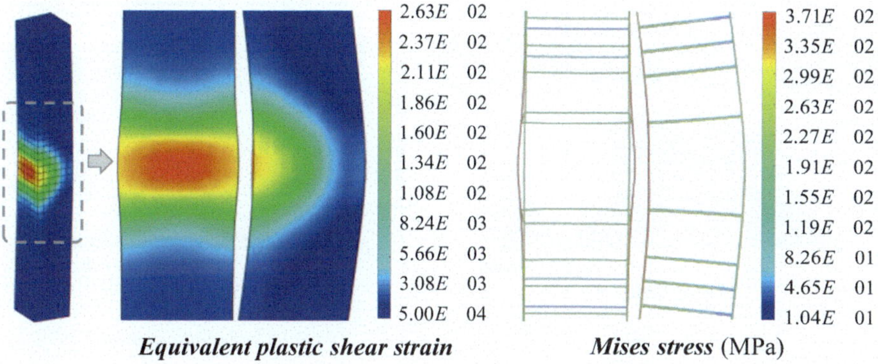

Fig. 10.21 Equivalent plastic shear strain and Mises stress distribution

LSM is about 42% of the Newton method. And the total CPU time of the LSM-based algorithm is about 44% of the time of the algorithm based on the Newton method.

10.4.5 3D Column Specimen Axial Compression Test

The fifth case study aims to verify the prediction results of the nonlocal model under conditions of pronounced softening behavior. As depicted in Fig. 10.24a a three-dimensional column specimen is presented, subjected to axial displacement loading, and simulated using three different mesh scales. The element type employed is C3D8R in Fig. 10.24b. The top and bottom surfaces are coupled with reference points RP-1 and RP-2, respectively, with RP-1 applying displacement-controlled loading.

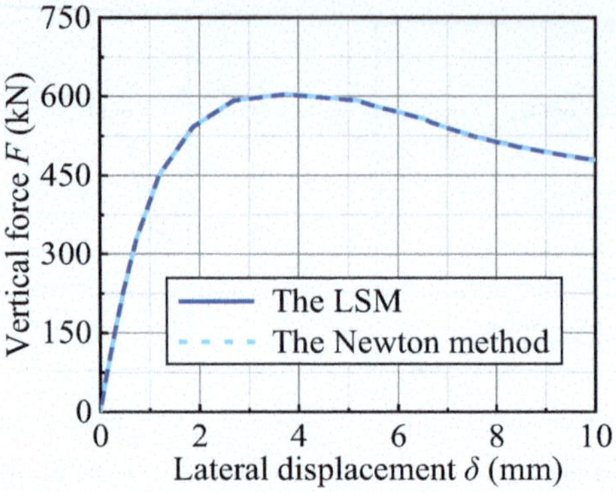

Fig. 10.22 F versus δ curves based on the LSM and the Newton method

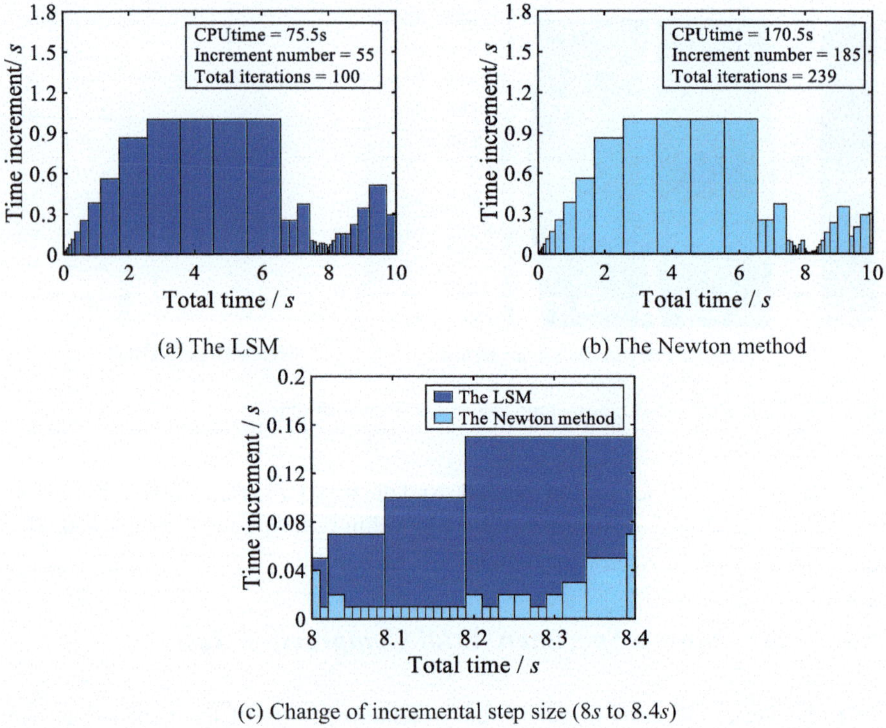

Fig. 10.23 Change of incremental step size for stress update algorithms

In the results presented in Fig. 10.25, simulation results are shown for the same specimen size with different numbers of elements: 279, 640, and 1250 elements. It should be noted that in situations where the softening behavior is exceptionally severe, not all local models can converge. Therefore, only the result of the nonlocal

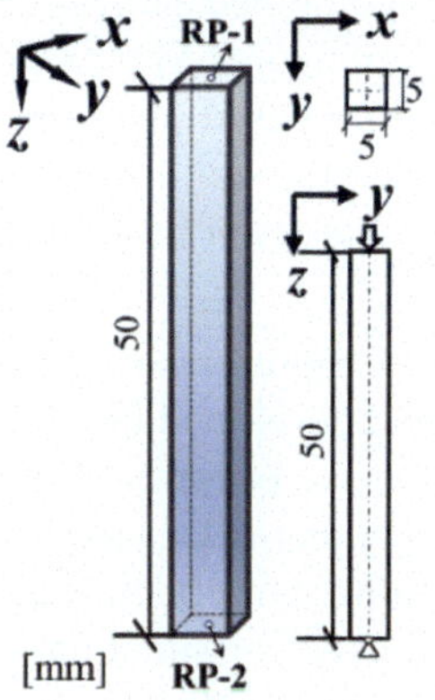

(a) Geometry and loading setup

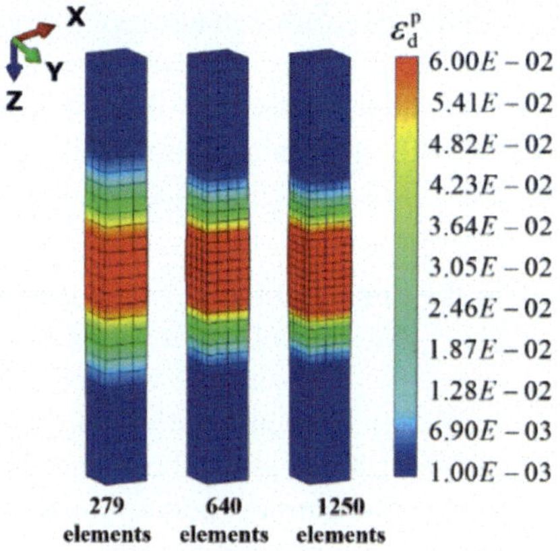

(b) Finite element mesh

Fig. 10.24 Numerical example information for the plain concrete column

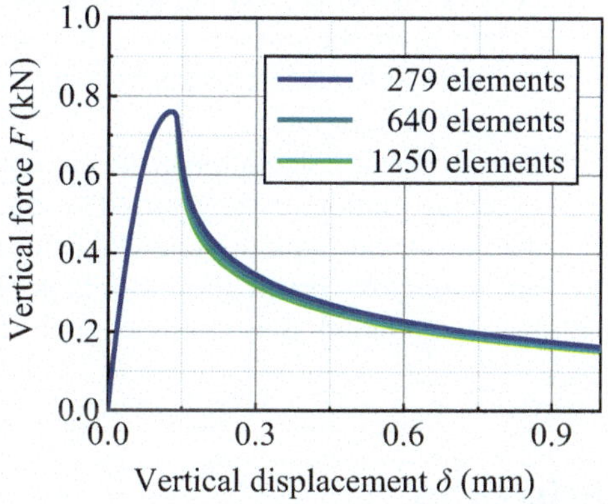

Fig. 10.25 F vs. δ curves of the plain concrete column under compressive condition

model is displayed here. It can be observed that the proposed nonlocal model exhibits regularization properties, even when faced with a more pronounced decay of the force–displacement curve. This demonstrates the efficacy of the nonlocal model in this study in handling challenging softening conditions.

10.5 Summary

In this chapter, we propose a nonlocal implicit stress update algorithm aimed at eliminating the mesh dependency of plastic damage models in structural failure analysis. Thanks to the characteristics of the effective stress modeling method, the implicit stress update algorithm of elastoplastic models is perfectly combined with the integral nonlocal method used for damage calculation. The non-orthogonal Mohr-Coulomb plasticity model is used as the application object of the proposed algorithm. The performance of the algorithm is comprehensively evaluated through five numerical examples under different load conditions. The main conclusions of this chapter are as follows:

(1) In the implementation of the model, the tensile and compressive damage variables that control strain softening behavior have been developed into nonlocal variables to ensure the well-posedness of PDEs after the occurrence of strain localization phenomenon. The analysis results of numerical examples with different mesh sizes show that the proposed algorithm can effectively eliminate the mesh dependency of the softening model, even for situations where softening is particularly severe.
(2) When calculating nonlocal damage variables, only the damage variables of the current integration point are the values at the end of the increment step, and the damage variables of other integration points are frozen at the beginning of the increment step. This approach simplifies the computation of the consistent tangent operator, avoiding its dependence on the quantities in the neighboring points. From the analysis of residual force, it can be seen that this simplification is successful and has almost no impact on the quadratic convergence speed of the global iteration.
(3) For the plastic calculation in the effective stress space, a fully implicit backward Euler integration strategy is used to ensure the stability and accuracy of solutions. The nonlinear implicit integral equations are solved using the LSM. Numerical analysis shows that it allows for larger load step sizes compared to the standard Newton method, thus saving overall computational costs.

References

Bažant, Z. P., & Jirásek, M. (2002). Nonlocal integral formulations of plasticity and damage: Survey of progress. *Journal of Engineering Mechanics, 128*(11), 1119–1149.

Candappa, D. C., Sanjayan, J. G., & Setunge, S. (2001). Complete triaxial stress-strain curves of high-strength concrete. *Journal of Materials in Civil Engineering, 13*(3), 209–215.

Chow, C. L., Mao, J. H., & Shen, J. (2011). Nonlocal damage gradient model for fracture characterization of aluminum alloy. *International Journal of Damage Mechanics, 20*(7), 1073–1093.

Faria, R., Oliver, J., & Cervera, M. (1998). A strain-based plastic viscous-damage model for massive concrete structures. *International Journal of Solids and Structures, 35*(14), 1533–1558.

Galavi, V., & Schweiger, H. F. (2010). Nonlocal multilaminate model for strain softening analysis. *International Journal of Geomechanics, 10*(1), 30–44.

Gao, Z. W., Zhao, J. D., & Li, X. (2021). The deformation and failure of strip footings on anisotropic cohesionless sloping grounds. *International Journal for Numerical and Analytical Methods in Geomechanics, 45*(10), 1526–1545.

Gao, Z. W., Li, X., & Lu, D. C. (2022). Nonlocal regularization of an anisotropic critical state model for sand. *Acta Geotechnica, 17*(2), 427-439.

Howard, M. P., Anderson, J. A., Nikoubashman, A., Glotzer, S. C., & Panagiotopoulos, A. Z. (2016). Efficient neighbor list calculation for molecular simulation of colloidal systems using graphics processing units. *Computer Physics Communications, 203*, 45–52.

Huang, M. S., Qu, X., & Lü, X. L. (2018). Regularized finite element modeling of progressive failure in soils within nonlocal softening plasticity. *Computational Mechanics, 62*(3), 347–358.

Imran, I., & Pantazopoulou, S. J. (1996). Experimental study of plain concrete under triaxial stress. *ACI Materials Journal, 93*(6), 589–601.

Jirasek, M., & Rolshoven, S. (2003). Comparison of integral-type nonlocal plasticity models for strain-softening materials. *International Journal of Engineering Science, 41*(13–14), 1553–1602.

Ju, J. W. (1989). On energy-based coupled elastoplastic damage theories: Constitutive modeling and computational aspects. *International Journal of Solids and Structures, 25*(7), 803–833.

Jukić, M., Brank, B., & Ibrahimbegović, A. (2013). Embedded discontinuity finite element formulation for failure analysis of planar reinforced concrete beams and frames. *Engineering Structures, 50*, 115–125.

Kamaludin, S., & Thamburaja, P. (2023). Efficient neighbour search algorithm for nonlocal-based simulations—application to failure mechanics. *Journal of Failure Analysis and Prevention, 23*(2), 540–547.

Kupfer, H. B., Hilsdorf, K. H., & Rush, H. (1969). Behavior of concrete under biaxial stresses. *ACI Journal Proceedings, 66*(8), 656–666.

Lagioia, R., & Panteghini, A. (2016). On the existence of a unique class of yield and failure criteria comprising Tresca, von Mises, Drucker–Prager, Mohr–Coulomb, Galileo–Rankine, Matsuoka–Nakai and Lade–Duncan. *Proceedings of the Royal Society a: Mathematical, Physical and Engineering Sciences, 472*(2185), 20150713.

Lai, M. H., Song, W., Ou, X. L., Chen, M. T., Wang, Q., & Ho, J. C. M. (2020). A path dependent stress-strain model for concrete-filled-steel-tube column. *Engineering Structures, 211*, 110312.

Lü, X. L., Xue, D. W., & Lim, K. (2020). Implicit gradient softening plasticity for the modeling of strain localization in soils. *Computer Methods in Applied Mechanics and Engineering, 364*, 112934.

Lu, X., & Hsu, C. T. T. (2007). Tangent poisson's ratio of high-strength concrete in triaxial compression. *Magazine of Concrete Research, 59*(1), 69–77.

Mánica, M. A., Gens, A., Vaunat, J., & Ruiz, D. F. (2018). Nonlocal plasticity modelling of strain localisation in stiff clays. *Computers and Geotechnics, 103*, 138–150.

Mazzucco, G., Xotta, G., Salomoni, V. A., & Majorana, C. (2019). Integral-type regularization of non associated softening plasticity for quasi brittle materials. *Computers & Structures, 224*, 106120.

Němeček, J., Padevět, P., Patzák, B., & Bittnar, Z. (2005). Effect of transversal reinforcement in normal and high strength concrete columns. *Materials and Structures, 38*(7), 665.

Ouyang, Y., Kwan, A. K. H., Lo, S. H., & Ho, J. C. M. (2017). Finite element analysis of concrete-filled steel tube (CFST) columns with circular sections under eccentric load. *Engineering Structures, 148*, 387–398.

Potts, D. M., Cui, W. J., & Zdravković, L. (2021). A coupled THM finite element formulation for unsaturated soils and a strategy for its nonlinear solution. *Computers and Geotechnics, 136*, 104221.

Ren, H. L., Zhuang, X. Y., Nguyen-Thoi, T., & Rabczuk, T. (2021). A nonlocal operator method for finite deformation higher-order gradient elasticity. *Computer Methods in Applied Mechanics and Engineering, 384*, 113963.

Rubin, M. B., & Cardon, A. H. (2002). Cosserat theories: shells, rods and points. Solid Mechanics and Its Applications, vol 79. *Applied Mechanics Reviews, 55*(6), B109–B110.

Sakino, K., Nakahara, H., Morino, S., & Nishiyama, I. (2004). Behavior of centrally loaded concrete-filled steel-tube short columns. *Journal of Structural Engineering, 130*(2), 180–188.

Sarkar, S., Singh, I. V., & Mishra, B. K. (2022). A localizing gradient plasticity model for ductile fracture. *Computer Methods in Applied Mechanics and Engineering, 388*, 114205.

Shi, G., Wang, M., Bai, Y., Wang, F., Shi, Y. J., & Wang, Y. Q. (2012). Experimental and modeling study of high-strength structural steel under cyclic loading. *Engineering Structures, 37*, 1–13.

Simo, J. C., & Hughes, T. J. R. (2006). Computational inelasticity. Springer Science & Business Media, New York.

Singh, V., Stanier, S., Bienen, B., & Randolph, M. F. (2021). Modelling the behaviour of sensitive clays experiencing large deformations using non-local regularisation techniques. *Computers and Geotechnics, 133*, 104025.

Strömberg, L., & Ristinmaa, M. (1996). FE-formulation of a nonlocal plasticity theory. *Computer Methods in Applied Mechanics and Engineering, 136*(1–2), 127–144.

Su, C. C., Lu, D. C., Zhou, X., Wang, G. S., Zhuang, X. Y., & Du, X. L. (2023). An implicit stress update algorithm for the plastic nonlocal damage model of concrete. *Computer Methods in Applied Mechanics and Engineering, 414*, 116189.

Yu, H. T., Wang, Z. B., Yuan, Y., & Li, W. T. (2016). Numerical analysis of internal blast effects on underground tunnel in soils. *Structure and Infrastructure Engineering, 12*(9), 1090–1105.

Appendix A
Definitions and Properties of Fractional Derivatives

The fractional derivative is an extension of the integer order derivative. There are usually two types of fractional derivatives commonly used in the non-orthogonal plastic flow rule, i.e., Riemann–Liouville (RL), and Caputo definitions. The RL and Caputo definitions are both in the form of integration and derivative, but the sequence of calculation is different. In the following content, the definitions and properties of these two fractional derivatives, as well as the Gamma function closely related to their definitions, are briefly described.

A.1 Gamma Function

Gamma function $\Gamma(x)$, also known as Euler's second integral, is an extension of the factorial function on real and complex numbers. Its expression is:

$$\Gamma(x) = \int_0^\infty e^{-\tau}\tau^{x-1}\mathrm{d}\tau \quad \mathrm{Re}(x) > 0 \tag{A.1}$$

where $\mathrm{Re}(x)$ represents the real part of x. τ is the integration variable.

The commonly used properties of gamma functions include:

(1) Recursive property: $\Gamma(x+1) = x\Gamma(x)$.
(2) For any positive integer n: $\Gamma(n) = (n-1)!$.
(3) Some special cases: $\Gamma(1) = \Gamma(2) = 1$, $\Gamma(0) = \pm\infty$, $\Gamma(-n) = \infty$.

D. Lu et al., *Multi-physical and Non-orthogonal Geotechnical Plasticity Modeling and Algorithms*, https://doi.org/10.1007/978-981-96-2369-3

A.2 Riemann–Liouville Fractional Derivative

The RL fractional derivative is the most important extension of ordinary calculus. Almost all other definitions for the fractional derivative represent a special case of the RL fractional derivative. Its expression is:

$$ {}_{a}^{\mathrm{RL}}D_{x}^{\mu}f(x) = \frac{1}{\Gamma(n-\mu)}\frac{\mathrm{d}^{n}}{\mathrm{d}x^{n}}\left(\int_{a}^{x}(x-\tau)^{n-\mu-1}f(\tau)d\tau\right) \tag{A.2} $$

where RL means the type of fractional derivative; a is the lower limit of the integral that is chosen as 0 herein; x is an independent variable that represents the loading stress in the fractional plastic flow rule or the loading time in the fractional rheology model; μ is the fractional order. Obviously, the fractional derivative is defined on the interval that is consist of the initial and current values of the variable, which is different from the integer derivative defined on the single point. Therefore, the value of the fractional derivative is related to the evolution history of the variable from the initial state to the current state. This implies that some information about the loading history can be included in the fractional gradient when the fractional gradient of yield function is used to determine the plastic flow direction of the material.

The RL fractional derivative has the following properties:

$$ {}_{0}^{RL}D_{x}^{\mu}C = \frac{Cx^{-\mu}}{\Gamma(1-\mu)} \tag{A.3} $$

$$ {}_{0}^{RL}D_{x}^{\mu}x^{m} = \frac{\Gamma(m+1)x^{m-\mu}}{\Gamma(m-\mu+1)} \tag{A.4} $$

$$ {}_{0}^{RL}D_{x}^{\mu}[\lambda_{1}f_{1}(x)+\lambda_{2}f_{2}(x)] = \lambda_{1}{}_{0}^{RL}D_{x}^{\mu}f_{1}(x)+\lambda_{2}{}_{0}^{RL}D_{x}^{\mu}f_{2}(x) \tag{A.5} $$

where C in Eq. (A.3) is an any constant; Eq. (A.4) is the RL fractional derivative of the power function; Eq. (A.5) indicates the that RL fractional derivative satisfies the linear superposition principle, where $\lambda_1, \lambda_2 \in R$.

A.3 Caputo Fractional Derivative

Caputo fractional derivatives have advantages such as weak singularity and taking the derivative of a constant to 0, and are therefore often used in mechanical modeling. The definition of Caputo fractional derivative is expressed as:

$$ {}_{0}^{C}D_{t}^{\mu}f(t) = \frac{1}{\Gamma(n-\mu)}\int_{a}^{t}(t-\tau)^{n-\mu-1}f^{(n)}(\tau)\mathrm{d}\tau \tag{A.6} $$

where C means the type of fractional derivative. It can be seen from the expressions of fractional derivative definition that Caputo fractional derivative operator is the derivative first and then integral, and RL fractional derivative operator is integral first and then derivative.

The Caputo fractional derivative satisfies the following properties:

$$ {}_0^C D_x^{\mu} C = 0 \tag{A.7} $$

$$ {}_0^C D_x^{\mu} x^m = \begin{cases} \frac{\Gamma(m+1)x^{m-\mu}}{\Gamma(m-\mu+1)} & n-1 < \mu < n, m > n-1, m \in R \\ 0 & n-1 < \mu < n, m \le n-1, m \in N \end{cases} \tag{A.8} $$

$$ {}_0^C D_x^{\mu} [\lambda_1 f_1(x) + \lambda_2 f_2(x)] = \lambda_1 {}_0^C D_x^{\mu} f_1(x) + \lambda_2 {}_0^C D_x^{\mu} f_2(x) \tag{A.9} $$

According to the definitions of two fractional derivatives, they have the following transformation relationship:

$$ {}_0^{RL} D_t^{\mu} f(t) = \sum_{k=0}^{n-1} \frac{f^{(k)}(0) t^{k-\mu}}{\Gamma(k+1-\mu)} + {}_0^C D_t^{\mu} f(t) \tag{A.10} $$

Appendix B
Operation Rules and Numerical Implementation of Hyper-Dual Numbers

The HDNs can be utilized for achieving high-precision numerical differentiation, yet the mathematical libraries of mainstream computing languages do not typically include a data type for hyper-dual numbers. Consequently, when adopting HDSDA, it is necessary to custom-define the data type for hyper-dual numbers along with their operational rules. In this chapter, we first provide a detailed introduction to the basic operational rules of HDNs. Subsequently, by utilizing the Operator overload and Function overload functionality of the Fortran language, we present an implementation scheme for the HDSDA, which involves defining the data structure and operational rules for HDNs in Fortran.

B.1 Operational Rules of Hyper-Dual Numbers

For any two HDNs $a = a_0 + a_1\vec{\epsilon}_1 + a_2\vec{\epsilon}_2 + a_3\vec{\epsilon}_1\vec{\epsilon}_2$ and $b = b_0 + b_1\vec{\epsilon}_1 + b_2\vec{\epsilon}_2 + b_3\vec{\epsilon}_1\vec{\epsilon}_2$, a is equal to b when their corresponding real and "imaginary" parts are equal

$$a = b \Leftrightarrow a_i = b_i (i = 0, 1, 2, 3) \tag{B.1}$$

The addition operation of HDNs a and b is defined by

$$a + b = (a_0 + b_0) + (a_1 + b_1)\vec{\epsilon}_1 + (a_2 + b_2)\vec{\epsilon}_2 + (a_3 + b_3)\vec{\epsilon}_1\vec{\epsilon}_2 \tag{B.2}$$

The multiplication operation of HDNs a and b is defined by

$$\begin{aligned} a \cdot b = &(a_0 b_0) + (a_0 b_1 + a_1 b_0)\vec{\epsilon}_1 + (a_0 b_2 + a_2 b_0)\vec{\epsilon}_2 \\ &+ (a_0 b_3 + a_3 b_0 + a_1 b_2 + a_2 b_1)\vec{\epsilon}_1\vec{\epsilon}_2 \end{aligned} \tag{B.3}$$

The division operation of HDNs a and b is defined by

D. Lu et al., *Multi-physical and Non-orthogonal Geotechnical Plasticity Modeling and Algorithms*, https://doi.org/10.1007/978-981-96-2369-3

$$\frac{a}{b} = \frac{a_0}{b_0} + \frac{a_1 b_0 - a_0 b_1}{b_0^2}\vec{\epsilon}_1 + \frac{a_2 b_0 - a_0 b_2}{b_0^2}\vec{\epsilon}_2 + \frac{-a_0 b_0 b_3 + 2a_0 b_1 b_2 - a_2 b_0 b_1 - a_1 b_0 b_2}{b_0^3}\vec{\epsilon}_1\vec{\epsilon}_2 \quad \text{(B.4)}$$

where the division operation holds under the condition of $b_0 \neq 0$.

The HDNs version of elementary functions is derived from its Taylor series expansion.

$$f(a) = f(a_0) + a_1 f'(a_0)\vec{\epsilon}_1 + a_2 f'(a_0)\vec{\epsilon}_2 + \left(a_3 f'(a_0) + a_1 a_2 f''(a_0)\right)\vec{\epsilon}_1\vec{\epsilon}_2 \quad \text{(B.5)}$$

Finally, the HDNs versions of several common functions are given as follows:

$$\cos(a) = \cos(a_0) - a_1 \sin(a_0)\vec{\epsilon}_1 - a_2 \sin(a_0)\vec{\epsilon}_2 - [a_3 \sin(a_0) + a_1 a_2 \cos(a_0)]\vec{\epsilon}_1\vec{\epsilon}_2 \quad \text{(B.6)}$$

$$\arccos(a) = \arccos(a_0) - \frac{a_1\vec{\epsilon}_1}{\sqrt{1-a_0^2}} - \frac{a_2\vec{\epsilon}_2}{\sqrt{1-a_0^2}} - \frac{a_3\vec{\epsilon}_1\vec{\epsilon}_2}{\sqrt{1-a_0^2}} - \frac{a_0 a_1 a_2\vec{\epsilon}_1\vec{\epsilon}_2}{\left(1-a_0^2\right)^{1.5}} \quad \text{(B.7)}$$

$$\sin(a) = \sin(a_0) + a_1 \cos(a_0)\vec{\epsilon}_1 + a_2 \cos(a_0)\vec{\epsilon}_2 + [a_3 \cos(a_0) - a_1 a_2 \sin(a_0)]\vec{\epsilon}_1\vec{\epsilon}_2 \quad \text{(B.8)}$$

$$\arcsin(a) = \arcsin(a_0) + \frac{a_1\vec{\epsilon}_1}{\sqrt{(1-a_0^2)}} + \frac{a_2\vec{\epsilon}_2}{\sqrt{(1-a_0^2)}} + \frac{a_3\vec{\epsilon}_1\vec{\epsilon}_2}{\sqrt{(1-a_0^2)}} + \frac{a_0 a_2 a_3\vec{\epsilon}_1\vec{\epsilon}_2}{\left(1-a_0^2\right)^{1.5}} \quad \text{(B.9)}$$

$$(a)^n = a_0^n + \left(a_1 n a_0^{n-1}\right)\vec{\epsilon}_1 + \left(a_2 n a_0^{n-1}\right)\vec{\epsilon}_2 + \left[a_3 n a_0^{n-1} + n(n-1)a_1 a_2 a_0^{n-2}\right]\vec{\epsilon}_1\vec{\epsilon}_2 \quad \text{(B.10)}$$

$$\exp(a) = \exp(a_0) + a_1 \exp(a_0)\vec{\epsilon}_1 + a_2 \exp(a_0)\vec{\epsilon}_2 + (a_3 + a_1 a_2)\exp(a_0)\vec{\epsilon}_1\vec{\epsilon}_2 \quad \text{(B.11)}$$

$$\log(a) = \log(a_0) + \frac{a_1}{a_0}\vec{\epsilon}_1 + \frac{a_2}{a_0}\vec{\epsilon}_2 + \left(\frac{a_3}{a_0} - \frac{a_0 a_1}{a_0^2}\right)\vec{\epsilon}_1\vec{\epsilon}_2 \quad \text{(B.12)}$$

B.2 Numerical Implementation of Hyper-Dual Number

The data type of HDNs can be defined by the derived data type of Fortran. In this user-defined data type, the information of one real part and three imaginary parts of the HDNs can be stored by four double-precision variables (i.e., *a*, *b*, *c*, and *d*). The *a*, *b*, *c*, and *d* are also called members of the derived data type. Then, members of the derived data type can be accessed using the percentage symbol %. For example, for a hyper-dual number $x = a + b\vec{\epsilon}_1 + c\vec{\epsilon}_2 + d\vec{\epsilon}_1\vec{\epsilon}_2$, executing commands $y = x\%b$ and $y = x\%d$ will respectively return the first and third "imaginary" parts of x. In practice, we write the above commands into custom functions to avoid direct operations on the real and imaginary parts of HDNs in the main program.

The source code of the example in Sect. 9.2.3 is provided in this appendix which contains not only the implementation details of HDSDA but also that of other numerical differentiation methods. It is important to note that mainstream programming languages have not yet adopted the definition of HDNs. Therefore, the code utilizes operator and function overloading to define the operation rules of HDNs and is implemented in Fortran 77. After the completion of the main program "ex01", the error data from different numerical differentiation methods are written to two files, namely "erorr1st.csv" and "erorr2nd.csv". These files can be utilized to generate the error curves of different numerical differentiation methods.

```
! ---------------------------------------------------------------------------
! Copyright (c)   2023   Dechun Lu, Xin Zhou, Anyu Shi
! Licensed under the MIT License. See LICENSE for details.
! This source code aims to illustrate the implementation details of
! different numerical differentiation methods
! ---------------------------------------------------------------------------
! Module for operator overloading of HDNs
        module hyper_dual_module
        implicit none
! definition of structure of HDNs
        type HyperDualNum
! a + b*e_1 + c*e_2 + d*e_1*e_2
        real*8 :: a = 0,b = 0,c = 0,d = 0
```

```
        end type HyperDualNum
! operator loading
        interface assignment(=);module procedure equal;end interface    ! HDNs =
(real,0,0,0)
        interface operator(+);module procedure add_dd; end interface   ! HDNs + HDNs
        interface operator(*);module procedure mul_dd; end interface   ! HDNs * HDNs
        interface operator(/);module procedure div_dd; end interface   ! HDNs / HDNs
        interface operator(**);module procedure pow; end interface     ! HDNs ** real
Number
! function loading
        interface dcos;module procedure dcos; end interface                    ! cosine
function
        interface dsin;module procedure dsin; end interface                 ! sine function
        interface dexp;module procedure dexp; end interface                   ! exponential
function
        contains
! subroutine of assignment
        subroutine equal(lhs,rhs)
            real*8, intent(in)::rhs;type(HyperDualNum),intent(out)::lhs
            lhs%a=rhs;lhs%b=0.0d0;lhs%c=0.0d0;lhs%d=0.0d0
        end subroutine equal

! function of addition
        function add_dd(u,v)        ! HDN (u) + HDN (v)
            type(HyperDualNum),intent(in)::u,v;type(HyperDualNum)::add_dd
            add_dd%a=u%a+v%a;add_dd%b=u%b+v%b
            add_dd%c=u%c+v%c;add_dd%d=u%d+v%d
        end function add_dd

! function of multiplication
        function mul_dd(u,v)        ! HDN(u)*HDN(v)
           type(HyperDualNum),intent(in)::u,v;type(HyperDualNum)::mul_dd
           mul_dd%a=u%a*v%a
           mul_dd%b=u%a*v%b+u%b*v%a
```

```
        mul_dd%c=u%a*v%c+u%c*v%a
        mul_dd%d=u%a*v%d+u%d*v%a+u%b*v%c+u%c*v%b
    end function mul_dd

! function of division
    function div_dd(u,v)       ! HDN(u)/HDN(v)
    type(HyperDualNum),intent(in)::u,v;type(HyperDualNum)::div_dd
      div_dd%a=u%a/v%a;    div_dd%b=(u%b*v%a-u%a*v%b)/(v%a*v%a)
      div_dd%c=(u%c*v%a-u%a*v%c)/(v%a*v%a)
      div_dd%d=(u%a*(2.d0*v%b*v%c-v%a*v%d)+
   & u%d*v%a*v%a-u%b*v%a*v%c-u%c*v%a*v%b)/(v%a*v%a*v%a)
    end function div_dd

! function of power function
    function pow(y,u)          ! HDN(y)**real number(u)
        real*8,intent(in)::u;type(HyperDualNum),intent(in)::y
        type(HyperDualNum)::pow;double precision deriv, yval, tol1
        yval = y%a;tol1 = 1.d-15    ! y's value, tolerance(e^(-15))
        if (abs(yval) < tol1) then
            if(yval >= 0.d0) then
                yval = tol1
            end if
            if(yval < 0.d0) then
                yval = - tol1
            end if
        end if
        deriv = u * (yval**(u-1.d0))
        pow%a = (y%a)**u; pow%b = y%b * deriv; pow%c = y%c * deriv
        pow%d = y%d*deriv+u*(u-1.d0)*y%b*y%c*(yval**(u-2.d0))
    end function pow

! function loading
    function dcos(x)        ! cosine function
        type(HyperDualNum),intent(in)::x;type(HyperDualNum)::dcos
```

```
        dcos%a = dcos(x%a);dcos%b=-x%b*dsin(x%a);dcos%c=-x%c*dsin(x%a)
        dcos%d = -x%d*dsin(x%a)-x%b*x%c*dcos(x%a)
    end function dcos
    function dsin(x)        ! sine function
        type(HyperDualNum),intent(in)::x;type(HyperDualNum)::dsin
        dsin%a=dsin(x%a);dsin%b=x%b*dcos(x%a);dsin%c = x%c*dcos(x%a)
        dsin%d = x%d*dcos(x%a)-x%b*x%c*dsin(x%a)
    end function dsin
    function dexp(x)        ! exponential function
        type(HyperDualNum),intent(in)::x;type(HyperDualNum)::dexp
        dexp%a = dexp(x%a);                 dexp%b = x%b*dexp(x%a)
        dexp%c = x%c*dexp(x%a);             dexp%d = (x%d+x%b*x%c)*dexp(x%a)
    end function dexp

! coefficient of e_1
    function dual1(x)
        type(HyperDualNum),intent(in)::x;real*8::dual1;dual1=x%b
    end function dual1

! coefficient of e_1*e_2
    function dual2(x)
        type(HyperDualNum),intent(in)::x;real*8::dual2;dual2=x%d
    end function dual2
    end module hyper_dual_module
!-----------------------------------------------------------------
```

```
! Main program ------------------------------------------------------
        program ex01
        implicit none
        double precision pi,x,hsize,af0,af1,af2
        double precision cd1,cd2,fd1,fd2,cs1,cs2,hd1,hd2
        double precision err1(17,4),err2(17,4)
        integer             i
        pi = dacos(-1.d0); x = pi / 12.d0
        call analytic_func(x,pi,af0,af1,af2)
! af0: the analytic original function; af1, af2: the analytic 1st and 2nd derivatives

        do i = 1,17
           hsize = 10.d0**(1 - i)                  ! perturbation value
           call fd_func(x,hsize,fd1,fd2)           ! calculate 1st and 2nd derivatives using
FDM
           call cd_func(x,hsize,cd1,cd2)           ! calculate 1st and 2nd derivatives
using CDM
           call cs_func(x,hsize,cs1,cs2)           ! calculate 1st and 2nd derivatives using
CSDA
           call hdns_func(x,hsize,hd1,hd2)         ! calculate 1st and 2nd derivatives
using HDSDA
! calculate the relative errors for four numerical differentiation methods
           err1(i,1)=abs((fd1-af1)/af1);err2(i,2)=abs((fd2-af2)/af2)
           err1(i,2)=abs((cd1-af1)/af1);err2(i,1)=abs((cd2-af2)/af2)
           err1(i,3)=abs((cs1-af1)/af1);err2(i,3)=abs((cs2-af2)/af2)
           err1(i,4)=abs((hd1-af1)/af1);err2(i,4)=abs((hd2-af2)/af2)
        enddo
        open(12,file="d:\erorr1st.csv")            ! output data to "d:\erorr1st.csv"
        do i = 1,17
            write(12,"(4(E15.7,','))") err1(i,:)  ! define the output format of data
        enddo
        close(12)
        open(13,file="d:\erorr2nd.csv")            ! output data to "d:\erorr2nd.csv"
        do i = 1,17
            write(13,"(4(E15.7,','))") err2(i,:)  ! define the output format of data
        enddo
        close(13)
        end program ex01
```

```
subroutine analytic_func(x,pi,analy_f,analy_1st,analy_2nd)
implicit none
double precision x,pi,analy_f,analy_1st,analy_2nd,denom,
& tem1,tem2

  denom = dsqrt( ( dsin(x) )**3.d0 +dcos(x)**3.d0 )
  ! analy_f is the value of the original function
  analy_f = x * dexp(x) / denom

  ! analy_1st denotes the analytical 1st derivative
  analy_1st = -x*dexp(x)*(-3.d0*(dcos(x)**2.d0)*dsin(x)+3.d0
& *dcos(x)*(dsin(x)**2.d0))/(2.d0*denom**3.d0)+
& dexp(x) / denom + x*dexp(x) / denom

  tem1 = 3.d0*((-3.d0*(dcos(x)**2.d0)*dsin(x)+3.d0*dcos(x)*(dsin(x)
&    **2.d0))**2.d0)/(4.d0*(denom**5.d0))

  tem2 = ( -3.d0*(dcos(x)**3.d0) + 6.d0*(dcos(x)**2.d0)*dsin(x) +
& 6.d0*(dsin(x)**2.d0)*dcos(x) - 3.d0*(dsin(x)**3.d0)) /
& (2.d0*(denom**3.d0))

  ! analy_2nd denotes the analytical 2nd derivative
  analy_2nd = -(x*dexp(x) + dexp(x))*(-3.d0*(dcos(x)**2.d0)*dsin(x)
& +3.d0*dcos(x)*(dsin(x)**2.d0))/(denom**3.d0)+(x*dexp(x)
& +2.d0*dexp(x))/(denom)+ x*dexp(x)* (tem1 - tem2)
  end
```

```
! The original function
        subroutine func_fx(x,fun)
        implicit none
        double precision x,fun
        fun = x * dexp(x) / ( ( dsin(x)**3.d0+dcos(x)**3.d0 )**0.5d0)
        end

! Central difference method
        subroutine cd_func(x,hsize,cd_1st,cd_2nd)
        implicit none
        double precision x,cd_1st,cd_2nd,hsize,fa,fb,f_self
        call func_fx(x+hsize,fa); call func_fx(x-hsize,fb);
        call func_fx(x,f_self)                              ! the original function
        cd_1st = (fa - fb) / 2.d0 / hsize                   ! the 1st derivative
        cd_2nd = (fa - 2.d0*f_self + fb) / (hsize**2.d0)    ! the 2nd derivative
        end

! Forward difference method
        subroutine fd_func(x,hsize,fd_1st,fd_2nd)
        implicit none
        double precision x,fd_1st,fd_2nd,hsize,fa,fb,f_self
        call func_fx(x+hsize,fa); call func_fx(x+2.d0*hsize,fb)
        call func_fx(x,f_self)                              ! the original function
        fd_1st = (fa - f_self) / hsize                      ! the 1st derivative
        fd_2nd = (fb - 2.d0*fa + f_self) / (hsize**2.d0)    ! the 2nd derivative
        end
```

```
! Complex step derivative appoximation
    subroutine cs_func(x,hsize,cs_1st,cs_2nd)
    implicit none
    double precision x,cs_1st,cs_2nd,hsize,fa,fb,f_self
    Complex(16)          cs_x,cs_f
    cs_x = cmplx(x,hsize,kind=8)
    cs_f = cs_x*exp(cs_x) /
   & ( ( sin(cs_x)**3.d0+cos(cs_x)**3.d0 )**0.5d0)          ! the original function
    cs_1st = imag(cs_f) / hsize                               ! the 1st derivative
    call func_fx(x,f_self)
    cs_2nd = 2.d0*(f_self - imag(cs_f)) / (hsize**2.d0) ! the 2nd derivative
    end

! Hyper dual step derivative appoximation
    subroutine hdns_func(x,hsize,hd_1st,hd_2nd)
    use hyper_dual_module
    implicit none
    double precision x,hd_1st,hd_2nd,hsize
    type(HyperDualNum)    hd_x,hd_f
    hd_x = 0.d0; hd_x = x
    ! assign the perturbation value to the 1st and 2nd imaginary parts
    hd_x%b = hsize;          hd_x%c = hsize
    hd_f=hd_x*dexp(hd_x)/((dsin(hd_x)**3.d0+dcos(hd_x)**3.d0)**0.5d0)
    ! hd_f denote HDN version of original function
    hd_1st = dual1(hd_f)/hsize                              ! the 1st derivative
    hd_2nd = dual2(hd_f)/(hsize**2.d0)                      ! the 2nd derivative
    end
```

The manufacturer's authorised representative in the EU is Springer Nature Customer Service Centre GmbH, Europaplatz 3, 69115 Heidelberg, Germany. If you have any concerns regarding our products, please contact ProductSafety@springernature.com

Printed and bound by CPI Group (UK) Ltd, Croydon, CR0 4YY
01/10/2026
02225926-0010